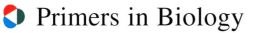 Primers in Biology

Protein Structure and Function

Primers in Biology:
Forthcoming titles

Immunity

Anthony DeFranco, Richard Locksley & Miranda Robertson

The Cell Cycle

David O Morgan

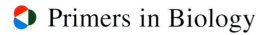
Primers in Biology

Protein Structure and Function

Gregory A Petsko

Dagmar Ringe

New Science Press Ltd

SINAUER ASSOCIATES, INC.
PUBLISHERS

Blackwell
Publishing

Editors: Eleanor Lawrence and Miranda Robertson
Managing Editor: Karen Freeland
Editorial Assistants: Amy Austin, Vivien Chen, Alexandra King
Production Manager: Liam Lynch
Design and Illustration: Matthew McClements, Blink Studio Ltd, London
Structure Graphics: Ezra Peisach
Indexer: Liza Furnival
Manufacturing: Adrienne Hanratty

Distributors:

Inside North America:

Sinauer Associates, Inc., Publishers,
23 Plumtree Road, PO Box 407, Sunderland, MA 01375, USA
orders@sinauer.com
www.sinauer.com

Outside North America:

Marston Book Services Ltd,
PO Box 269, Abingdon, Oxford, OX14 4YN, UK
direct.orders@marston.co.uk
www.blackwellpublishing.com

ISBN 0-9539181-4-9   (paperback) New Science Press Ltd
ISBN 1-4051-1922-5   (paperback) Blackwell Publishing Ltd
ISBN 0-87893-663-7   (paperback) Sinauer Associates, Inc.

British Library Cataloguing-in-Publication Data

A catalogue record for this title is available from the British Library
ISBN 0-9539181-4-9

Published by New Science Press Ltd
Middlesex House
34-42 Cleveland Street
London W1P 6LB
UK
www.new-science-press.com

in association with
Blackwell Publishing Ltd
and
Sinauer Associates, Inc., Publishers

Printed by Stamford Press PTE Singapore

15 14 13 12 11 10 9 8 7 6 5 4 3 2 1

The Authors

Gregory A Petsko studied chemistry and classics as an undergraduate at Princeton University before going to Oxford as a Rhodes scholar to work for his PhD with David Phillips. He is currently Director of the Rosenstiel Center at Brandeis University.

Dagmar Ringe graduated in chemistry from Barnard College, Columbia, and took her PhD in bioorganic chemistry from Boston University. She is currently Professor of Biochemistry and Chemistry at Brandeis University.

Primers in Biology:
a note from the publisher

section heading one-sentence subheading

bottom margin:
definitions and references

Protein Structure and Function is the first in a series of books constructed on a modular principle that is intended to make them easy to teach from, to learn from, and to use for reference, without sacrificing the synthesis that is essential for any text that is to be truly instructive. The diagram above illustrates the modular structure and special features of these books. Each chapter is broken down into two-page sections each covering a defined topic and containing all the text, illustrations, definitions and references relevant to that topic. Within each section, the text is divided into subsections under one-sentence headings that reflect the sequence of ideas and the global logic of the chapter.

The modular structure of the text, and the transparency of its organization, make it easy for instructors to choose their own path through the material and for students to revise; or for working scientists using the book as an up-to-date reference to find the topics they want, and as much of the conceptual context of any individual topic as they may need.

All of the definitions and references are collected together at the end of the book, with the section or sections in which they occur indicated in each case. Glossary definitions may sometimes contain helpful elaboration of the definition in the text, and references contain a full list of authors instead of the abbreviated list in the text.

From Sequence to Consequence

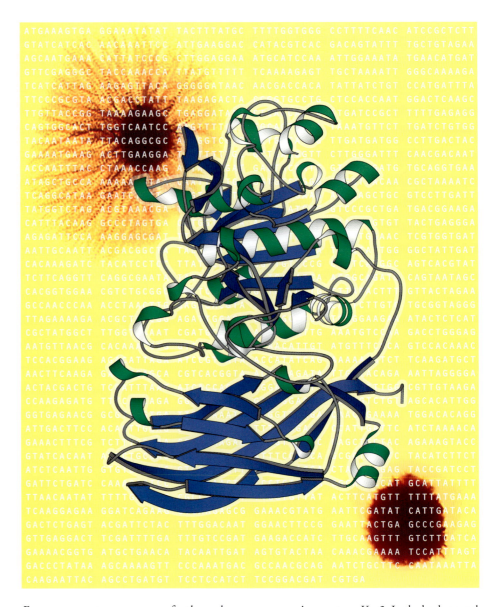

From sequence to consequence for the prohormone processing protease Kex2. In the background is the amino acid sequence for the yeast protease Kex2, the prototype of all the dibasic prohormone processing enzymes in higher eukaryotes. The sequence is the result of translating the information contained in the gene that codes for this protein. In the center is the three-dimensional structure of the protein as determined by Todd Holyoak, Mark Wilson and Dagmar Ringe in collaboration with Robert Fuller. The structure is specified by the information contained in the amino acid sequence. The protein has two domains; the upper domain contains the catalytic site where prohormones are cleaved; the lower domain is of unknown function. In the background are images (courtesy of Nina Agabian, UCSF) of two strains of the pathogenic fungus *Candida albicans* that illustrate the consequence for the organism of not having this protein in the genome. The upper strains are wild-type and contain functional Kex2 protein. They form hyphae, the stringy projections necessary for virulence. The lower strains are mutants in which both copies of the Kex2 gene have been deleted. They are unable to form hyphae and are greatly attenuated in virulence, establishing that the cellular and biochemical functions of this protein are important for pathogenicity. Based on a figure by Timothy Fenn.

Biology in the age of genomics is a journey from sequence to consequence

Dagmar Ringe

Preface

Sequence determines structure determines function - The real central dogma of biology

Two major revolutions have occurred within the last twenty years that have changed the way biologists think about their science and have influenced the way experiments are designed. These two revolutions are related, although they have a very different basis experimentally.

The first revolution was the structural revolution. The development of techniques for visualizing a model of a macromolecule in three dimensions and for manipulating that model *in silico* has provided a powerful tool for predicting the outcome of an experiment and for the interpretation of the results of a wide variety of experiments. It also allows the design of experiments that would never otherwise have been possible. Availability of structure has extended the insight possible from experiments as diverse as enzyme kinetics to the mechanics of signaling pathways. With the advent of structural enzymology, it is now possible to describe the course of an enzyme catalyzed reaction in detail, including the structures of intermediates that could only be surmised before. The connection between chemistry and biology at this level has been powerful and is now all-pervasive, forming the basis of structure-aided drug design.

The second revolution is the genomic revolution. The availability of the sequences of all genes and all intervening nucleic acids between them in the genomes of organisms ranging from bacteria to man is an entree to the description of the working of a cell as a whole and to its most intimate parts. The relationships of genes to each other in a functioning cell provide the basis to describe the machinery by which the cell lives and dies. It is now possible to determine the fate of a cell from a primary signal to the final ratchet which produces its demise. The relationships of cells to each other in organs and organisms will ultimately be understood in terms of this machinery as well.

It is not our purpose here to give a detailed account of the many sophisticated experimental methods on which our current understanding of protein structure and function is based: our objective is to convey an up-to-date picture of what is known, rather than how that knowledge is obtained. To provide some general perspective on the types of strategies used however, we have briefly summarized in Chapter 4 (4–4) some of the most important current techniques for exploring protein function; and the pragmatic art of structure determination is outlined in Chapter 5. We address in this book the questions to which these techniques are providing the answers. How does the chemical description of a protein relate to its actual three-dimensional structure? How does the three-dimensional structure relate to the machinery that brings about a chemical reaction or that recognizes another molecule? How does the sequence of a gene encode not only the sequence of a protein but, more importantly, the three-dimensional architecture of the protein, and finally, the function of that protein? We attempt in this book to give a concise and critical perspective on the current state of the art in answering these questions.

We apologize to those whose favorite protein or pathway has not been mentioned. The nature of a primer like this does not allow an exhaustive compendium. We have tried to pick representative examples that illustrate general principles. We recognize that in some cases exceptions exist, and that more than one example could usefully illustrate a particular point.

To our former student, Ezra Peisach, we express appreciation for his tireless efforts to provide beautiful and informative pictures of protein structures. We are particularly grateful to Eleanor Lawrence and Matthew McClements, and to the staff of New Science Press—Amy Austin, Vivien Chen, Karen Freeland and Liam Lynch—for their patience, helpfulness, and enthusiastic support for this project. Finally, a very special thanks to our publisher Miranda Robertson, who nurtured the project to fruition.

Gregory A Petsko

Dagmar Ringe

The Protein Data Bank: a note from the authors

Throughout this book, protein structures are given with their Protein Data Bank identity code, a combination of four letters and numbers that allows the atomic coordinates for the structure to be downloaded from the Protein Data Bank. The Protein Data Bank (PDB) is the single world-wide repository for the processing and distribution of 3-D biological macromolecular structure data. The PDB is operated by Rutgers, The State University of New Jersey; the San Diego Supercomputer Center at the University of California, San Diego; and the Center for Advanced Research in Biotechnology of the National Institute of Standards and Technology. The PDB is described in Berman, H.M., Westbrook, J., Feng, Z., Gilliland, G., Bhat, T.N., Weissig, H., Shindyalov, I.N. and Bourne, P.E.: **The Protein Data Bank.** *Nucleic Acids Research* 2000, **28**:235–242, and a new book that includes several chapters on the PDB and how to use it is: Bourne, P.E. and Weissig, H.: *Structural Bioinformatics* (Wiley, Hoboken, NJ, 2003).

The PDB is available via the World Wide Web and may be accessed at the URL: http://www.pdb.org/

In addition to files containing the atomic coordinates for macromolecular structures, the PDB site has simple programs that can be used via a web browser for viewing and manipulating such structures, as well as information on the structural genomics initiatives now underway.

Acknowledgements

The following individuals provided expert advice on entire chapters or parts of chapters:

Chapter 1   Walter Englander, University of Pennsylvania; Robert Fletterick, University of California, San Francisco; Stephen Harrison, Harvard University; Liisa Holm, European Bioinformatics Institute; Susan Marqusee, University of California, Berkeley; Robert Stroud, University of California, San Francisco; Janet Thornton, European Bioinformatics Institute

Chapter 2 Karen Allen, Boston University School of Medicine; Tom Bruice, University of California, Santa Barbara; Stephen Harrison, Harvard University; Dan Herschlag, Stanford University; Jack F. Kirsch, University of California, Berkeley; Jeremy Knowles, Harvard University; Jack Kyte, University of California, San Diego; Wendell Lim, University of California, San Francisco; Peter Moore, Yale University; Vernon Schramm, The Albert Einstein College of Medicine, University of Yeshiva

Chapter 3   Alex Bateman, The Wellcome Trust Sanger Institute; Henry Bourne, University of California, San Francisco; Mark Bedford, The University of Texas MD Anderson Cancer Center; Patrick Casey, Duke University; Chin Ha Chung, Seoul National University; Gary Felsenfeld, National Institutes of Heath; Alfred Goldberg, Harvard University; Colin Gordon, MRC Human Genetics Unit, Western General Hospital; Rachel Green, Johns Hopkins University School of Medicine; Steve Gross, Weill Medical College, Cornell University; Mark Hochstrasser, Yale University; Wendell Lim, University of California, San Francisco; John Lowesnstein, Brandeis University; Emil Pai, University of Toronto; Richard Silverman, Northwestern University; Steven Sprang, University of Texas South Western Medical Center, Dallas; Henry Paulus, Harvard Medical School; Tony Pawson, Samuel Lunenfeld Research Institute, Mount Sinai Hospital; Pauline Rudd, Oxford University; Elena Sablin, University of California, San Francisco; James Spudich, Stanford University; Ann Stock, The State University of New Jersey, Rutgers; Brian Strahl, University of North Carolina School of Medicine; Fyodor Urnov, Advanced Genomics Technologies, Sangamo Biosciences

Chapter 4   Charles Brenner, Thomas Jefferson University; George Church, Harvard Medical School; Alexander Johnson, University of California, San Francisco; Roman Laskowski, European Bioinformatics Institute; Timothy Richmond, Swiss Federal Institute of Technology; Andrej Sali, The Rockefeller University; Jeff Skolnick, Donald Danforth Plant Science Center; Michael Snyder, Yale University; Song Tan, Penn State University

We are grateful to the following for providing or permitting the use of illustrations:

Figure 1-20 Table of conformational preferences of the amino acids. Reprinted from *Biochimica et Biophysica Acta*, Volume **916**, Williams, R.W., Chang, A., Juretic, D. and Loughran, S.: **Secondary structure predictions and medium range interactions.** Pages 200–204. ©1987, with permission from Elsevier Science.

Figure 1-66 "Open-book" view of the complementary structural surfaces that form the interface between interleukin-4 and its receptor. Kindly provided by Walter Sebald and Peter Reineme.

Figure 1-67 Coiled-coil alpha-helical interactions. Kindly provided by Carolyn Cohen. Cohen, C. and Parry, D.A.: **Alpha-helical coiled coils and bundles: how to design an alpha-helical protein.** *Proteins* 1990, 7:1–15. Copyright ©1990 Wiley. Reprinted by permission of Wiley-Liss, Inc., a subsidiary of John Wiley & Sons, Inc.

Figure 1-71 Sickle-cell hemoglobin. Kindly provided by Stuart J. Edelstein.

Figure 2-2 Substrate binding to anthrax toxin lethal factor. Kindly provided by Robert Liddington. Pannifer, A.D., Wong, T.Y., Schwarzenbacher, R., Renatus, M., Petosa, C., Bienkowska, J., Lacy, D.B., Collier, R.J., Park, S., Leppla, S.H., Hanna, P. and Liddington, R.C.: **Crystal structure of the anthrax lethal factor.** *Nature* 2001, 414:229–233. With copyright permission from Nature.

Figure 2-12 Surface view of the heme-binding pocket of cytochrome c6. Kindly provided by Paul Roesch. Beissinger, M., Sticht, H., Sutter, M., Ejchart, A., Haehnel, W., and Roesch, P.: **Solution structure of cytochrome c6 from the thermophilic cyanobacterium *Synechococcus elongatus*.** *EMBO J.* 1998, 2:27–36, by permission of Oxford University Press.

Figure 2-16 Structure of the 50S (large) subunit of the bacterial ribosome. Kindly provided by Poul Nissen and Thomas Steitz.

Figure 2-22 The electrostatic potential around the enzyme Cu,Zn-superoxide dismutase. Kindly provided by Barry Honig and Emil Alexov.

Figure 2-45 Three consecutive reactions are catalyzed by the active sites of the enzyme carbamoyl phosphate synthetase. Kindly provided by Frank Raushel and Hazel Holden.

Figure 3-5 Cathepsin D conformational switching. Kindly provided by John Erickson. Lee, A.Y.,

Gulink, S.V. and Erickson, J.W.: **Conformational switching in an aspartic proteinase.** *Nat. Struct. Biol.* 1998, 5:866–871. With copyright permission from Nature.

Figure 3-17 Models for the motor actions of muscle myosin and kinesin. Kindly provided by Graham Johnson and Ron Milligan. Reprinted with permission from Vale, R.D. and Milligan, R.A.: **The way things move: looking under the hood of molecular motor proteins.** *Science* 2000, 288:88–95. ©2000 American Association for the Advancement of Science.

Figure 3-18 Structural and functional similarity between different families of molecular switches. Kindly provided by Elena P. Sablin. Reprinted from *Curr. Opin. Struct. Biol.*, Volume 11, Sablin, E.P., and Fletterick, R.J.: **Nucleotide switches in molecular motors: structural analysis of kinesins and myosins.** 716–724, ©2001, with permission from Elsevier Science.

Figure 3-19 The eukaryotic proteasome. Kindly provided by U.S. Department of Energy Genomes to Life Program, http://doegenomestolife.org.

Figure 3-24 The conserved protein kinase catalytic domain. Kindly provided by Ming Lei and Stephen Harrison.

Figure 3-28 The substrate-binding site of Cdk2. Kindly provided by Jane Endicott and Martin Noble.

Figure 3-30 Conserved features of RR regulatory domains. Kindly provided by Ann Stock. Reprinted from *Trends Biochem. Sci.*, Volume **26**, West, A.H. and Stock, A.M.: **Histidine kinases and response regulator proteins in two-component signaling systems.** Pages 369–376, ©2001, with permission from Elsevier.

Figure 3-42 The structure of Glc3Man9GlcNac2. Kindly provided by Mark Wormald.

Figure 3-43 Secreted immunoglobulin A. Kindly provided by Louise Royle and Pauline Rudd (Royle, L.M., Wormald, M.R., Merry, A.H., Dwek, R.A. and Rudd, P.M., unpublished).

Figure 4-5 Phylogenetic tree comparing the three major MAP kinase subgroups. Kindly provided by James E. Ferrell Jr.

Figure 4-16 The phenotype of a gene knockout can give clues to the role of the gene. Kindly provided by Ute Hochgeschwender. Yaswen, L., Diehl, N., Brennan, M.B. and Hochgeschwender, U.: **Obesity in the mouse model of pro-opiomelanocortin deficiency responds to peripheral melanocortin.** *Nat. Med.* 1999, 5:1066–1070. With copyright permission from Nature.

Figure 4-17 Protein localization in the cell. Kindly provided by Daniel Moore and Terry Orr-Weaver. Reprinted from *Cell*, Volume **83**, Kerrebrock, A.W., Moore, D.P., Wu, J.S. and Orr-Weaver, T.L.: **Mei-S332, a *Drosophila* protein required for sister-chromatid cohesion, can localize to meiotic centromere regions.** Pages 247–256, ©1995, with permission from Elsevier.

Figure 4-23 Evolutionary conservation and interactions between residues in the protein-interaction domain PDZ and in rhodopsin. Kindly provided by Rama Ranganathan. Reprinted with permission from Lockless, S.W. and Ranganathan, R.: **Evolutionarily conserved pathways of energetic connectivity in protein families.** *Science* 1999, 286:295–299. ©1999 American Association for the Advancement of Science.

Figure 4-26 Some decoy structures produced by the Rosetta method. Reprinted from *J. Mol. Biol.*, Volume **268**, Simons, K.T., Kooperberg, C., Huang, E. and Baker, D.: **Assembly of protein tertiary structures from fragments with similar local sequences using simulated annealing and Bayesian scoring functions.** Pages 209–225, ©1997, with permission from Elsevier.

Figure 4-27 Examples of the best-center cluster found by Rosetta for a number of different test proteins. Kindly provided by Richard Bonneau and David Baker. Bonneau, R., Tsai, J., Ruczinski, I., Chivian, D., Rohl, C., Strauss, C.E. and Baker, D.: **Rosetta in CASP4: Progress in *ab initio* protein structure prediction.** *Proteins* 2001, 45(S5):119–126. Copyright ©2001 Wiley. Reprinted by permission of Wiley-Liss, Inc., a subsidiary of John Wiley & Sons, Inc.

Figure 4-34 Ribbon representation showing the experimentally derived functionality map of thermolysin. Kindly provided by Roderick E. Hubbard. English, A.C., Groom, C.R. and Hubbard, R.E.: **Experimental and computational mapping of the binding surface of a crystalline protein.** *Protein Eng.* 2001, 14:47–59, by permission of Oxford University Press.

Figure 4-54 Structures of serine protease inhibitors. Adapted from Ye, S. and Goldsmith, E.J.: **Serpins and other covalent protease inhibitors.** *Curr. Opin. Struct. Biol.* 2001, 11:740–745.

Figure 5-1 Portion of a protein electron density map at three different resolutions. Kindly provided by Aaron Moulin.

Figure 5-5 Different ways of presenting a protein structure. Kindly provided by Todd Holyoak.

Contents summary

Contents in full

CHAPTER 3 Control of Protein Function

CHAPTER 5 Structure Determination

1

From Sequence to Structure

The genomics revolution is providing gene sequences in exponentially increasing numbers. Converting this sequence information into functional information for the gene products coded by these sequences is the challenge for post-genomic biology. The first step in this process will often be the interpretation of a protein sequence in terms of the three-dimensional structure into which it folds. This chapter summarizes the basic concepts that underlie the relationship between sequence and structure and provides an overview of the architecture of proteins.

1-0 Overview: Protein Function and Architecture

Binding

Specific recognition of other molecules is central to protein function. The molecule that is bound (the ligand) can be as small as the oxygen molecule that coordinates to the heme group of myoglobin, or as large as the specific DNA sequence (called the TATA box) that is bound—and distorted—by the TATA binding protein. Specific binding is governed by shape complementarity and polar interactions such as hydrogen bonding.

TATA binding protein

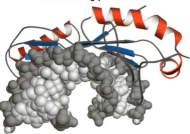

The TATA binding protein binds a specific DNA sequence and serves as the platform for a complex that initiates transcription of genetic information. (PDB 1tgh)

Myoglobin

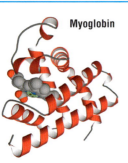

Myoglobin binds a molecule of oxygen reversibly to the iron atom in its heme group (shown in grey with the iron in green). It stores oxygen for use in muscle tissues. (PDB 1a6k)

Catalysis

Essentially every chemical reaction in the living cell is catalyzed, and most of the catalysts are protein enzymes. The catalytic efficiency of enzymes is remarkable: reactions can be accelerated by as much as 17 orders of magnitude over simple buffer catalysis. Many structural features contribute to the catalytic power of enzymes: holding reacting groups together in an orientation favorable for reaction (proximity); binding the transition state of the reaction more tightly than ground state complexes (transition state stabilization); acid-base catalysis, and so on.

DNA polymerase

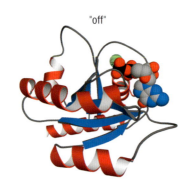

DNA replication is catalyzed by a specific polymerase that copies the genetic material and edits the product for errors in the copy. (PDB 1pbx)

HIV protease

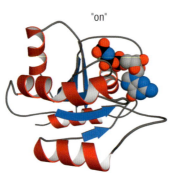

Replication of the AIDS virus HIV depends on the action of a protein-cleaving enzyme called HIV protease. This enzyme is the target for protease-inhibitor drugs (shown in grey). (PDB 1a8k)

Switching

Proteins are flexible molecules and their conformation can change in response to changes in pH or ligand binding. Such changes can be used as molecular switches to control cellular processes. One example, which is critically important for the molecular basis of many cancers, is the conformational change that occurs in the small GTPase Ras when GTP is hydrolyzed to GDP. The GTP-bound conformation is an "on" state that signals cell growth; the GDP-bound structure is the "off" signal.

Ras

"off"   "on"

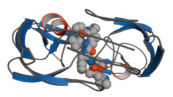

The GDP-bound ("off"; PDB 1pll) state of Ras differs significantly from the GTP-bound ("on"; PDB 121p) state. This difference causes the two states to be recognized by different proteins in signal transduction pathways.

Structural Proteins

Protein molecules serve as some of the major structural elements of living systems. This function depends on specific association of protein subunits with themselves as well as with other proteins, carbohydrates, and so on, enabling even complex systems like actin fibrils to assemble spontaneously. Structural proteins are also important sources of biomaterials, such as silk, collagen, and keratin.

Silk

Silk derives its strength and flexibility from its structure: it is a giant stack of antiparallel beta sheets. Its strength comes from the covalent and hydrogen bonds within each sheet; the flexibility from the van der Waals interactions that hold the sheets together. (PDB 1slk)

F-actin

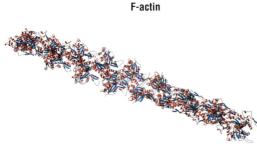

Actin fibers are important for muscle contraction and for the cytoskeleton. They are helical assemblies of actin and actin-associated proteins. (Courtesy of Ken Holmes)

Figure 1-1 Four examples of biochemical functions performed by proteins

Proteins are the most versatile macromolecules of the cell

This book is concerned with the functions that proteins perform and how these are determined by their structures. "Protein function" may mean the biochemical function of the molecule in isolation, or the cellular function it performs as part of an assemblage or complex with other molecules, or the phenotype it produces in the cell or organism.

Major examples of the biochemical functions of proteins include binding; catalysis; operating as molecular switches; and serving as structural components of cells and organisms (Figure 1-1). Proteins may bind to other macromolecules, such as DNA in the case of DNA polymerases or gene regulatory proteins, or to proteins in the case of a transporter or a receptor that binds a signaling molecule. This function exploits the ability of proteins to present structurally and chemically diverse surfaces that can interact with other molecules with high specificity. Catalysis requires not only specific binding, to substrates and in some cases to regulatory molecules, but also specific chemical reactivity. Regulated enzymes and switches, such as the signaling G proteins (which are regulated enzymes that catalyze the hydrolysis of GTP), require large-scale conformational changes that depend on a delicate balance between structural stability and flexibility. Structural proteins may be as strong as silk or as tough and durable as keratin, the protein component of hair, horn and feathers; or they may have complex dynamic properties that depend on nucleotide hydrolysis, as in the case of actin and tubulin. This extraordinary functional diversity and versatility of proteins derives from the chemical diversity of the side chains of their constituent amino acids, the flexibility of the polypeptide chain, and the very large number of ways in which polypeptide chains with different amino acid sequences can fold.

There are four levels of protein structure

Proteins are polymers of 20 different amino acids joined by peptide bonds. At physiological temperatures in aqueous solution, the polypeptide chains of proteins fold into a form that in most cases is globular (see Figure 1-2c). The sequence of the different amino acids in a protein, which is directly determined by the sequence of nucleotides in the gene encoding it, is its *primary structure* (Figure 1-2a). This in turn determines how the protein folds into higher-level structures. The *secondary structure* of the polypeptide chain can take the form either of alpha helices or of beta strands, formed through regular hydrogen-bonding interactions between N–H and C=O groups in the invariant parts of the amino acids in the polypeptide **backbone** or main chain (Figure 1-2b). In the globular form of the protein, elements of either alpha helix, or beta sheet, or both, as well as loops and links that have no secondary structure, are folded into a *tertiary structure* (Figure 1-2c). Many proteins are formed by association of the folded chains of more than one polypeptide; this constitutes the *quaternary structure* of a protein (Figure 1-2d).

For a polypeptide to function as a protein, it must usually be able to form a stable tertiary structure (or *fold*) under physiological conditions. On the other hand, the demands of protein function require that the folded protein should not be too rigid. Presumably because of these constraints, the number of folds adopted by proteins, though large, is limited. Whether the limited number of folds reflects physical constraints on the number of stable folds, or simply the expedience of divergent evolution from an existing stable fold, is not known, but it is a matter of some practical importance: if there are many possible stable folds not represented in nature, it should be possible to produce completely novel proteins for industrial and medical applications.

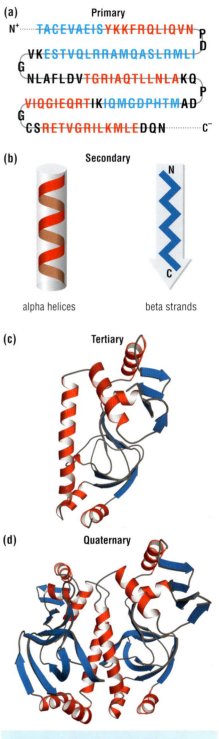

(a) Primary

N+ TACEVAEISYKKFRQLIQVN P
D
VKESTVQLRRAMQASLRMLI
G
NLAFLDVTGRIAQTLLNLAKQ P
VIQGIEQRTIKIQMGDPHTMAD P
G
CSRETVGRILKMLEDQN C−

(b) Secondary

alpha helices   beta strands

N

C

(c) Tertiary

(d) Quaternary

Figure 1-2 Levels of protein structure illustrated by the catabolite activator protein (a) The amino-acid sequence of a protein (primary structure) contains all the information needed to specify **(b)** the regular repeating patterns of hydrogen-bonded backbone conformations (secondary structure) such as alpha helices (red) and beta sheets (blue), as well as **(c)** the way these elements pack together to form the overall fold of the protein (tertiary structure) (PDB 2cgp). **(d)** The relative arrangement of two or more individual polypeptide chains is called quaternary structure (PDB 1cgp).

Definitions

backbone: the regularly repeating part of a polymer. In proteins it consists of the amide –N–H, alpha carbon –C–H and the carbonyl –C=O groups of each amino acid.

References

Alberts, B. *et al.*: *Molecular Biology of the Cell* 4th ed. Chapter 3 (Garland, New York, 2002).

Jansen, R. and Gerstein, M.: **Analysis of the yeast transcriptome with structural and functional cate-** gories: characterizing highly expressed proteins. *Nucleic Acids Res.* 2000, **28**:1481–1488.

Michal, G., ed.: *Boehringer Mannheim Biochemical Pathways Wallcharts*, Roche Diagnostics Corporation, Roche Molecular Biochemicals, P.O. Box 50414, Indianapolis, IN 46250-0414, USA.

Voet, D. and Voet, J.G.: *Biochemistry* 2nd ed. Chapters 4 to 7 (Wiley, New York, 1995).

http://www.expasy.ch/cgi-bin/search-biochem-index (searchable links to molecular pathways and maps).

The chemical characters of the amino-acid side chains have important consequences for the way they participate in the folding and functions of proteins

The amino-acid **side chains** (Figure 1-3) have different tendencies to participate in interactions with each other and with water. These differences profoundly influence their contributions to protein stability and to protein function.

Hydrophobic amino-acid **residues** engage in *van der Waals* interactions only. Their tendency to avoid contact with water and pack against each other is the basis for the *hydrophobic effect*. Alanine and leucine are strong helix-favoring residues, while proline is rarely found in helices because its backbone nitrogen is not available for the hydrogen bonding required for helix formation. The aromatic side chain of phenylalanine can sometimes participate in weakly polar interactions.

Hydrophilic amino-acid residues are able to make *hydrogen bonds* to one another, to the peptide backbone, to polar organic molecules, and to water. This tendency dominates the interactions in which they participate. Some of them can change their charge state depending on their pH or the microenvironment. Aspartic acid and glutamic acid have pK_a values near 5 in aqueous solution, so they are usually unprotonated and negatively charged at pH 7. But in the hydrophobic interior of a protein molecule their pK_a may shift to 7 or even higher (the same effect occurs if a negative charge is placed nearby), allowing them to function as proton donors at physiological pH. The same considerations apply to the behavior of lysine, which has a pK_a greater than 10 in water and so is usually depicted as positively charged. But in a nonpolar environment, or in the presence of a neighboring positive charge, its pK_a can shift to less than 6, and the resulting neutral species can be a proton acceptor. Histidine is perhaps the most versatile of all the amino acids in this regard, which explains why it is also the residue most often found in enzyme active sites. It has two titratable –N–H groups, each with pK_a values around 6. When one of these –N–H groups loses a proton, however, the pK_a of the other one becomes much greater than 10. When both are protonated, the residue as a whole is positively charged. When only one is protonated (usually it is the one farthest from the main chain of the protein) the side chain is neutral and has the ability both to donate and to accept a proton. The fully deprotonated form is negatively charged, and occurs rarely. Arginine is always completely protonated at neutral pH; its positive charge is localized primarily at the carbon atom of the guanidium head. Serine, threonine, glutamine and asparagine do not ionize but are able both to donate and to accept hydrogen bonds simultaneously. Cysteine, like histidine, is commonly found in enzyme active sites, because the thiolate anion is the most powerful *nucleophile* available from the naturally occurring amino acids.

Amphipathic residues have both polar and nonpolar character, making them ideal for forming interfaces. It may seem surprising to consider the charged side chain of lysine as amphipathic, but its long hydrophobic region is often involved in van der Waals interactions with hydrophobic side chains. Tyrosine does not usually ionize at physiological pH (its pK_a is about 9) but in some enzyme active sites it can participate in acid-base reactions because the environment can lower this pK_a. The –O–H group is able both to donate and to accept hydrogen bonds, and the aromatic ring can form weakly polar interactions. Tryptophan behaves similarly, but the indole –N–H group does not ionize. Methionine is the least polar of the amphipathic amino acids, but the thioether sulfur is an excellent ligand for many metal ions.

Definitions

amphipathic: having both polar and nonpolar character and therefore a tendency to form interfaces between **hydrophobic** and **hydrophilic** molecules.

hydrophilic: tending to interact with water. Hydrophilic molecules are polar or charged and, as a consequence, are very soluble in water. In polymers, hydrophilic **side chains** tend to associate with other hydrophilic side chains, or with water molecules, usually by means of hydrogen bonds.

hydrophobic: tending to avoid water. Hydrophobic molecules are nonpolar and uncharged and, as a consequence, are relatively insoluble in water. In polymers, hydrophobic **side chains** tend to associate with each other to minimize their contact with water or polar side chains.

residue: the basic building block of a polymer; the fragment that is released when the bonds that hold the polymer segments together are broken. In proteins, the residues are the amino acids.

side chain: a chemical group in a polymer that protrudes from the repeating backbone. In proteins, the side chain, which is bonded to the alpha carbon of the backbone, gives each of the 20 amino acids its particular chemical identity.

References

Creighton, T.E.: *Proteins: Structure and Molecular Properties* 2nd ed. Chapter 1 (Freeman, New York, 1993).

A website summarizing the physical-chemical properties of the standard amino acids may be found at: http://prowl.rockefeller.edu/aainfo/contents.htm

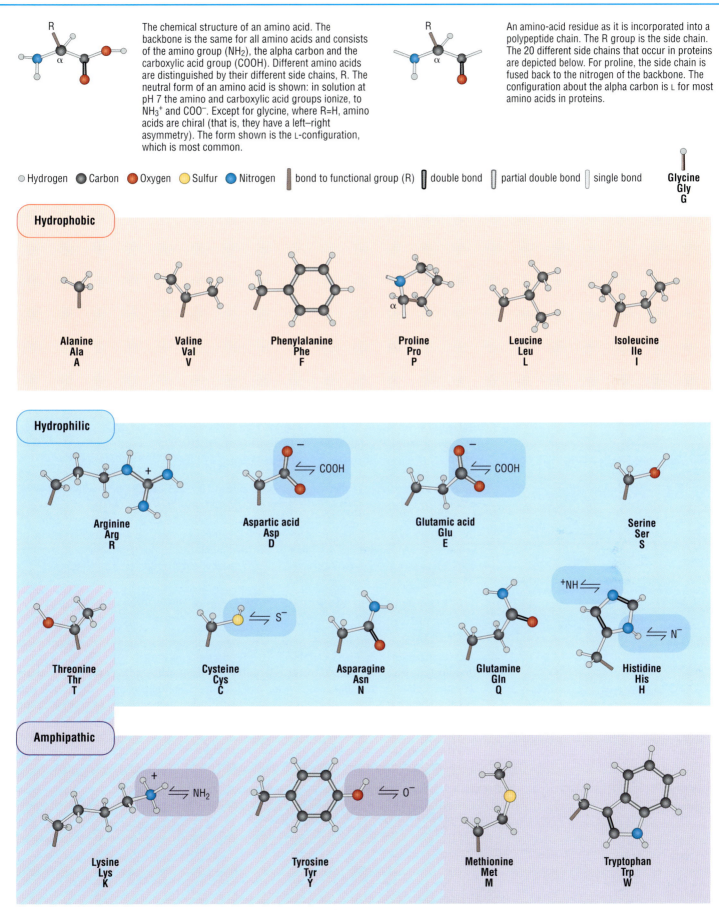

The chemical structure of an amino acid. The backbone is the same for all amino acids and consists of the amino group (NH₂), the alpha carbon and the carboxylic acid group (COOH). Different amino acids are distinguished by their different side chains, R. The neutral form of an amino acid is shown: in solution at pH 7 the amino and carboxylic acid groups ionize, to NH$_3^+$ and COO$^-$. Except for glycine, where R=H, amino acids are chiral (that is, they have a left–right asymmetry). The form shown is the L-configuration, which is most common.

An amino-acid residue as it is incorporated into a polypeptide chain. The R group is the side chain. The 20 different side chains that occur in proteins are depicted below. For proline, the side chain is fused back to the nitrogen of the backbone. The configuration about the alpha carbon is L for most amino acids in proteins.

○ Hydrogen ● Carbon ● Oxygen ● Sulfur ● Nitrogen ▮ bond to functional group (R) ▮ double bond ▮ partial double bond ▯ single bond

Glycine Gly G

Hydrophobic

Alanine Ala A

Valine Val V

Phenylalanine Phe F

Proline Pro P

Leucine Leu L

Isoleucine Ile I

Hydrophilic

Arginine Arg R

Aspartic acid Asp D ⇌ COOH

Glutamic acid Glu E ⇌ COOH

Serine Ser S

Threonine Thr T

Cysteine Cys C ⇌ S$^-$

Asparagine Asn N

Glutamine Gln Q

Histidine His H $^+$NH ⇌ / ⇌ N$^-$

Amphipathic

Lysine Lys K $^+$ ⇌ NH₂

Tyrosine Tyr Y ⇌ O$^-$

Methionine Met M

Tryptophan Trp W

Figure 1-3 Amino-acid structure and the chemical characters of the amino-acid side chains Charged side chains are shown in the form that predominates at pH 7. For proline, the nitrogen and alpha carbon are shown because the side chain is joined to the nitrogen atom to form a ring that includes these atoms.

1st position (5' end)	2nd position				3rd position (3' end)
	U	C	A	G	
U	Phe	Ser	Tyr	Cys	U
	Phe	Ser	Tyr	Cys	C
	Leu	Ser	STOP	STOP	A
	Leu	Ser	STOP	Trp	G
C	Leu	Pro	His	Arg	U
	Leu	Pro	His	Arg	C
	Leu	Pro	Gln	Arg	A
	Leu	Pro	Gln	Arg	G
A	Ile	Thr	Asn	Ser	U
	Ile	Thr	Asn	Ser	C
	Ile	Thr	Lys	Arg	A
	Met	Thr	Lys	Arg	G
G	Val	Ala	Asp	Gly	U
	Val	Ala	Asp	Gly	C
	Val	Ala	Glu	Gly	A
	Val	Ala	Glu	Gly	G

Amino acids	Abbreviations		Codons
Alanine	Ala	**A**	GCA GCC GCG GCU
Cysteine	Cys	**C**	UGC UGU
Aspartic acid	Asp	**D**	GAC GAU
Glutamic acid	Glu	**E**	GAA GAG
Phenylalanine	Phe	**F**	UUC UUU
Glycine	Gly	**G**	GGA GGC GGG GGU
Histidine	His	**H**	CAC CAU
Isoleucine	Ile	**I**	AUA AUC AUU
Lysine	Lys	**K**	AAA AAG
Leucine	Leu	**L**	UUA UUG CUA CUC CUG CUU
Methionine	Met	**M**	AUG
Asparagine	Asn	**N**	AAC AAU
Proline	Pro	**P**	CCA CCC CCG CCU
Glutamine	Gln	**Q**	CAA CAG
Arginine	Arg	**R**	AGA AGG CGA CGC CGG CGU
Serine	Ser	**S**	AGC AGU UCA UCC UCG UCU
Threonine	Thr	**T**	ACA ACC ACG ACU
Valine	Val	**V**	GUA GUC GUG GUU
Tryptophan	Trp	**W**	UGG
Tyrosine	Tyr	**Y**	UAC UAU

Figure 1-4 The genetic code Each of the 64 possible three-base codons codes for either an amino acid or a signal for the end of the coding portion of a gene (a stop codon). Amino acids shaded pink have nonpolar (hydrophobic) side chains; those shaded blue have polar or charged side chains. Those shaded mauve are amphipathic. Glycine has no side chain. Almost all of the amino acids can be specified by two or more different codons that differ only in the third position in the codon. Single-base changes elsewhere in the codon usually produce a different amino acid but with similar physical-chemical properties.

There is a linear relationship between the DNA base sequence of a gene and the amino-acid sequence of the protein it encodes

The **genetic code** is the formula that converts hereditary information from genes into proteins. Every amino acid in a protein is represented by a **codon** consisting of three consecutive **nucleotides** in the gene. DNA contains four different nucleotides, with the **bases** adenine (A), guanine (G), thymidine (T) and cytosine (C), whose sequence in a gene spells out the sequence of the amino acids in the protein that it specifies: this is the **primary structure** of the protein. The nucleotide sequence of the DNA is **transcribed** into **messenger RNA** (**mRNA**), with uridine (U) replacing thymine (T). Figure 1-4 shows the correspondence between the 64 possible three-base codons in mRNA and the 20 naturally occurring amino acids. Some amino acids are specified by only one codon, whereas others can be specified by as many as six different codons: the genetic code is **degenerate**. There are three codons that do not code for amino acids, but signal the termination of the polypeptide chain (**stop codons**). The process by which the nucleotide sequence of the DNA is first transcribed into RNA and then **translated** into protein is outlined in Figure 1-5.

In bacteria and other lower organisms, the relationship between the base sequence of the gene and the amino acid sequence of the corresponding protein is strictly linear: the protein sequence can be read directly from the gene sequence (Figure 1-5 left-hand side). In higher organisms, however, genes are typically segmented into coding regions (**exons**) that are interrupted by non-coding stretches (**introns**). These non-coding introns are transcribed into RNA, but are enzymatically excised from the resulting transcript (the **primary transcript**), and the exons are then spliced together to make the mature mRNA (Figure 1-5 right-hand side).

The process of intron removal and exon ligation has been exploited in the course of evolution through **alternative splicing**, in which exon segments as well as intron segments may be differentially excised from the primary transcript to give more than one mRNA and thus more than one protein. Depending on the arrangement of the introns, alternative splicing can lead to truncated proteins, proteins with different stretches of amino acids in the middle, or frameshifts in which the sequence of a large part of the protein is completely different from that specified by an in-frame reading of the gene sequence. Coding sequences can also be modified by **RNA editing**. In this process, some nucleotides are changed to others, and stretches of additional nucleotides can be inserted into the mRNA sequence before translation occurs. Modification of the coding sequences by RNA processing in these ways complicates the interpretation of genomic sequences in terms of protein structure, though this complication does not apply to cDNA sequences, which are artificially copied by reverse transcription from mRNA.

The organization of the genetic code reflects the chemical grouping of the amino acids

The amino acids fall into groups according to their physical-chemical properties (see Figure 1-3). The organization of the genetic code reflects this grouping, as illustrated in Figure 1-4. Note that single-base changes (**single-nucleotide polymorphism**) in the third position in a codon will often produce the same amino acid. Single-base changes elsewhere in the codon will usually produce a different amino acid, but with the same physical-chemical properties: for example, the second base specifies if the amino acid is polar or hydrophobic. Changes of this sort are known as **conservative substitutions** and when they are found in comparisons of protein sequences they are taken to indicate conservation of structure between two proteins. Examination of protein sequences for the same gene product over a large evolutionary distance

Definitions

alternative splicing: the selection of different coding sequences from a gene by the removal during RNA processing of portions of the RNA containing or affecting coding sequences.

base: the aromatic group attached to the sugar of a **nucleotide**.

codon: three consecutive **nucleotides** in a strand of DNA or RNA that represent either a particular amino acid or a signal to stop translating the transcript of the gene.

conservative substitution: replacement of one amino acid by another that has similar chemical and/or physical properties.

degenerate: having more than one **codon** for an amino acid.

exon: coding segment of a gene (compare **intron**).

genetic code: the relationship between each of the 64 possible three-letter combinations of A, U (or T), G and C and the 20 naturally occurring amino acids that make up proteins.

intron: noncoding DNA within a gene.

messenger RNA (mRNA): the RNA molecule transcribed from a gene after removal of **introns** and editing.

nucleotide: the basic repeating unit of a nucleic acid polymer. It consists of a **base** (A, U [in RNA, T in DNA], G or C), a sugar (ribose in RNA, deoxyribose in DNA) and a phosphate group.

primary structure: the amino-acid sequence of a polypeptide chain.

primary transcript: the RNA molecule directly

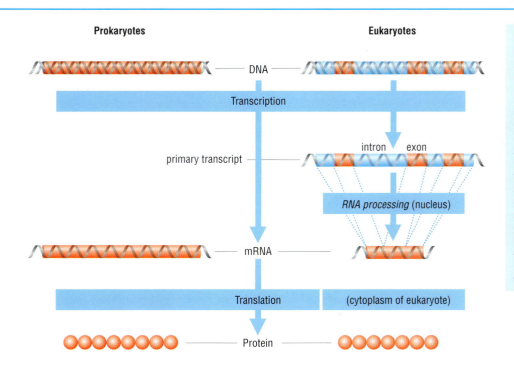

Prokaryotes   Eukaryotes

DNA
Transcription
intron   exon
primary transcript
RNA processing (nucleus)
mRNA
Translation   (cytoplasm of eukaryote)
Protein

Figure 1-5 The flow of genetic information in prokaryotes (left) and eukaryotes (right) The amino-acid sequences of proteins are coded in the base sequence of DNA. This information is transcribed into a complementary base sequence in messenger RNA (mRNA). In prokaryotes, the mRNA is generated directly from the DNA sequence (left-hand side of diagram). Eukaryotic genes (right-hand side) are often interrupted by one or more noncoding intervening segments called introns. These are transcribed along with the exons to produce a primary transcript, from which the introns iare excised in the nucleus and the coding segments, the exons, joined together to generate the mRNA. Finally, the mRNA base sequence is translated into the corresponding amino-acid sequence on the ribosome, a process that occurs in the cytoplasm of eukaryotic cells. (Diagram not to scale.)

illustrates this principle (Figure 1-6). An amino acid that is altered from one organism to another in a given position in the protein sequence is most often changed to a residue of similar physical-chemical properties, exactly as predicted by the organization of the code.

	Gly	Ala	Val	Leu	Ile	Met	Cys	Ser	Thr	Asn	Gln	Asp	Glu	Lys	Arg	His	Phe	Tyr	Trp	Pro
Gly																				
Ala	58																			
Val	10	37																		
Leu	2	10	30																	
Ile		7	66	25																
Met	1	3	8	21	6															
Cys	1	3	3		2															
Ser	45	77	4	3	2	2	12													
Thr	5	59	19	5	13	3	1	70												
Asn	16	11	1	4	4			43	17											
Gln	3	9	3	8	1	2		5	4	5										
Asp	16	15	2		1			10	6	53	8									
Glu	11	27	4	2	4	1		9	3	9	42	83								
Lys	6	6	2	4	4	9		17	20	32	15		10							
Arg	1	3	2	2	3	2	1	14	2	2	12	9		48						
His	1	2	3	4			1	3	1	23	24	4	2	2	10					
Phe	2	2	1	17	9	2		4	1	1					1	2				
Tyr		2	2	2	1		3	2	2	4				1	1		4	26		
Trp			1					2							3		1	1		
Pro	5	35	5	4	1		1	27	7	3	9	1	4	4	7	5	1			

Figure 1-6 Table of the frequency with which one amino acid is replaced by others in amino-acid sequences of the same protein from different organisms The larger the number, the more common a particular substitution. For example, glycine is commonly replaced by alanine and vice versa; this makes chemical sense because these are the amino acids with the smallest side chains. Similarly, aspartic acid and glutamic acid, the two negatively charged residues, frequently substitute for one another. There are some surprises: for example, serine and proline often substitute for each other, as do glutamic acid and alanine. Serine may substitute for proline because the side-chain OH can receive a hydrogen bond from its own main-chain NH, mimicking the fused ring of proline.

transcribed from a gene, before processing.

RNA editing: enzymatic modification of the RNA **base** sequence.

single-nucleotide polymorphism (SNP): a mutation of a single **base** in a **codon**.

stop codon: a **codon** that signals the end of the coding sequence and usually terminates **translation**.

transcription: the synthesis of RNA from the coding strand of DNA by DNA-dependent RNA polymerase.

translation: the transfer of genetic information from the sequence of **codons** in **mRNA** into a sequence of amino acids in a polypeptide chain.

References

Alberts, B. et al.: Molecular Biology of the Cell 4th ed. Chapters 3 and 6 (Garland, New York, 2002).

Argyle, E.: **A similarity ring for amino acids based on their evolutionary substitution rates.** Orig. Life 1980, **10**:357–360.

Dayhoff, M.O. et al.: **Establishing homologies in protein sequences.** Methods Enzymol. 1983, **91**:524–545.

Jones, D.T. et al.: **The rapid generation of mutation data matrices from protein sequences.** Comput. Appl. Biosci. 1992, **8**:275–282.

Topham, C.M. et al.: **Fragment ranking in modelling of protein structure. Conformationally constrained environmental amino acid substitution tables.** J. Mol. Biol. 1993, **229**:194–220.

1-3 The Peptide Bond

Proteins are linear polymers of amino acids connected by amide bonds

Amino acids are crucial components of living cells because they are easy to polymerize. α-Amino acids are preferable to β-amino acids because the latter are too flexible to form spontaneously folding polymers. The amino acids of a protein chain are covalently joined by amide bonds, often called **peptide bonds**: for this reason, proteins are also known as **polypeptides**. Proteins thus have a repeating **backbone** from which 20 different possible kinds of side chains protrude (see Figure 1-8). On rare occasions, nonstandard side chains are found. In plants, a significant number of unusual amino acids have been found in proteins. In mammals, however, they are largely confined to small hormones. Sometimes, post-translational modification of a conventional amino acid may convert it into a nonstandard one. Examples are the nonenzymatic carbamylation of lysine, which can produce a metal-ion ligand, thereby activating an enzyme; and the deamidation of asparagine, which alters protein stability and turnover rate.

Chemically, the peptide bond is a covalent bond that is formed between a carboxylic acid and an amino group by the loss of a water molecule (Figure 1-7). In the cell, the synthesis of peptide bonds is an enzymatically controlled process that occurs on the ribosome and is directed by the mRNA template. Although peptide bond formation can be reversed by the addition of water (**hydrolysis**), **amide bonds** are very stable in water at neutral pH, and the hydrolysis of peptide bonds in cells is also enzymatically controlled.

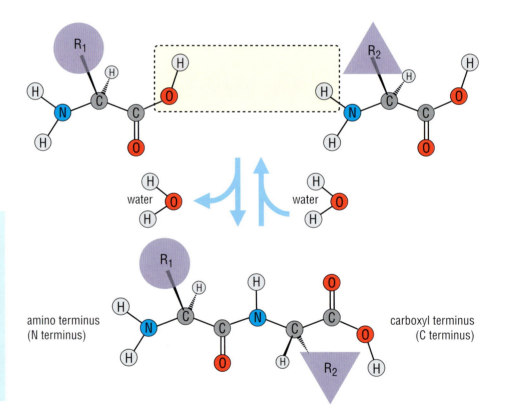

Figure 1-7 Peptide bond formation and hydrolysis Formation (top to bottom) and hydrolysis (bottom to top) of a peptide bond requires, conceptually, loss and addition, respectively, of a molecule of water. The actual chemical synthesis and hydrolysis of peptide bonds in the cell are enzymatically controlled processes that in the case of synthesis nearly always occurs on the ribosome and is directed by an mRNA template. The end of a polypeptide with the free amino group is known as the amino terminus (N terminus), that with the free carboxyl group as the carboxyl terminus (C terminus).

Definitions

amide bond: a chemical bond formed when a carboxylic acid condenses with an amino group with the expulsion of a water molecule.

backbone: the repeating portion of a **polypeptide** chain, consisting of the N–H group, the alpha-carbon C–H group, and the C=O of each amino-acid residue. Residues are linked to each other by means of **peptide bonds**.

dipole moment: an imaginary vector between two

separated charges that may be full or partial. Molecules or functional groups having a dipole moment are said to be polar.

hydrolysis: breaking a covalent bond by addition of a molecule of water.

peptide bond: another name for **amide bond**, a chemical bond formed when a carboxylic acid condenses with an amino group with the expulsion of a water molecule. The term peptide bond is used only when both groups come from amino acids.

phi torsion angle: see **torsion angle**.

polypeptide: a polymer of amino acids joined together by **peptide bonds**.

psi torsion angle: see **torsion angle**.

resonance: delocalization of bonding electrons over more than one chemical bond in a molecule. Resonance greatly increases the stability of a molecule. It can be represented, conceptually, as if the properties of the molecule were an average of several structures in which the chemical bonds differ.

The properties of the peptide bond have important effects on the stability and flexibility of polypeptide chains in water

The properties of the amide bond account for several important properties of polypeptide chains in water. The stability of the peptide bond, as well as other properties important for the behavior of polypeptides, is due to **resonance**, the delocalization of electrons over several atoms. Resonance has two other important consequences. First, it increases the polarity of the peptide bond: the **dipole moment** of each peptide bond is shown in Figure 1-8. The polarity of the peptide bond can make an important contribution to the behavior of folded proteins, as discussed later in section 1-6.

Second, because of resonance, the peptide bond has partial double-bond character, which means that the three non-hydrogen atoms that make up the bond (the carbonyl oxygen O, the carbonyl carbon C and the amide nitrogen N) are coplanar, and that free rotation about the bond is limited (Figure 1-9). The other two bonds in the basic repeating unit of the polypeptide backbone, the N–C$_\alpha$ and C$_\alpha$–C bonds (where C$_\alpha$ is the carbon atom to which the side chain is attached), are single bonds and free rotation is permitted about them provided there is no steric interference from, for example, the side chains. The angle of the N–C$_\alpha$ bond to the

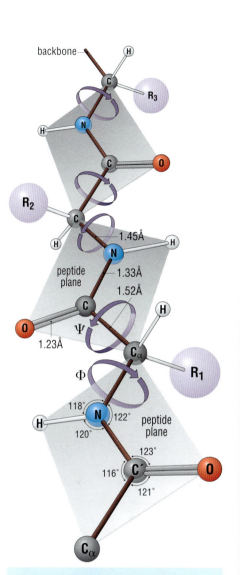

Figure 1-9 Extended polypeptide chain showing the typical backbone bond lengths and angles The planar peptide groups are indicated as shaded regions and the backbone torsion angles are indicated with circular arrows, with the phi and psi torsion angles marked. The omega torsion angle about the C–N peptide bond is usually restricted to values very close to 180° (*trans*), but can be close to 0° (*cis*) in rare cases. X–H bond lengths are all about 1 Å.

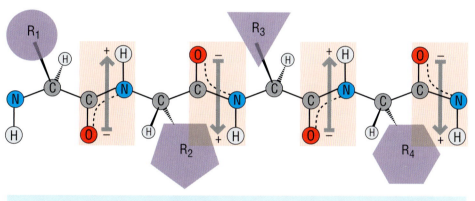
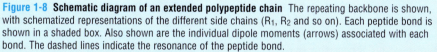

Figure 1-8 Schematic diagram of an extended polypeptide chain The repeating backbone is shown, with schematized representations of the different side chains (R$_1$, R$_2$ and so on). Each peptide bond is shown in a shaded box. Also shown are the individual dipole moments (arrows) associated with each bond. The dashed lines indicate the resonance of the peptide bond.

adjacent peptide bond is known as the **phi torsion angle**, and the angle of the C–C$_\alpha$ bond to the adjacent peptide bond is known as the **psi torsion angle** (see Figure 1-9). Thus a protein is an unusual kind of polymer, with rotatable covalent bonds alternating with rigid planar ones. This combination greatly restricts the number of possible conformations that a polypeptide chain can adopt and makes it possible to determine from simple steric considerations the most likely backbone conformation angles for polypeptide residues other than glycine.

torsion angle: the angle between two groups on either side of a rotatable chemical bond. If the bond is the C$_\alpha$–N bond of a peptide backbone the torsion angle is called **phi**. If the bond is the C$_\alpha$–C backbone bond, the angle is called **psi**.

References

Martin, R.B.: **Peptide bond characteristics**. *Met. Ions Biol. Syst.* 2001, **38**:1–23.

Pauling, L.C.: *The Nature of the Chemical Bond and the Structure of Molecules and Crystals* 3rd ed. Chapter 8 (Cornell Univ. Press, Ithaca, New York, 1960).

Voet, D. and Voet, J.G.: *Biochemistry* 2nd ed. (Wiley, New York, 1995), 67–68.

1-4 Bonds that Stabilize Folded Proteins

Folded proteins are stabilized mainly by weak noncovalent interactions

The amide bonds in the backbone are the only covalent bonds that hold the residues together in most proteins. In proteins that are secreted, or in the extracellular portions of cell-surface proteins, which are not exposed to the **reducing environment** in the interior of the cell, there may be additional covalent linkages present in the form of **disulfide bridges** between the side chains of cysteine residues. Except for cross-links like these, however, the remainder of the stabilization energy of a folded protein comes not from covalent bonds but from noncovalent weakly polar interactions. The properties of all the interactions that hold folded proteins together are listed in Figure 1-10. Weakly polar interactions depend on the electrostatic attraction between opposite charges. The charges may be permanent and full, or fluctuating and partial. In general, the term **electrostatic interaction** is reserved for those interactions due to full charges, and this convention is observed in Figure 1-10. But in principle, all polar interactions are electrostatic and the effect is the same: positively polarized species will associate with negatively polarized ones. Such interactions rarely contribute even one-tenth of the enthalpy contributed by a single covalent bond (see Figure 1-10), but in any folded protein structure there may be hundreds to thousands of them, adding up to a very large contribution. The two most important are the **van der Waals interaction** and the **hydrogen bond**.

Van der Waals interactions occur whenever the fluctuating electron clouds on an atom or group of bonded atoms induce an opposite fluctuating dipole on a non-bonded neighbor, resulting in a very weak electrostatic interaction. The effect is greatest with those groups that are the most polarizable; in proteins these are usually the methyl groups and methylene groups of hydrophobic side chains such as leucine and valine. Van der Waals interactions diminish rapidly as the interacting species get farther apart, so only atoms that are already close together (about 5 Å apart or less) have a chance to participate in such interactions. A given van der Waals interaction is extremely weak (see Figure 1-10), but in proteins they sum up to a substantial energetic contribution.

Hydrogen bonds are formed when a hydrogen atom has a significant partial positive charge by virtue of being covalently bound to a more electronegative atom, such as oxygen, and is attracted to a neighboring atom that has a significant partial negative charge (see Figure 1-10). This electrostatic interaction draws the two non-hydrogen atoms closer together than the sum of their atomic radii would normally allow. So, if two polar non-hydrogen atoms in a protein, one of which has a hydrogen attached, are found to be less than 3.5 Å apart, a hydrogen bond is assumed to exist between them. It is thought that the hydrogen-bonding effect is energetically most favorable if the three-atom system is roughly linear. The atom to which the hydrogen is covalently attached is called the **donor atom**; the non-bonded one is termed the **acceptor atom**. If the donor, the acceptor or both are fully charged, the hydrogen bond is stronger than when both are uncharged. When both the donor and acceptor are fully charged, the bonding energy is significantly higher and the hydrogen-bonded ion pair is called a **salt bridge** (see Figure 1-10).

The strengths of all polar weak interactions depend to some extent on their environment. In the case of hydrogen bonding, the strength of the interaction depends critically on whether the groups involved are exposed to water.

Figure 1-10 Table of the typical chemical interactions that stabilize polypeptides Values for the interatomic distances and free energies are approximate average values; both can vary considerably. Any specific number is highly dependent on the context in which the interaction is found. Therefore values such as these should only be taken as indicative of the approximate value.

Definitions

disulfide bridge: a covalent bond formed when the reduced –S–H groups of two cysteine residues react with one another to make an oxidized –S–S– linkage.

electrostatic interaction: noncovalent interaction between atoms or groups of atoms due to attraction of opposite charges.

hydrogen bond: a noncovalent interaction between the **donor atom**, which is bound to a positively polarized hydrogen atom, and the **acceptor atom**, which is negatively polarized. Though not covalent, the hydrogen bond holds the donor and acceptor atom close together.

reducing environment: a chemical environment in which the reduced states of chemical groups are favored. In a reducing environment, free –S–H groups are favored over –S–S– bridges. The interior of most cells is a highly reducing environment.

salt bridge: a **hydrogen bond** in which both donor and acceptor atoms are fully charged. The bonding energy of a salt bridge is significantly higher than that of a hydrogen bond in which only one participating atom is fully charged or in which both are partially charged.

van der Waals interaction: a weak attractive force between two atoms or groups of atoms, arising from the fluctuations in electron distribution around the nuclei. Van der Waals forces are stronger between less electronegative atoms such as those found in hydrophobic groups.

The hydrogen-bonding properties of water have important effects on protein stability

Water, which is present at 55 M concentration in all aqueous solutions, is potentially both a donor and an acceptor of hydrogen bonds. Water molecules hydrogen bond to one another, which is what makes water liquid at ordinary temperatures (a property of profound biological significance) and has important energetic consequences for the folding and stability of proteins. The ability of water molecules to hydrogen-bond to the polar groups of proteins has important effects on the energy, or strength, of the hydrogen bonds formed between such groups. This is most clearly seen by comparing hydrogen bonds made by polar groups on the surface and in the interior of proteins.

The strengths of polar weak interactions depend to some extent on their environment. A polar group on the surface of a protein can make interactions with water molecules that are nearly equivalent in energy to those it can make with other surface groups of a protein. Thus, the difference in energy between an isolated polar group and that of the same species when involved in a hydrogen bond with another polar group from that protein, is small. If, however, the interaction occurs in the interior of the protein, away from bulk solvent, the net interaction energy reflects the difference between the group when hydrogen-bonded and when not.

It is energetically very unfavorable not to make a hydrogen bond, because that would leave one or more uncompensated partial or full charges. Thus, in protein structure nearly all potential hydrogen-bond donors and acceptors are participating in such interactions, either between polar groups of the protein itself or with water molecules. In a polypeptide chain of indeterminate sequence the most common hydrogen-bond groups are the peptide C=O and N–H; in the interior of a protein these groups cannot make hydrogen bonds with water, so they tend to hydrogen bond with one another, leading to the secondary structure which stabilizes the folded state.

Chemical Interactions that Stabilize Polypeptides

Interaction	Example	Distance dependence	Typical distance	Free energy (bond dissociation enthalpies for the covalent bonds)
Covalent bond	$-C_\alpha-C-$	-	1.5 Å	356 kJ/mole (610 kJ/mole for a C=C bond)
Disulfide bond	$-Cys-S-S-Cys-$	-	2.2 Å	167 kJ/mole
Salt bridge		Donor (here N), and acceptor (here O) atoms <3.5 Å	2.8 Å	12.5–17 kJ/mole; may be as high as 30 kJ/mole for fully or partially buried salt bridges (see text), less if the salt bridge is external
Hydrogen bond	$N-H \cdots O=C$	Donor (here N), and acceptor (here O) atoms <3.5 Å	3.0 Å	2–6 kJ/mole in water; 12.5–21 kJ/mole if either donor or acceptor is charged
Long-range electrostatic interaction		Depends on dielectric constant of medium. Screened by water. $1/r$ dependence	Variable	Depends on distance and environment. Can be very strong in nonpolar region but very weak in water
Van der Waals interaction		Short range. Falls off rapidly beyond 4 Å separation. $1/r^6$ dependence	3.5 Å	4 kJ/mole (4–17 in protein interior) depending on the size of the group (for comparison, the average thermal energy of molecules at room temperature is 2.5 kJ/mole)

References

Burley, S.K. and Petsko, G.A.: **Weakly polar interactions in proteins.** *Adv. Prot. Chem.* 1988, **39**:125–189.

Dunitz, J.D.: **Win some, lose some: enthalpy-entropy compensation in weak intermolecular interactions.** *Chem. Biol.* 1995, **2**:709–712.

Fersht, A.R.: **The hydrogen bond in molecular recognition.** *Trends Biochem. Sci.* 1987, **12**:301–304.

Jaenicke, R.: **Stability and stabilization of globular proteins in solution.** *J. Biotechnol.* 2000, **79**:193–203.

Pauling, L.C.: *The Nature of the Chemical Bond and the Structure of Molecules and Crystals* 3rd ed. Chapter 8 (Cornell Univ. Press, Ithaca, New York, 1960).

Sharp, K.A. and Englander, S.W.: **How much is a stabilizing bond worth?** *Trends Biochem. Sci.* 1994, **19**:526–529.

Spearman, J. C.: *The Hydrogen Bond and Other Intermolecular Forces* (The Chemical Society, London, 1975).

1-5 Importance and Determinants of Secondary Structure

Folded proteins have segments of regular conformation

Although proteins are linear polymers, the structures of most proteins are not the random coils found for synthetic non-natural polymers. Most soluble proteins are globular and have a tightly packed core consisting primarily of hydrophobic amino acids. This observation can be explained by the tendency of hydrophobic groups to avoid contact with water and interact with one another. Another striking characteristic of folded polypeptide chains is that segments of the chain in nearly all proteins adopt conformations in which the phi and psi torsion angles of the backbone repeat in a regular pattern. These regular segments form the elements of the **secondary structure** of the protein. Three general types of secondary structure elements have been defined (see section 1-0): helices, of which the most common by far is the **alpha helix**; **beta sheets** (sometimes called **pleated sheets**), of which there are two forms, parallel and antiparallel; and beta turns, in which the chain is forced to reverse direction and which make the compact folding of the polypeptide chain possible.

Secondary structure contributes significantly to the stabilization of the overall protein fold. Helices and pleated sheets consist of extensive networks of hydrogen bonds in which many consecutive residues are involved as we shall see in the next two sections. The hydrogen bonding in these elements of structure provides much of the enthalpy of stabilization that allows the polar backbone groups to exist in the hydrophobic core of a folded protein.

The arrangement of secondary structure elements provides a convenient way of classifying types of folds

Prediction of the location of secondary structure elements from the amino-acid sequence alone is accurate to only about 70% (see section 1-8). Such prediction is sometimes useful because the pattern of secondary structure elements along the chain can be characteristic of certain overall protein folds. For example, a beta-sheet strand followed by an alpha helix, repeated eight times, usually signifies a type of fold called a TIM barrel. All TIM barrels known to date are enzymes, so recognition of a TIM-barrel fold in a sequence suggests that the protein has a catalytic function. However, it is a general rule that while classification of a protein may suggest function it cannot define it, and TIM barrels are known that catalyze many different reactions, so prediction of a more specific function cannot be made from recognition of the fold alone. Moreover, relatively few folds can be recognized in this way. Individual secondary structure elements are rarely associated with specific functions, although there are some interesting exceptions such as the binding of alpha helices in the major groove of DNA in two families of DNA-binding proteins.

Steric constraints dictate the possible types of secondary structure

The physical size of atoms and groups of atoms limits the possible phi and psi torsion angles (see Figure 1-9) that the backbone of a polypeptide chain can adopt without causing protruding groups like the carbonyl and side chains to bump into each other. These allowed values can be plotted on a phi, psi diagram called a **Ramachandran plot** (Figure 1-11). Two broad regions of phi, psi space are permitted by steric constraints: the regions that include the torsion angles of the right-handed alpha helix and of the extended beta or pleated sheet. Residues may have phi, psi values that lie outside the allowed regions in cases where the protein fold stabilizes a locally strained conformation.

Definitions

alpha helix: a coiled conformation, resembling a right-handed spiral staircase, for a stretch of consecutive amino acids in which the backbone –N–H group of every residue n donates a hydrogen bond to the C=O group of every residue n+4.

beta sheet: a **secondary structure** element formed by backbone hydrogen bonding between segments of extended polypeptide chain.

beta turn: a tight turn that reverses the direction of the

polypeptide chain, stabilized by one or more backbone hydrogen bonds. Changes in chain direction can also occur by loops, which are peptide chain segments with no regular conformations.

hairpin turn: another name for **beta turn**.

pleated sheet: another name for **beta sheet**.

Ramachandran plot: a two-dimensional plot of the values of the backbone torsion angles phi and psi, with allowed regions indicated for conformations where there is no steric interference. Ramachandran plots are

used as a diagnosis for accurate structures: when the phi and psi torsion angles of an experimentally determined protein structure are plotted on such a diagram, the observed values should fall predominantly in the allowed regions.

reverse turn: another name for **beta turn**.

secondary structure: folded segments of a polypeptide chain with repeating, characteristic phi, psi backbone torsion angles, that are stabilized by a regular pattern of hydrogen bonds between the peptide –N–H and C=O groups of different residues.

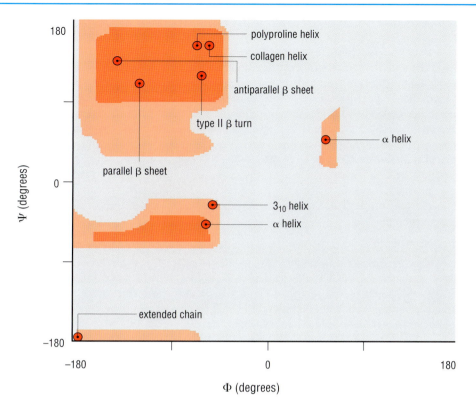

Figure 1-11 **Ramachandran plot** Shown in red are those combinations of the backbone torsion angles phi and psi (see Figure 1-9) that are "allowed" because they do not result in steric interference. The pink regions are allowed if some relaxation of steric hindrance is permitted. Common protein secondary structure elements are marked at the positions of their average phi, psi values. The isolated pink alpha-helical region on the right is actually for a left-handed helix, which is only rarely observed in short segments in proteins. The zero values of phi and psi are defined as the *trans* configuration.

The simplest secondary structure element is the beta turn

The simplest secondary structure element usually involves four residues but sometimes requires only three. It consists of a hydrogen bond between the carbonyl oxygen of one residue (n) and the amide N–H of residue n+3, reversing the direction of the chain (Figure 1-12). This pattern of hydrogen bonding cannot ordinarily continue because the turn is too tight. This tiny element of secondary structure is called a **beta turn** or **reverse turn** or, sometimes, a **hairpin turn** based on its shape. In a few cases, this interaction can be made between residue n and n+2, but such a turn is strained. Although the reverse turn represents a simple way to satisfy the hydrogen-bonding capability of a peptide group, inspection of this structure reveals that most of the C=O and N–H groups in the four residues that make up the turn are not making hydrogen bonds with other backbone atoms (Figure 1-12). Water molecules can donate and accept hydrogen bonds to these groups if the turn is not buried. Therefore, beta turns are found on the surfaces of folded proteins, where they are in contact with the aqueous environment, and by reversing the direction of the chain they can limit the size of the molecule and maintain a compact state.

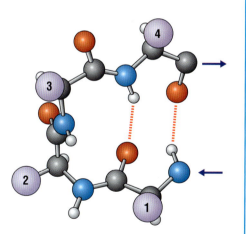

Figure 1-12 **Typical beta turn** Schematic diagram showing the interresidue backbone hydrogen bonds that stabilize the reversal of the chain direction. Side chains are depicted as large light-purple spheres. The tight geometry of the turn means that some residues, such as glycine, are found more commonly in turns than others.

References

Deane, C.M. *et al.*: **Carbonyl-carbonyl interactions stabilize the partially allowed Ramachandran conformations of asparagine and aspartic acid.** *Protein Eng.* 1999, **12**:1025–1028.

Mattos, C. *et al.*: **Analysis of two-residue turns in proteins.** *J. Mol. Biol.* 1994, **238**:733–747.

Ramachandran, G.N. *et al.*: **Stereochemistry of polypeptide chain configurations.** *J. Mol. Biol.* 1963, **7**:95–99.

Richardson, J.S. and Richardson, D.C.: **Principles and patterns of protein conformation** in *Prediction of Protein Structure and the Principles of Protein Conformation* 2nd ed. Fasman, G.D. ed. (Plenum Press, New York, 1990), 1–98.

Alpha helices are versatile cylindrical structures stabilized by a network of backbone hydrogen bonds

Alpha helices are the commonest secondary structural elements in a folded polypeptide chain, possibly because they are generated by local hydrogen bonding between C=O and N–H groups close together in the sequence. In an alpha helix, the carbonyl oxygen atom of each residue (n) accepts a hydrogen bond from the amide nitrogen four residues further along (n+4) in the sequence (Figure 1-13c), so that all of the polar amide groups in the helix are hydrogen bonded to one another except for the N–H group of the first residue in the helical segment (the amino-terminal end) and the C=O group of the last one (the carboxy-terminal end). The result is a cylindrical structure where the wall of the cylinder is formed by the hydrogen-bonded backbone, and the outside is studded with side chains. The protruding side chains determine the interactions of the alpha helix both with other parts of a folded protein chain and with other protein molecules.

The alpha helix is a compact structure, with approximate phi, psi values of –60° and –50° respectively: the distance between successive residues along the helical axis (translational rise) is only 1.5 Å (Figure 1-13a). It would take a helix 20 residues long to span a distance of 30 Å, the thickness of the hydrophobic portion of a **lipid bilayer** (alpha helices are common in the trans-membrane portions of proteins that span the lipid bilayer in cell membranes; see section 1-11). Alpha helices can be right-handed (clockwise spiral staircase) or left-handed (counterclockwise), but because all amino acids except glycine in proteins have the L-configuration, steric constraints favor the right-handed helix, as the Ramachandran plot indicates (see Figure 1-11), and only a turn or so of left-handed alpha helix has ever been observed in the structure of a real protein. There appears to be no practical limit to the length of an alpha helix; helices hundreds of Ångstroms long have been observed, such as in the keratin fibers that make up human hair. There are variants of the alpha helix with slightly different **helical parameters** (Figure 1-14), but they are much less common and are not very long because they are slightly less stable.

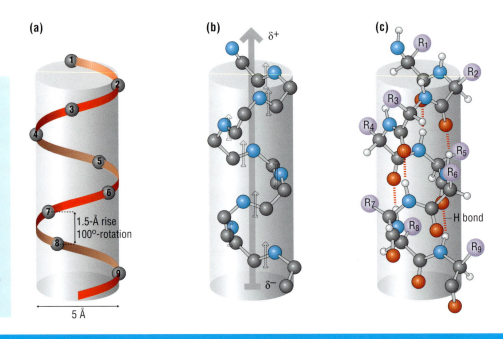

(a) **(b)** **(c)**

1.5-Å rise
100°-rotation

5 Å

–H bond

R_1 R_2 R_3 R_4 R_5 R_6 R_7 R_8 R_9

Figure 1-13 The alpha helix The chain path with average helical parameters is indicated showing **(a)** the alpha carbons only, **(b)** the backbone fold with peptide dipoles and **(c)** the full structure with backbone hydrogen bonds in red. All three chains run from top to bottom (that is, the amino-terminal end is at the top). Note that the individual peptide dipoles align to produce a macrodipole with its positive end at the amino-terminal end of the helix. Note also that the amino-terminal end has unsatisfied hydrogen-bond donors (N–H groups) whereas the carboxy-terminal end has unsatisfied hydrogen-bond acceptors (C=O groups). Usually a polar side chain is found at the end of the helix, making hydrogen bonds to these donors and acceptors; such a residue is called a helix cap.

Definitions

amphipathic alpha helix: an alpha helix with a hydrophilic side and a hydrophobic side.

helical parameters: set of numerical values that define the geometry of a helix. These include the number of residues per turn, the translational rise per residue, and the main-chain torsional angles.

helix dipole: the macrodipole that is thought to be formed by the cumulative effect of the individual peptide dipoles in an alpha helix. The positive end of

the dipole is at the beginning (amino terminus) of the helix; the negative end is at the carboxyl terminus of the helical rod.

lipid bilayer: the structure of cellular membranes, formed when two sheets of lipid molecules pack against each other with their hydrophobic tails forming the interior of the sandwich and their polar head-groups covering the outside.

References

Hol, W.G.: **The role of the alpha helix dipole in protein function and structure.** *Prog. Biophys. Mol. Biol.* 1985, **45**:149–195.

Pauling, L. *et al.*: **The structure of proteins: two hydrogen-bonded helical configurations of the polypeptide chain.** *Proc. Natl Acad. Sci. USA* 1951, **37**:205–211.

Scott, J.E.: **Molecules for strength and shape.** *Trends Biochem. Sci.* 1987, **12**:318–321.

Average Conformational Parameters of Helical Elements

Conformation	Phi	Psi	Omega	Residues per turn	Translation per residue
Alpha helix	−57	−47	180	3.6	1.5
3-10 helix	−49	−26	180	3.0	2.0
Pi-helix	57	−70	180	4.4	1.15
Polyproline I	−83	+158	0	3.33	1.9
Polyproline II	−78	+149	180	3.0	3.12
Polyproline III	−80	+150	180	3.0	3.1

Figure 1-14 Table of helical parameters
Average conformational parameters of the most commonly found helical secondary structure elements.

In a randomly coiled polypeptide chain the dipole moments of the individual backbone amide groups point in random directions, but in an alpha helix the hydrogen-bonding pattern causes all of the amides—and their dipole moments—to point in the same direction, roughly parallel to the helical axis (Figure 1-13b). It is thought that, as a result, the individual peptide dipoles in a helix add to make a macrodipole with the amino-terminal end of the helix polarized positively and the carboxy-terminal end polarized negatively. The magnitude of this **helix dipole** should increase with increasing length of the helix, provided the cylinder remains straight. Because favorable electrostatic interactions could be made between oppositely charged species and the ends of the helix dipole, one might expect to find, at frequencies greater than predicted by chance, negatively charged side chains and bound anions at the amino-terminal ends of helices, and positively charged side chains and cations interacting with the carboxy-terminal ends. Experimentally determined protein structures and studies of model peptides are in accord with these predictions. Indeed, the helix dipole in some cases contributes significantly to the binding of small charged molecules by proteins.

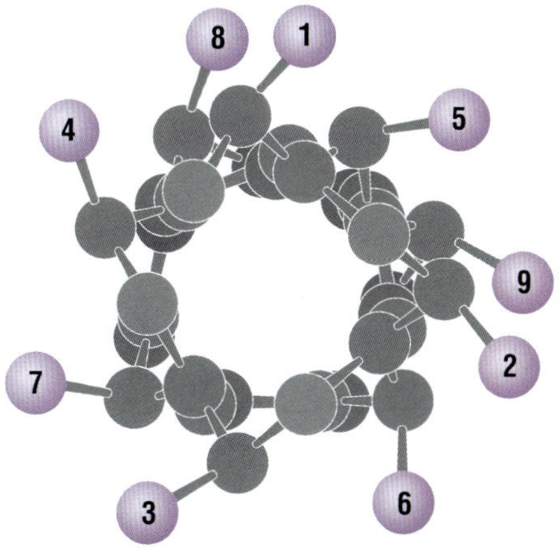

Alpha helices can be amphipathic, with one polar and one nonpolar face

The alpha helix has 3.6 residues per turn, corresponding to a rotation of 100° per residue, so that side chains project out from the helical axis at 100° intervals, as illustrated in Figure 1-15, which shows the view down the helix axis. This periodicity means that, broadly speaking, residues 3-4 amino acids apart in the linear sequence will project from the same face of an alpha helix. In many alpha helices, polar and hydrophobic residues are distributed 3-4 residues apart in the sequence, to produce an alpha helix with one hydrophilic face and one hydrophobic face; such a helix is known as an **amphipathic alpha helix**, which can stabilize helix–helix packing. Helices with this character frequently occur on the surfaces of proteins, where their polar faces are in contact with water, or at interfaces where polar residues interact with one another: the distribution of polar and hydrophobic residues in a sequence is therefore useful in positioning alpha helices in structure prediction, and in predicting their positions at interfaces.

Figure 1-15 View along the axis of an idealized alpha-helical polypeptide The view is from the amino-terminal end. Side chains project outward from the helical axis at 100° intervals. Note that side chains four residues apart in the sequence tend to cluster on the same face of the helix, for shorter helices. For long helices any such pattern would slowly coil about the helix axis, so if two long helices had a pattern of hydrophobic groups four residues apart they would interact by forming a coiled coil (see Figure 1-67).

Collagen and polyproline helices have special properties

Although the amino acid proline, which lacks an N–H group, is not frequently found in an alpha helix, two interesting helical structures can be formed from sequences rich in proline residues. The first is the collagen triple helix (Figure 1-16). Collagen is the main constituent of the bones, tendons, ligaments and blood vessels of higher organisms and consists of a repeating tripeptide in which every third residue is a glycine (GlyXY)n. X and Y are usually proline residues, although lysine occurs sometimes. Many of the proline residues are hydroxylated post-translationally. Each collagen strand forms a (left-handed) helical conformation and three such strands coil around each other like those of a rope. The effect is to create a fibrous protein of great tensile strength. Collagen molecules more than 2 μm in length have been observed. Denaturing the collagen triple helix by heating converts it to a disordered, dissociated, random mass that we call gelatin.

The second proline-rich conformation is that formed by polyproline sequences. When the peptide bonds in a polyproline sequence are all *trans* it forms a left-handed helix with three residues per turn. Such a conformation is easily recognized by other proteins, and helical polyproline sequences often serve as docking sites for protein recognition modules, such as SH3 domains in signal transduction pathways.

collagen triple helix

Figure 1-16 The structure of collagen Collagen is a three-chain fibrous protein in which each chain winds round the others. The rise per residue is much larger than in an alpha helix.

Beta sheets are extended structures that sometimes form barrels

In contrast to the alpha helix, the beta pleated sheet, whose name derives from the corrugated appearance of the extended polypeptide chain (Figure 1-17), involves hydrogen bonds between backbone groups from residues distant from each other in the linear sequence. In beta sheets, two or more strands that may be widely separated in the protein sequence are arranged side by side, with hydrogen bonds between the strands (Figure 1-17). The strands can run in the same direction (**parallel beta sheet**) or **antiparallel** to one another; **mixed sheets** with both parallel and antiparallel strands are also possible (Figure 1-17).

Nearly all polar amide groups are hydrogen bonded to one another in a beta-sheet structure, except for the N–H and C=O groups on the outer sides of the two edge strands. Edge strands may make hydrogen bonds in any of several ways. They may simply make hydrogen bonds to water, if they are exposed to solvent; or they may pack against polar side chains in, for example, a neighboring alpha helix; or they may make hydrogen bonds to an edge strand in another protein chain, forming an extended beta structure that spans more than one subunit and thereby stabilizes quaternary structure (Figure 1-18). Or the sheet may curve round on itself to form a barrel structure, with the two edge strands hydrogen bonding to one another to complete the closed cylinder (Figure 1-19). Such **beta barrels** are a common feature of protein architecture.

Parallel sheets are always buried and small parallel sheets almost never occur. Antiparallel sheets by contrast are frequently exposed to the aqueous environment on one face. These observations suggest that antiparallel sheets are more stable, which is consistent with their hydrogen bonds being more linear (see Figure 1-17). Silk, which is notoriously strong, is made up of stacks of antiparallel beta sheets. Antiparallel sheets most commonly have beta turns connecting the strands, although sometimes the strands may come from discontiguous regions of the linear sequence, in which case the connections are more complex and may include segments of alpha

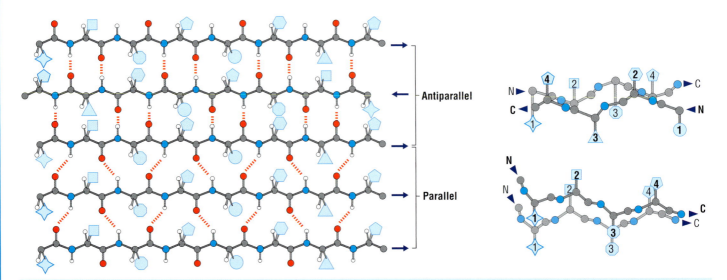

Figure 1-17 The structure of the beta sheet The left figure shows a mixed beta sheet, that is one containing both parallel and antiparallel segments. Note that the hydrogen bonds are more linear in the antiparallel sheet. On the right are edge-on views of antiparallel (top) and parallel sheets (bottom). The corrugated appearance gives rise to the name "pleated sheet" for these elements of secondary structure. Consecutive side chains, indicated here as numbered geometric symbols, point from alternate faces of both types of sheet.

Definitions

antiparallel beta sheet: a beta sheet, often formed from contiguous regions of the polypeptide chain, in which each strand runs in the opposite direction from its immediate neighbors.

beta barrel: a beta sheet in which the last strand is hydrogen bonded to the first strand, forming a closed cylinder.

mixed beta sheet: beta sheet containing both parallel and antiparallel strands.

parallel beta sheet: a beta sheet, formed from non-contiguous regions of the polypeptide chain, in which every strand runs in the same direction.

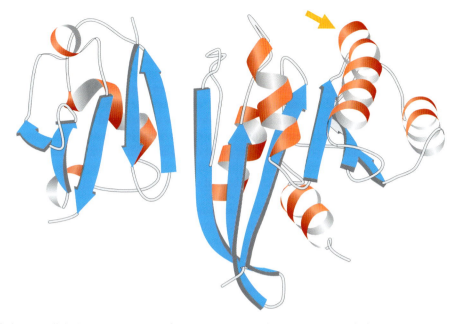

Figure 1-18 Two proteins that form a complex through hydrogen bonding between beta strands (the Rap–Raf complex, PDB 1gua) Two antiparallel edge strands of individual beta sheets hydrogen bond to each other at the protein–protein interface, forming a continuous mixed sheet that stabilizes the complex. The protein on the right contains a parallel beta sheet where each strand is connected to the next by an alpha helix, such as the one indicated with the yellow arrow. These helices pack against the faces of the sheet.

helix. Parallel sheet strands are of necessity always discontiguous, and the most common connection between them is an alpha helix that packs against a face of the beta sheet (see for example the helix indicated in Figure 1-18).

The polypeptide chain in a beta sheet is almost fully extended. The distance between consecutive residues is 3.3 Å and the phi and psi angles for peptides in beta sheets are approximately –130° and +125° respectively. Beta strands usually have a pronounced right-handed twist (see for example the sheets in Figure 1-19), due to steric effects arising from the L-amino acid configuration. Parallel strands are less twisted than antiparallel ones. The effect of the strand twist is that sheets consisting of several long strands are themselves twisted.

Because the polypeptide chain in a beta sheet is extended, amino-acid side chains such as those of valine and isoleucine, which branch at the beta carbon, can be accommodated more easily in a beta structure than in a tightly coiled alpha helix where side chains are crowded more closely together. Although unbranched side chains can fit in beta structures as well, branched side chains appear to provide closer packing so they are found more frequently in sheets than other residues. However, it is generally easier to identify helical stretches in sequences than to identify sections of beta structure, and locating the ends of beta sections from sequence alone is particularly difficult.

Amphipathic beta sheets are found on the surfaces of proteins

Like alpha helices (see section 1-6), beta strands can be amphipathic. Because nearly all peptide bonds are *trans* (that is, the C=O and N–H groups point in opposite directions to avoid collision between them), as one proceeds along a beta strand the side chains point in opposite directions (see Figure 1-17). Thus, a stretch of sequence with alternating hydrophobic and hydrophilic residues could have one hydrophobic and one hydrophilic face, forming an amphipathic beta strand (or, depending on its length, several strands of amphipathic antiparallel sheet). Such strands and sheets are found on the surface of proteins.

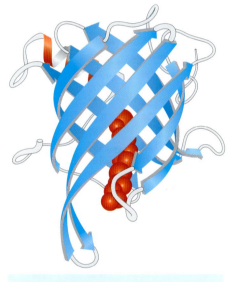

Figure 1-19 Beta barrel In this retinol-binding protein (PDB 1rlb), a large antiparallel beta sheet curves all the way around so that the last strand is hydrogen bonded to the first, forming a closed cylinder. The interior of this beta barrel is lined with hydrophobic side chains; nonpolar molecules such as retinol (shown in red) can bind inside.

References

Gellman, S.H.: **Minimal model systems for beta sheet secondary structure in proteins.** *Curr. Opin. Chem. Biol.* 1998, **2**:717–725.

Pauling, L. and Corey, R.B.: **Configurations of polypeptide chains with favored orientations around single bonds: two new pleated sheets.** *Proc. Natl Acad. Sci. USA* 1951, **37**:729–740.

Richardson, J.S.: **The anatomy and taxonomy of protein structure.** *Adv. Prot. Chem.* 1981, **34**:167–339.

Tutorial for using Protein Explorer to view structures:

http://www.umass.edu/microbio/chime/explorer/pe_tut.htm

1-9 Folding

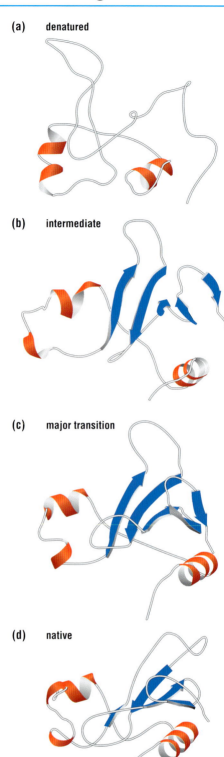

(a) denatured

(b) intermediate

(c) major transition

(d) native

The folded structure of a protein is directly determined by its primary structure

The three-dimensional or tertiary structure of a protein is determined by the sequence of amino acids encoded by the gene that specifies the protein. Translation of the mRNA produces a linear polymer of amino acids that usually folds spontaneously into a more compact, stable structure. Sometimes folding is assisted by other proteins called **chaperones**, but most proteins can be unfolded and refolded in dilute solution, demonstrating that the primary structure contains all the information necessary to specify the folded state. Protein folding can occur quite rapidly, but there is evidence that one or more partially folded intermediate states often exist, transiently, along the path to the final structure (Figure 1-22). The structures of these intermediates are not as well characterized as the native structures, but have many of the secondary structure elements of the fully folded protein without the closely packed interior and full complement of weak interactions that characterize what is termed the **native state**.

Competition between self-interactions and interactions with water drives protein folding

Consider a protein of arbitrary sequence emerging from the ribosome. If the chain is made up of only polar and charged amino acids, nearly every chemical group in it can hydrogen bond to water whether the chain is folded up or not, so there will be no driving force to form a compact or regular structure. Many such sequences are known in nature, and, as expected, they have no stable folded structure on their own in solution. The amino-acid sequences of soluble proteins tend to be mixtures of polar and nonpolar residues, sometimes in patches, but most often distributed along the chain with no discernible pattern. When such a sequence is synthesized in water, it cannot remain as a fully extended polymer. True, the polar and charged side chains, and the polar peptide groups, will be able to form hydrogen bonds with water; but the nonpolar side chains cannot. Their physical presence will disrupt the hydrogen-bonded structure of water without making any compensating hydrogen bonds with the solvent. To minimize this effect on the water structure, these side chains will tend to clump together the way oil droplets do when dispersed in water. This **hydrophobic effect**—the clustering of hydrophobic side chains from diverse parts of the polypeptide sequence—causes the polypeptide to become compact (Figure 1-23). From an energetic point of view the compactness produces two favorable results: it minimizes the total hydrophobic surface area in contact with water, and it brings the polarizable hydrophobic groups close to each other, allowing van der Waals interactions between them. Polar side chains do not need to be shielded from the solvent because they can hydrogen bond to water, so they will tend to be distributed on the outside of this "oil drop" of hydrophobic residues.

Figure 1-22 Folding intermediates Structures of **(a)** denatured, **(b)** intermediate, **(c)** major transition and **(d)** native states of barnase. Structures were determined from molecular dynamics calculations and NMR experiments, illustrating a possible folding pathway. Note that during folding, segments of secondary structure form that do not completely coincide with their final positions in the sequence, and that the non-native states are considerably expanded and more flexible relative to the final folded form. These characteristics appear to be common to most, if not all, protein-folding pathways. (Bond, C.J. *et al.*: *Proc. Natl Acad. Sci. USA* 1997, **94**:13409–13413.)

Definitions

chaperone: a protein that aids in the folding of another protein by preventing the unwanted association of the unfolded or partially folded forms of that protein with itself or with others.

hydrophobic effect: the tendency of nonpolar groups in water to self-associate and thereby minimize their contact surface area with the polar solvent.

native state: the stably folded and functional form of a biological macromolecule.

But this hydrophobic collapse has a negative consequence. When the nonpolar side chains come together to form a hydrophobic core, they simultaneously drag their polar backbone amide groups into the greasy interior of the protein (see Figure 1-23). These polar groups made hydrogen bonds to water when the chain was extended, but now they are unable to do so. Leaving these groups with unsatisfied hydrogen bonds would lead to a significant energy penalty. Yet they cannot make hydrogen bonds to polar side chains because most such side chains are on the surface interacting with water. The result is that most peptide N–H and carbonyl groups of folded proteins hydrogen bond to each other. It is this tendency of the amide groups of polypeptide chains to satisfy their hydrogen-bonding potential through self-interactions that gives rise to secondary structure, as described in section 1-4.

Although most hydrophobic side chains in a protein are buried, some are found on the surface of the folded polypeptide chain in contact with water. Presumably, this unfavorable situation is offset by the many favorable interactions that provide a net stability for the folded protein. As a rule, such residues occur in isolation; when hydrophobic side chains cluster on the surface they are usually part of a specific binding site for other molecules, or form a patch of mutually interacting nonpolar groups.

Computational prediction of folding is not yet reliable

Recently there have been a number of efforts to fold amino-acid sequences into the correct three-dimensional structures *ab initio* purely computationally. Such methods vary in detail (for example, some start with secondary structure prediction, others do not) but in the end all depend on several assumptions. First, it is assumed that the equilibrium conformation is the global free-energy minimum on a folding pathway. This assumption is likely to be correct, but no one knows for sure. Second, it is assumed that the current empirical potential energy parameters used to compute the contributions of hydrogen bonds, van der Waals interactions and so forth to the overall stabilization energy are sufficiently accurate. This is uncertain. Third, very many globular proteins are oligomeric, and many will not fold as monomers; but only monomeric proteins are treated by these methods. The problem of recognizing that a given sequence will produce a dimeric or tetrameric protein, and how to treat oligomerization in computational approaches to folding, has not even begun to be addressed.

Helical membrane proteins may fold by condensation of preformed secondary structure elements in the bilayer

The hydrophobic environment of a membrane interior allows formation of the same secondary structure elements as does aqueous solution, but the range of protein architectures appears to be much more limited. Thus far, only all-helical and all-beta-barrel integral membrane proteins have been observed (see section 1-11).

Much less is known about the mechanism of folding of proteins whose structure is largely embedded in the hydrophobic interior of a lipid bilayer. Because water molecules do not occupy stable positions in this region of a membrane, the polar N–H and C=O groups of a peptide backbone have no option but to hydrogen bond to one another. Thus, it is thought that transmembrane segments of integral membrane proteins form secondary structure (usually alpha helices) very early in the folding process, and that these elements then assemble to give the final structure by diffusional motion in the bilayer until the most favorable set of side-chain interactions is found. The folding pathway of all-beta-sheet membrane proteins is unclear.

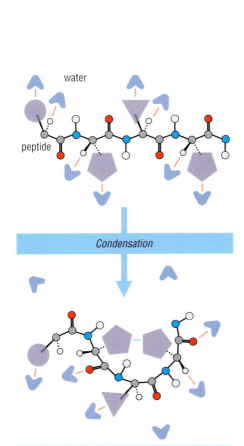

Figure 1-23 Highly simplified schematic representation of the folding of a polypeptide chain in water In the unfolded chain, side-chain and main-chain groups interact primarily with water, even if they are hydrophobic and the interaction is unfavorable. Burying the hydrophobic groups in the interior of a compact structure enables them to interact with each other (blue line), which is favorable, and leaves polar side chains on the surface where they can interact with water (red lines). The polar backbone groups that are buried along with the hydrophobic side chains must make hydrogen bonds to each other (not shown), as bulk water is no longer available.

References

Bond, C.J. *et al.*: **Characterization of residual structure in the thermally denatured state of barnase by simulation and experiment: description of the folding pathway.** *Proc. Natl Acad. Sci. USA* 1997, **94**:13409–13413.

Cramer, W.A. *et al.*: **Forces involved in the assembly and stabilization of membrane proteins.** *FASEB J.* 1992, **6**:3397–3402.

Dinner, A.R. *et al.*: **Understanding protein folding via free-energy surfaces from theory and experiment.** *Trends Biochem. Sci.* 2000, **25**:331–339.

Eaton, W.A. *et al.*: **Fast kinetics and mechanisms in protein folding.** *Annu. Rev. Biophys. Biomol. Struct.* 2000, **29**:327–359.

Fersht, A.: *Structure and Mechanism in Protein Science. A Guide to Enzyme Catalysis and Protein Folding* (Freeman, New York, 1999).

The condensing of multiple secondary structural elements leads to tertiary structure

In a folded protein, the secondary structure elements fold into a compact and nearly solid object stabilized by weak interactions involving both polar and nonpolar groups. The resulting compact folded form is called the **tertiary structure** of the protein. Because tertiary structure is not regular, it is hard to describe it simply. One way to characterize tertiary structure is by the topological arrangement of the various secondary structure elements as they pack together. In fact, the same types of secondary structure elements can come together in many different ways depending on the sequence (Figure 1-24). Tertiary structure is sometimes classified according to the arrangement of secondary structure elements in the linear sequence and in space. One effect of tertiary structure is to create a complex surface topography that enables a protein to interact specifically either with small molecules that may bind in clefts, or with other macromolecules, with which it may have regions of complementary topology and charge. These recognition sites are often formed from the stretches of amino acids joining secondary structure elements.

Figure 1-24 Comparison of the structures of triosephosphate isomerase and dihydrofolate reductase Two proteins with similar secondary structure elements but different tertiary structures. Approximately the same secondary structure elements can be arranged in more than one way. Both TIM (left) and DHFR (right) consist of eight beta strands with connecting alpha helices, yet the former is a singly wound parallel alpha/beta barrel whereas the latter is a doubly wound alpha/beta domain with a mixed sheet. (PDB 1tim and 1ai9)

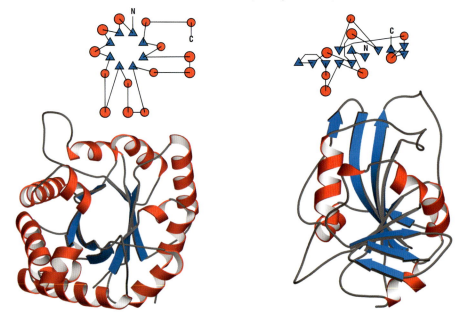

Although helical segments and beta strands are often connected by tight turns, more often there are long stretches of amino acids in between secondary structural elements that do not adopt regular backbone conformations. Such loops are found at the surface of proteins and typically protrude into the solvent. Consequently they provide convenient sites for protein recognition, ligand binding and membrane interaction. For example, the antigen-binding site in immunoglobulins is made up of a series of loops that project up from the core beta structure like the fingers of a cupped hand (Figure 1-25). Because these protruding loops often contribute little to the stabilization of the overall fold, they can tolerate mutations more readily than can the core of the protein. Since

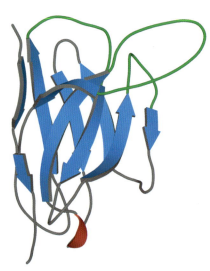

Figure 1-25 Variable loops Three-dimensional structure of the V domain of an immunoglobulin light chain showing the hypervariable loops (green) protruding from the ends of a sandwich formed by two antiparallel beta sheets. The structure resembles a cupped hand, with the hypervariable loops forming the fingers. These loops form the antigen-binding site. (PDB 1ogp)

Definitions

packing motif: an arrangement of secondary structure elements defined by the number and types of such elements and the angles between them. The term motif is used in structural biology in a number of contexts and thus can be confusing.

tertiary structure: the folded conformation of a protein, formed by the condensation of the various secondary elements, stabilized by a large number of weak interactions.

References

Barlow, D.J. and Thornton, J.M.: **Helix geometry in proteins.** *J. Mol. Biol.* 1988, **201**:601–619.

Eilers, M. *et al.*: **Internal packing of helical membrane proteins.** *Proc. Natl Acad. Sci. USA* 2000, **97**:5796–5801.

Lesk, A.M. and Chothia, C.: **Solvent accessibility, protein surfaces and protein folding.** *Biophys. J.* 1980, **32**:35–47.

Richards, F.M. and Richmond, T.: **Solvents, interfaces and protein structure.** *Ciba. Found. Symp.* 1997, **60**:23–45.

Rose, G.D. and Roy, S.: **Hydrophobic basis of packing in globular proteins.** *Proc. Natl Acad. Sci. USA* 1980, **77**:4643–4647.

Walther, D. *et al.*: **Principles of helix-helix packing in proteins: the helical lattice superposition model.** *J. Mol. Biol.* 1996, **255**:536–553.

they are also often involved in function, their mutability provides a mechanism for the evolution of new functions. Although surface loops are often drawn as being open, like a lariat, in reality their side chains frequently pack together so that the loop is nearly solid. This means that when loops undergo conformational changes they often move as rigid bodies.

Bound water molecules on the surface of a folded protein are an important part of the structure

When the polar backbone groups of a polypeptide chain become involved in secondary and tertiary structure interactions, the water molecules that were interacting with them in the unfolded protein are freed to rejoin the structure of liquid water. But there are many polar groups, both backbone and side-chain, on the surface of a folded protein that must remain in contact with water. Atomic-resolution structures of proteins show a layer of bound water molecules on the surfaces of all folded soluble proteins (Figure 1-26); these waters are making hydrogen bonds with polar backbone and side-chain groups and also with one another. There are several such water molecules per residue. Some are in fixed positions and are observed every time the structure is determined. However, others are in non-unique positions and reflect an ensemble of water–protein interactions that hydrate the entire surface. A few additional water molecules are trapped inside the protein in internal cavities. Because bound water molecules make important interactions with groups that would otherwise make none, the waters in fixed positions should be considered as part of the tertiary structure, and any detailed structure description that does not include them is incomplete.

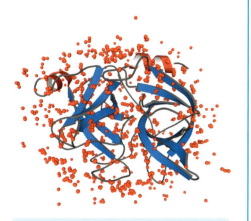

Figure 1-26 Porcine pancreatic elastase showing the first hydration shell surrounding the protein In any one structure determination, only a subset of these water molecules is seen. This picture is a composite of the results of parallel structure determinations of the same protein.

Tertiary structure is stabilized by efficient packing of atoms in the protein interior

The individual secondary structure elements in a protein pack together in part to bury the hydrophobic side chains, forming a compact molecule with very little empty space in the interior (Figure 1-27). The interactions that hold these elements together are the weak interactions described earlier: polar interactions between hydrophilic groups and van der Waals interaction between nonpolar groups. Close packing of atoms maximizes both the probability that these interactions will occur and their strength.

Maintaining a close-packed interior can be accomplished by many different modes of packing of helices with each other and of sheets with each other, and between helices and sheets. These various types of packing arrangements can be described in terms of a set of **packing motifs** that have been used to classify protein tertiary structures in general terms. For example, in helix–helix interactions, the protruding side chains of one helix fit into grooves along the cylindrical surface of the other helix in what has been described as a "ridges and grooves" arrangement. This principle is best illustrated in alpha-helical dimers and we return to it later (see Figure 1-67). These steric considerations permit several different interhelical crossing angles, each set of which constitutes a distinct packing motif (Figure 1-28).

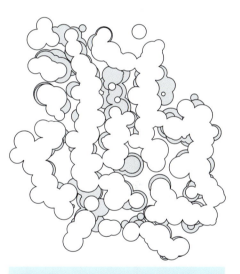

Figure 1-27 Cut-away view of the interior of a folded protein The atoms in the interior of a protein are packed almost as closely as in a solid. Note that there are a few cavities and small channels in some parts of the structure. These packing defects provide room for neighboring atoms to move, allowing the structure to have some flexibility.

Although the density of atoms in the hydrophobic core of a folded protein is high, the packing is not perfect. There are many cavities that range in size from subatomic packing defects to ones large enough to accommodate several water molecules. If the cavity walls are lined with hydrophobic side chains, the cavity is usually found to contain no ordered waters, but more commonly there are some polar groups lining the cavity and these interact with buried water molecules that fill the space.

(a) **(b)**

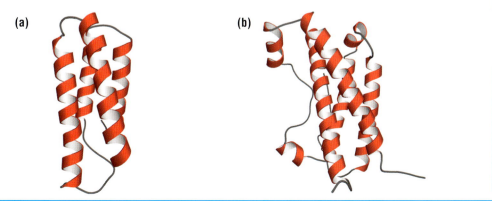

Figure 1-28 Packing motifs of a helical structure When two helices pack together, their side chains interdigitate. Because several interhelical crossing angles allow good interdigitation, a number of distinct arrangements of helical bundles are possible. The two examples illustrated here are **(a)** cytochrome b562 (PDB 256b) and **(b)** human growth hormone (PDB 3hhr).

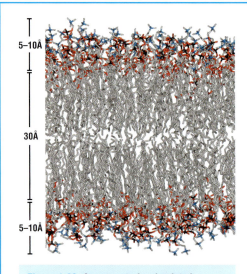

Figure 1-29 A segment of a simulated membrane bilayer
http://www.lrz-muenchen.de/~heller/membrane/membrane.html

The principles governing the structures of integral membrane proteins are the same as those for water-soluble proteins and lead to formation of the same secondary structure elements

Not all proteins in the cell exist in an aqueous environment. Some are embedded in the hydrophobic interior of the membranes that form the surfaces of cells, organelles and vesicles. Most biological membranes are bilayers of lipid molecules (derived from fatty acids) with polar or charged head-groups (Figure 1-29). The bilayer resembles a sandwich with the head-groups as the bread and the lipid tails as an almost completely hydrophobic filling. The nonpolar interior of the membrane is approximately 30 Å across; the head-group layers contribute an additional 5–10 Å on each side to the total thickness of the membrane.

A protein that is inserted into a membrane is exposed to an almost completely nonpolar environment. The side chains of amino acids forming transmembrane segments of proteins are usually hydrophobic, and can be accommodated with no energetic cost; but the polar backbone carbonyl and amide groups will all have unfavorable interactions with the nonpolar lipid tails. There will thus be the same strong driving force for these groups to hydrogen bond with one another as there is in the hydrophobic interior of a soluble protein when it folds up in water, and with the same results. Formation of alpha-helical and beta-sheet secondary structure elements is thus strongly favored in the membrane interior. Because hydrogen bonds in a completely nonpolar environment are considered stronger than if the same groups were exposed to solvent, an isolated alpha helix can exist stably in a membrane, whereas non-interacting helices are rare in water-soluble proteins. Any polar side chains will be found either on the protein surface that protrudes out of the membrane, interacting with the polar head-groups of the lipids, or in the core of the membrane-embedded part of the protein, where they can interact with each other or form a polar surface that often constitutes a pore or ion channel through the bilayer.

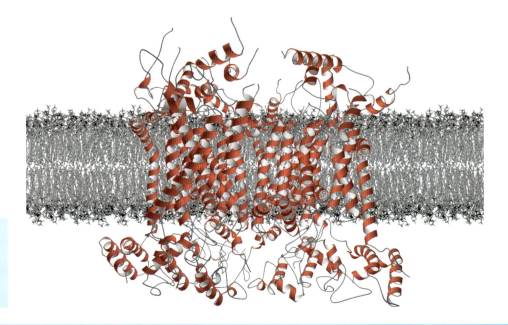

Figure 1-30 The three-dimensional structure of part of the cytochrome bc1 complex The protein (PDB 1bgy) is shown with a simulated lipid bilayer showing the transmembrane parts and some of the cytosolic segments both above and below the membrane.

References

Doyle, D.A. *et al.*: **The structure of the potassium channel: molecular basis of K+ conduction and selectivity.** *Science* 1998, **280**:69–77.

Ferguson, A.D. *et al.*: **Siderophore-mediated iron transport: crystal structure of FhuA with bound lipopolysaccharide.** *Science* 1998, **282**:2215–2220.

Heller, H. *et al.*: **Molecular dynamics simulation of a bilayer of 200 lipids in the gel and in the liquid crystal phases.** *J. Phys. Chem.* 1993, **97**:8343–8360.

Koebnik, R. *et al.*: **Structure and function of bacterial outer membrane proteins: barrels in a nutshell.** *Mol. Microbiol.* 2000, **37**:239–253.

Kyte, J. and Doolittle, R.F.: **A simple method for displaying the hydropathic character of a protein.** *J. Mol. Biol.* 1982, **157**:105–132.

Popot, J.L. and Engelman, D.M.: **Helical membrane protein folding, stability and evolution.** *Annu. Rev. Biochem.* 2000, **69**:881–922.

Popot, J.L. and Engelman, D.M.: **Membrane protein folding and oligomerization: the two-stage model.** *Biochemistry* 1990, **29**:4031–4037.

von Heijne, G.: **Recent advances in the understanding of membrane protein assembly and structure.** *Q. Rev. Biophys.* 1999, **32**:285–307.

Xia, D. *et al.*: **Crystal structure of the cytochrome bc1 complex from bovine heart mitochondria.** *Science* 1997, **277**:60–66.

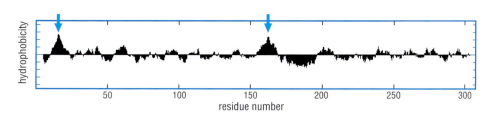

Figure 1-31 Hydropathy plot of the *Rhizobium meliloti* protein DctB The plot shows that two membrane-spanning alpha-helical regions are predicted. The plot represents the average hydrophobicity of an eight-residue moving window that slides along the sequence.

Because the backbone hydrogen bonds of an alpha helix are local, alpha helices are by far the most common secondary structure element in membrane proteins (Figure 1-30). As the translation per residue in a helix is 1.5 Å, a stretch of about 20 consecutive hydrophobic residues can form a helix that spans the bilayer if the helix axis is not tilted with respect to the membrane plane. Such stretches are easily recognized in protein sequences and are considered diagnostic for internal membrane proteins in analysis of genome sequences, because they do not occur frequently in soluble proteins. Figure 1-31 illustrates a hydropathy plot (plot of mean residue hydrophobicity) as a function of sequence for a carboxylic acid transport sensor (DctB) in the nitrogen-fixing bacterium *Rhizobium meliloti*. Two membrane-spanning alpha-helical regions are predicted. Many membrane-associated proteins are embedded in the lipid bilayer via only one or two membrane-spanning segments. These are always helical.

Beta sheets also occur in membrane proteins, but they are harder to recognize in the sequence. A beta strand 8–9 residues long would span the membrane (the translation per residue is about 3.5 Å) if the chain were perpendicular to the plane of the bilayer, but such stretches occur in soluble proteins and the variable twist of beta sheets makes it likely that the strand will be tilted. In those membrane proteins in which beta sheets have been found so far, they are antiparallel sheets with short polar turns. Because the edge strands in a beta sheet that is embedded in a membrane would have many unsatisfied hydrogen-bond donor and acceptor groups in their backbones, all such sheets examined to date form closed barrels with the first and last strands hydrogen bonded to each other (Figure 1-32). These beta sheets will have hydrophobic side chains covering their exterior surface, but can have polar or charged side chains lining the interior of the barrel. Such barrels seem to be used primarily as channels to permit water or ions to diffuse across the membrane. Channels can also be made from primarily helical proteins, as in the case of the potassium channel (Figure 1-33).

No integral membrane proteins with both helical and beta-sheet secondary structure have yet been found. There is good reason to expect that these are less common than all-helical or all-beta types: the need to hydrogen bond the polar groups on the edge strands of a beta sheet would be difficult to satisfy in a mixed structure. At present there are too few membrane protein structures determined to permit us to generalize with confidence on this point, or to allow creation of a detailed taxonomy of membrane protein fold families.

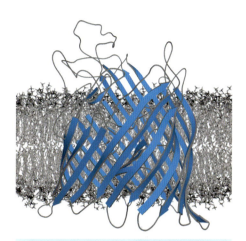

Figure 1-32 The three-dimensional structure of the all-beta transport protein FhuA The protein (PDB 1by3) is shown with a simulated lipid bilayer in the position it would occupy in the bilayer. The beta strands form a barrel that serves as a pore in the membrane, with hydrophobic side chains on the outside of the barrel and polar side chains lining the pore.

Figure 1-33 Three-dimensional structure of the bacterial potassium channel The protein is a homotetramer with a single channel formed at the interface of the four subunits. Although the protein is primarily helical, the pore of the channel is formed, in part, by two extended strands (grey) from each subunit. Shown here is the view looking down the channel from the extracellular side of the membrane: the outer ends of the helices in the figure are at the top of the chanel looking down, and the pore-forming strands can be seen looping inward. Two potassium ions are depicted as green spheres within the pore. (PDB 1bl8)

Tool for producing hydropathy plots on the Internet:

http://arbl.cvmbs.colostate.edu/molkit/hydropathy/index.html

Membrane models on the Internet:

http://www.lrz-muenchen.de/~heller/membrane/membrane.html

The folded protein is a thermodynamic compromise

Protein tertiary structure is maintained by the sum of many weak forces, some of which are stabilizing and some of which are destabilizing, some of which are internal to the protein and some of which are between the protein and its environment. The net effect is a folded structure that is only marginally stable in water at room temperature.

The contributions of the forces to protein stability are usually quantified in terms of the energy associated with any one of them. The heat released when such an interaction is formed in an isolated system is the **enthalpy** of the bond. However, bond enthalpies do not give a complete picture of the energetics of interactions in biological systems, in part because they neglect the contributions of water. Water plays two major roles in modulating the strengths of weak interactions. First, interactions between polar groups contribute only the difference in enthalpy between the groups when they are bonded to each other and the same groups when they are bonded to water. Because the interactions of water molecules with, for example, hydrogen-bond donors and acceptors are often similar, and of nearly equivalent enthalpy, to those that these groups can make to one another, the net enthalpy term is small.

Second, the contribution of water to the **entropy** of a weak interaction is also considerable. Entropy is a measure of randomness or disorder. The second law of thermodynamics states that spontaneous processes such as protein folding tend to increase the total entropy of a system plus its surroundings. An example of the importance of entropic contributions from water is found in the hydrophobic effect. Nonpolar groups in water tend to be surrounded by a cluster of water molecules that are more ordered than in the normal structure of liquid water (Figure 1-34). When such hydrophobic groups clump together, expelling water, the water molecules that are released undergo an increase in entropy. Although there will be a shell of ordered water around the clump, the total number of these ordered water molecules will be smaller than if all the hydrophobic groups were exposed to solvent individually. The gain in solvent entropy that results from the association of hydrophobic groups together is the driving force behind the hydrophobic effect. Thus, in evaluating the energetic consequences of a weak interaction, the changes in entropy of the interacting groups and the water around them all need to be considered simultaneously.

Stability is defined as a net loss of **free energy**, a function of the combined effects of entropy and enthalpy. Such a loss may result predominantly from a loss of enthalpy when a bond forms, or predominantly from a gain in entropy when the disorder of a system (protein) plus its surroundings (water) increases, or from a balance between enthalpy and entropy changes. Most weak interactions release about 4–13 kJ/mole of free energy when they occur in water and therefore contribute only a small amount to the total stability of a protein. However, there are a large number of them, adding up to a very large free-energy decrease when secondary and tertiary structures form.

Finally, even though many hundreds of hydrogen bonds and van der Waals interactions occur in a folded protein, the net free energy of stabilization of most folded proteins—the difference in free energy between the folded and unfolded states—is actually rather small, about 21–42 kJ/mole, or only about 10 times the average thermal energy available at physiological temperature. Most folded proteins are marginally stable because the free energy released when hundreds of weak interactions form is almost exactly counterbalanced by the enormous loss of conformational flexibility (loss of entropy) that occurs when the unfolded chain folds into a compact, ordered structure. A folded protein is a thermodynamic compromise.

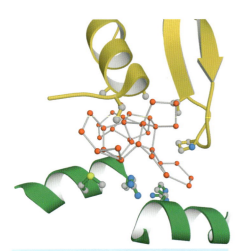

Figure 1-34 Illustration of the ordered arrays of water molecules surrounding exposed hydrophobic residues in bovine pancreatic ribonuclease A Such waters (red) often form pentagonal arrays. It is thought that this ordering of water around exposed polar groups is the driving force for the hydrophobic effect. (PDB 1dyg)

Definitions

denaturant: a chemical capable of unfolding a protein in solution at ordinary temperatures.

denatured state: the partially or completely unfolded form of a biological macromolecule in which it is incapable of carrying out its biochemical and biological functions.

enthalpy: a form of energy, equivalent to work, that can be released or absorbed as heat at constant pressure.

entropy: a measure of the disorder or randomness in a molecule or system.

free energy: a function, designed to produce a criterion for spontaneous change, that combines the **entropy** and **enthalpy** of a molecule or system. Free energy decreases for a spontaneous process, and is unchanged at equilibrium.

mesophilic: favoring moderate temperatures. Mesophilic organisms normally cannot tolerate extremes of heat or cold. Mesophilic enzymes typically denature at moderate temperatures (over 40 °C or so).

temperature-sensitive: losing structure and/or function at temperatures above physiological or room temperature. A temperature-sensitive mutation is a change in the amino-acid sequence of a protein that causes the protein to inactivate or fail to fold properly at such temperatures.

thermophilic: favoring high temperatures. A thermophilic organism is one that requires high temperatures (above approximately 50 °C) for survival. A thermophilic enzyme is one that functions optimally and is stable at temperatures at which **mesophilic** proteins denature.

Protein structure can be disrupted by a variety of agents

High temperatures break the weak interactions that stabilize the folded or native form of a protein and eventually convert the structure to a largely unfolded or denatured one, in which these interactions are replaced by hydrogen bonds with water. The **denatured state** is usually defined empirically, either by loss of biological or biochemical activity, or by spectroscopic signals characteristic of an unfolded polypeptide (Figure 1-35). Because the free-energy difference between the native and denatured states is small, loss of a single interaction in the native state can sometimes bring the free-energy difference close to the thermal energy available at ordinary temperatures. A mutation that causes a normally stable protein to unfold at relatively low temperatures is called a **temperature-sensitive** (ts) mutation. These mutations are widely used in experimental biology to test the function of a protein in cells by raising the temperature and thereby disabling the mutant protein. Similarly, just a few additional interactions can greatly increase the stability of a protein at elevated temperatures, producing a protein that is more able to withstand heating, prolonged storage or shipping for industrial applications. One example is the thermostable Taq DNA polymerase used in PCR.

Another way to unfold a protein is by the use of chemical **denaturants** such as urea or guanidinium hydrochloride, or detergents like SDS. In contrast to thermal denaturation, these compounds are thought to unfold proteins in large part by competing for hydrogen bonds with the polar groups of the backbone and side chains.

Some proteins are naturally very stable to thermal or chemical denaturation. One important class of very stable proteins consists of those from microorganisms that normally live at high temperatures. These so-called **thermophilic** proteins sometimes retain their structure—and activity—at temperatures approaching the boiling point of water. No single type of interaction or effect accounts for such hyperthermostability. Comparisons of structures of proteins with similar sequences and functions isolated from thermostable microbes and their **mesophilic** counterparts show a variety of differences: some thermophilic proteins have more salt bridges, while others appear to have more hydrophobic interactions and shorter protruding loops, and so forth. There seem to be many ways to achieve the same effect, and when this is the case it is usual in biology to find all of them.

The marginal stability of protein tertiary structure allows proteins to be flexible

Above absolute zero, all chemical bonds have some flexibility: atoms vibrate and chemical groups can rotate relative to each other. In proteins, because most of the forces that stabilize the native state are noncovalent, there is enough thermal energy at physiological temperatures for weak interactions to break and reform frequently. Thus a protein molecule is more flexible than a molecule in which only covalent forces dictate the structure. Protein structures continuously fluctuate about the equilibrium conformation observed by techniques such as X-ray diffraction and nuclear magnetic resonance (NMR) (Figure 1-36). Thermally driven atomic fluctuations range in magnitude from a few hundredths of an Ångstrom for a simple atomic vibration to many Ångstroms for the movement of a whole segment of a protein structure relative to the rest. These fluctuations are large enough to allow small molecules such as water to penetrate into the interior of the protein. They are essential for protein functions such as ligand binding and catalysis, for they allow the structure to adjust to the binding of another molecule or to changes in the structure of a substrate as a reaction proceeds.

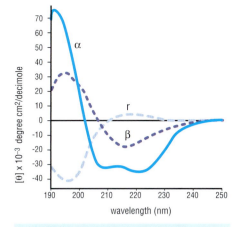

Figure 1-35 Computed circular dichroism spectra for the evaluation of protein conformation Circular dichroism spectrum of poly(Lys) in the alpha-helical (α), anti-parallel beta sheet (β) and random coil (r) conformations. (From Greenfield, N.J. and Fasman, G.D.: *Biochemistry* 1969, **8**:4108–4116.)

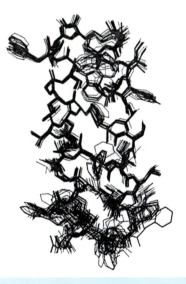

Figure 1-36 Results of a molecular dynamics simulation of two interacting alpha helices The diagram shows fluctuations of portions of the structure. Some parts of the protein seem to be more mobile than others.

References

Dill, K.A and Bromberg, S.: *Molecular Driving Forces: Statistical Thermodynamics in Chemistry and Biology* (Garland, New York and London, 2003).

Ferreira, S.T. and De Felice, F.G.: **Protein dynamics, folding and misfolding: from basic physical chemistry to human conformational diseases.** *FEBS Lett.* 2001, **498**:129–134.

Greenfield, N.J. and Fasman, G.D.: **Computed circular dichroism spectra for the evaluation of protein**

conformation. *Biochemistry* 1969, **8**:4108–4116.

Jaenicke, R.: **Stability and stabilization of globular proteins in solution.** *J. Biotechnol.* 2000, **79**:193–203.

Kauzmann, W.: **Some factors in the interpretation of protein denaturation.** *Adv. Protein Chem.* 1959, **14**:1–63.

Sharp, K.A. and Englander, S.W.: **How much is a stabilizing bond worth?** *Trends Biochem. Sci.* 1994, **19**:526–529.

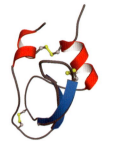

Figure 1-37 The structure of the small protein bovine pancreatic trypsin inhibitor, BPTI The three disulfide bonds are yellow, beta strands are blue, and alpha helices are red. If these disulfide bonds are reduced this small protein unfolds, presumably because there is not enough secondary structure to stabilize the fold without them. (PDB 1bpi)

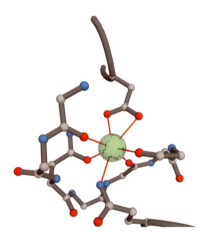

Figure 1-38 Stabilization by coordinate covalent bonds Close-up of one of the three calcium ion binding sites in the bacterial protein subtilisin, showing the coordination of the metal ion by the protein. This site is used only for protein stability, not for catalysis. Removal of this metal ion significantly destabilizes the protein. (PDB 1sca)

Covalent bonds can add stability to tertiary structure

Noncovalent forces are the principal interactions that stabilize protein tertiary structure but they are not the only ones. Many proteins also are stabilized by additional, covalent interactions that provide a form of cross-linking between segments of secondary structure in the native state. The most common of these covalent bonds is the disulfide bridge that can form between two cysteine side chains that are brought close together by the tertiary structure (Figure 1-37). Formation of a disulfide bridge (also called an S–S bridge or a disulfide bond) involves the oxidation of the two sulfhydryl groups as a coupled redox reaction in the endoplasmic reticulum. Conversely, the bridge can be broken by reduction. Thus, S–S bridges are uniquely sensitive to their environment. They are not found in most intracellular proteins, because the environment inside the cell is highly reducing, but they are common in proteins that are secreted from that environment into the oxidizing conditions found outside the cell.

The second most common cross-linking interaction in proteins is the coordination of a metal ion to several protein side chains; the **coordinate covalent bonds** between the protein and the metal ion form a type of internal metal chelate (Figure 1-38). The strength of the binding of the metal ion to the protein varies from very loose (K_d of mM) to very tight (K_d of nM) depending on the nature of the metal ion and the protein ligands. Not all of the ligands are contributed by the protein; one or more water molecules can also occur in the coordination sphere. A given protein can have more than one stabilizing metal ion binding site. The metal ions that most commonly form such chelates are calcium (Ca^{2+}) and zinc (Zn^{2+}), although monovalent cations such as potassium and sodium can also function in this way. These stabilizing metal ions carry out no chemistry and are distinct from metal ions in active sites of metalloproteins which carry out the biochemical function of a protein (see below). Sometimes when these metal ions are removed by chelating agents such as EDTA, the resulting protein remains folded, although less stable, under physiological conditions. In other cases, removal of the metal ions from the protein leads to denaturation.

Finally, some proteins are stabilized by the covalent binding of a dissociable organic or organometallic **cofactor** at the active site, or by the formation of a covalent cross-link between amino-acid side chains that is different from a disulfide bridge (Figure 1-39a). So far, these cross-links have always been found at the active site, where they contribute critically to the chemical function of the protein. The covalent bond between cofactor and protein may be formed with the organic part of some cofactors as in the case of D-amino acid aminotransferase (DaAT) (Figure 1-39a), or with a metal ion that is an integral part of some cofactors as in the case of vitamin B12, chlorophyll, and the heme group in some heme-containing proteins (Figure 1-39b), or with both, as in the case of the heme group in cytochrome c (Figure 1-39c). However, in some cases the cofactor is not a separable molecule, but is created by the chemical cross-linking of two amino-acid side chains, as in the case of the redox active cofactor PQQ (Figure 1-39d) and the bioluminescent chromophore in green fluorescent protein. Although many proteins that are stabilized in this way remain folded when the cofactor is dissociated, some do not.

Post-translational modification can alter both the tertiary structure and the stability of a protein

Proteins in eukaryotic cells that are destined to be placed on the cell surface or secreted into the environment are often modified by the covalent attachment of one or more chains of carbohydrate molecules at specific serine, threonine or asparagine residues. This is known as **glycosylation**, and along with the covalent attachment of lipids, is among the most important

Definitions

cofactor: an organic or organometallic molecule that binds to a protein and provides an essential chemical function for that protein.

coordinate covalent bond: a bond formed when a lone pair of electrons from an atom in a ligand is donated to a vacant orbital on a metal ion.

glycosylation: the post-translational covalent addition of sugar molecules to asparagine, serine or threonine residues on a protein molecule. Glycosylation can add a single sugar or a chain of sugars at any given site and is usually enzymatically catalyzed.

K_d: the dissociation constant for the binding of a ligand to a macromolecule. Typical values range from 10^{-3} M to 10^{-10} M. The lower the K_d, the tighter the ligand binds.

limited proteolysis: specific cleavage by a protease of a limited number of the peptide bonds in a protein substrate. The fragments thus produced may remain associated or may dissociate.

N-acetylation: covalent addition of an acetyl group

from acetyl-CoA to a nitrogen atom at either the amino terminus of a polypeptide or in a lysine side chain. The reaction is catalyzed by *N*-acetyltransferase.

phosphorylation: covalent addition of a phosphate group, usually to one or more amino-acid side chains on a protein, catalyzed by protein kinases.

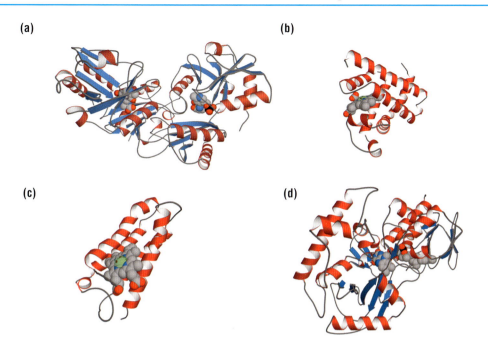

(a) (b) (c) (d)

Figure 1-39 Examples of stabilization by cofactor binding **(a)** DaAT is a dimeric protein with covalently bound pyridoxal phosphate bound in each subunit. (PDB 3daa) **(b)** The heme group of myoglobin is attached to the protein by a coordinate covalent bond between a histidine side chain and the heme iron. (PDB 1a6k) **(c)** In cytochrome c the heme group is attached to the protein in two places: by a coordinate covalent bond between the sulfur of a methionine residue and the heme iron, and by covalent attachment of the porphyrin cofactor to the protein. (PDB 1a7v) **(d)** The cofactor PQQ in polyamine oxidases is formed by the reaction of two side chains with each other, leaving it attached to the protein. (PDB 1b37) In all cases, the apoprotein (the protein without the cofactor) is thermally less stable and/or more susceptible to proteolytic digestion than the protein with the covalently attached cofactor.

of the post-translational modifications. Both are enzymatically catalyzed by specific enzymes. Probably the most important for protein stability is glycosylation. The roles of the attached sugars are not known precisely for most proteins, but on proteins expressed on the surface of blood cells, for example, they are believed to be important in preventing the cells from sticking to one another or to vessel walls and obstructing blood flow. Some protein glycosylation sites are involved in protein–protein recognition. In many instances the removal of carbohydrates from glycosylated proteins leads either to unfolding or to aggregation. This characteristic limits the use of prokaryotic systems in the application of recombinant DNA technology to such proteins, since prokaryotes do not carry out this post-translational modification. It is generally believed that glycosylation does not alter the tertiary structure of a protein, but can significantly influence thermal stability, stability to degradation and quaternary structure.

There are other forms of post-translational modification that can also alter the stability—and in some cases the tertiary structure—of folded proteins. In contrast to glycosylation, these modifications usually alter the function of the protein. Some of these modifications, such as **phosphorylation** and *N*-**acetylation**, are reversible and thus can act as conformational switches. The functional consequences of post-translational modification will be discussed in Chapter 3. Others, such as **limited proteolysis**, the cleavage of the polypeptide chain at one or more sites, are irreversible, and thereby change the structure and function of a protein permanently. Limited proteolysis, for example, sometimes generates an active protein from an inactive precursor. Common kinds of post-translational modification that influence the stability of proteins are summarized in Figure 1-40. Although many types of post-translational modification can increase or decrease the stability of a protein, such modifications often also serve additional functions, such as signaling or activation of catalysis.

Figure 1-40 Table of post-translational modifications affecting protein stability

Most Common Post-translational Modifications

Reversible	Irreversible
disulfide bridge	cofactor binding
cofactor binding	proteolysis
glycosylation	ubiquitination
phosphorylation	peptide tagging
acylation	lysine hydroxylation
ADP-ribosylation	methylation
carbamylation	
N-acetylation	

References

Imperiali, B. and O'Connor, S.E.: **Effect of *N*-linked glycosylation on glycopeptide and glycoprotein structure.** *Curr. Opin. Chem. Biol.* 1999, **3**:643–649.

Morris, A.J. and Malbon, C.C.: **Physiological regulation of G protein-linked signaling.** *Physiol. Rev.* 1999, **79**:1373–1430.

Palade, G.E.: **Protein kinesis: the dynamics of protein trafficking and stability.** *Cold Spring Harbor Symp. Quant. Biol.* 1995, **60**:821–831.

Rattan, S.I. *et al.*: **Protein synthesis, posttranslational modification, and aging.** *Annls N.Y. Acad. Sci.* 1992, **663**:48–62.

Tainer, J.A. *et al.*: **Protein metal-binding sites.** *Curr. Opin. Biotechnol.* 1992, **3**:378–387.

Zhang, T. *et al.*: **Entropic effects of disulfide bonds on protein stability.** *Nat. Struct. Biol.* 1994, **1**:434–438.

For a database of all known post-translational modifications, organized by type of amino acid modified, see:

http://pir.georgetown.edu/pirwww/dbinfo/resid.html

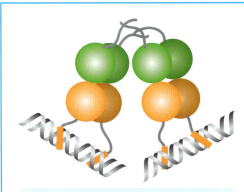

Figure 1-41 Schematic diagram of the Lac repressor tetramer binding to DNA Each monomer of the Lac repressor is made up of a tetramerization domain (green) and a DNA-binding domain (orange).

Globular proteins are composed of structural domains

Some proteins, such as the keratin of hair, are fibrous: their polypeptide chains are stretched out in one direction. Most proteins, however, are globular: their polypeptide chains are coiled up into compact shapes. Since proteins range in molecular weight from a thousand to over a million, one might have thought that the size of these globular folds would increase with molecular weight, but this is not the case. Proteins whose molecular weights are less than about 20,000 often have a simple globular shape, with an average molecular diameter of 20 to 30 Å, but larger proteins usually fold into two or more independent globules, or structural domains. A **domain** is a compact region of protein structure that is often, but not always, made up of a continuous segment of the amino-acid sequence, and is often capable of folding stably enough to exist on its own in aqueous solution. The notion that the domains of large proteins are independently stable has been verified by cloning the corresponding DNA sequences and expressing them independently. Not only do many of them form stable, folded structures in solution, they often retain part of the biochemical function of the larger protein from which they are derived. The bacterial Lac repressor, which is a tetrameric protein that binds tightly to a specific DNA sequence, is a good example (Figure 1-41). One of the two domains in the monomer can dimerize by itself, and binds to DNA with an affinity that nearly matches that of the intact protein. The function of the other domain is to form the tetramer by making protein–protein interactions; by itself it tetramerizes but does not bind to DNA.

Not all domains consist of continuous stretches of polypeptide. In some proteins, a domain is interrupted by a block of sequence that folds into a separate domain, after which the original domain continues. The enzyme alanine racemase has an interrupted domain of this type (Figure 1-42).

Domains vary in size but are usually no larger than the largest single-domain protein, about 250 amino acids, and most are around 200 amino acids or less. Forty-nine per cent of all domains are in the range 51 to 150 residues. The largest single-chain domain so far has 907 residues, and the largest number of domains found in a protein to date is 13. As the domains in a protein associate with one another by means of the same interactions that stabilize their internal structures, what is true for domains is true for whole proteins, and vice versa: the same structural principles apply to both.

Domains have hydrophobic cores

Hydrophobic cores appear to be essential for the stability of domains. Concentrating hydrophobic groups in the core is energetically favorable because it minimizes the number of unfavorable interactions of hydrophobic groups with water, and maximizes the number of van der Waals interactions the hydrophobic groups make with each other. All the different polypeptide folding patterns presented in this book can be thought of as alternative solutions to a single problem: how to fold a polypeptide chain so as to maximize the exposure of its hydrophilic groups to water while minimizing the exposure of its hydrophobic groups. It is the presence of a hydrophobic core that usually allows protein domains to fold stably when they are expressed on their own.

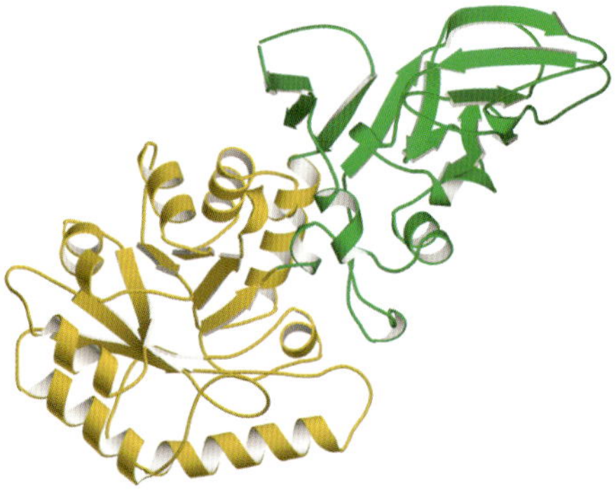

Figure 1-42 Structure of alanine racemase The diagram shows that one of its structural domains (yellow) is interrupted by insertion of another domain (green).

Definitions

domain: a compact unit of protein structure that is usually capable of folding stably as an independent entity in solution. Domains do not need to comprise a contiguous segment of peptide chain, although this is often the case.

Multidomain proteins probably evolved by the fusion of genes that once coded for separate proteins

There are many examples of proteins with two or more domains of nearly identical structure. The *Escherichia coli* thioesterase, for example, is organized into two equal-sized domains of almost identical structure (Figure 1-43a). The domains can be overlaid on top of one another with almost perfect overlap in the paths of their polypeptide chains, except for a few of the external loops and the polypeptide that links the domains.

It is likely that this protein, and others that have internal similarity of structure, evolved by gene duplication. A single gene coding for a protein resembling one domain is assumed to have been duplicated in tandem, and the two genes to have fused so that their sequences are expressed as a single polypeptide. In some proteins the duplicated domains retain some sequence identity, but in other proteins they do not. Whether duplicated domains display sequence identity depends on how long ago the duplication occurred, and the nature of the functional constraints that guided their divergence. The more ancient the gene duplication, the more time for mutation to obscure the sequence relationship. In thioesterase, for example, completely different sequences give rise to the same overall fold. The original sequence identity between the two thioesterase domains has been largely obliterated by random mutations over millennia. Sometimes the original gene that is duplicated can be identified. In the case of thioesterase, the protein thioester dehydrase, which carries out a similar function to thioesterase, is a homodimeric protein. Each monomer has the same fold as one of the domains of thioesterase (Figure 1-43b).

Sometimes, gene duplication can occur within a single structural domain. An example is shown in Figure 1-44, which depicts the fold of the eye-lens protein gamma-crystallin. The protein as a whole is made up of two similar domains, which are 40% identical in sequence. Closer inspection of the two domains reveals that they, too, are made up of two essentially identical halves. Each eight-stranded beta-sheet domain is composed of two four-stranded antiparallel sheets of the same topology. Internal symmetry such as this does not prove tandem duplication of a smaller gene, as this arrangement of beta strands could simply be a stable configuration for two antiparallel beta sheets to pack against one another. For the gamma-crystallins, however, there is enough residual sequence identity to justify the conclusion that the individual domain did indeed evolve by gene duplication and fusion. Additional evidence for this evolutionary history can be found in the sequences of the genes for other crystallins. The gene for mouse beta-crystallin, a protein closely related in amino-acid sequence to gamma-crystallin, is divided into four exons, and each exon codes for one four-stranded beta-sheet segment. In other words, the positions of the three introns correspond to the junctions between each of the subdomains that were presumably encoded by the primordial gene from which the modern genes arose by duplication. This is persuasive evidence for gene duplication in crystallin evolution.

If the fusion of tandem genes can account for proteins with internal symmetry, then it is likely that this mechanism also explains the origin of multidomain proteins where the domains are structurally unrelated, as in the Lac repressor (see Figure 1-41).

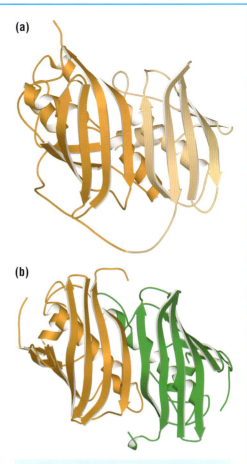

Figure 1-43 Structures of thioesterase and thioester dehydrase (a) Structure of *E. coli* thioesterase, a protein composed of two nearly identical domains (dark and light orange) fused together. (PDB 1c8u) Each domain resembles the subunit of thioester dehydrase (PDB 1mkb) **(b)**, a protein composed of two identical subunits.

Figure 1-44 Structure of gamma-crystallin Gamma-crystallin is composed of two nearly identical domains. Each domain is also made up of two nearly identical halves. (PDB 1gcs)

References

Burchett, S.A.: **Regulators of G protein signaling: a bestiary of modular protein binding domains.** *J. Neurochem.* 2000, **75**:1335–1351.

Campbell, I.D. and Downing, A.K.: **Building protein structure and function from modular units.** *Trends Biotechnol.* 1994, **12**:168–172.

Dengler, U. *et al.*: **Protein structural domains: analysis of the 3Dee domains database.** *Proteins* 2001, **42**:332–344.

Hawkins, A.R. and Lamb, H.K.: **The molecular biology of multidomain proteins. Selected examples.** *Eur. J. Biochem.* 1995, **232**:7–18.

Hegyi, H. and Bork, P.: **On the classification and evolution of protein molecules.** *J. Prot. Chem.* 1997, **16**:545–551.

Richardson, J.S.: **The anatomy and taxonomy of protein structure.** *Adv. Protein Chem.* 1981, **34**:167–339.

Thornton, J.W. and DeSalle, R.: **Gene family evolution and homology: genomics meets phylogenetics.** *Annu. Rev. Genomics Hum. Genet.* 2000, **1**:41–73.

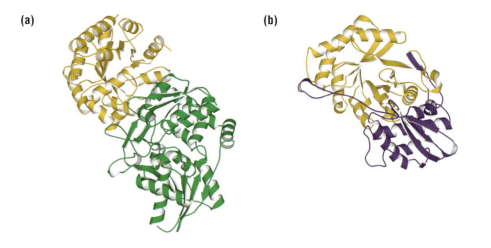

(a)   (b)

Figure 1-45 Structures of tryptophan synthase and galactonate dehydratase (a) Tryptophan synthase (PDB 1ttp). **(b)** Galactonate dehydratase. Both proteins have one domain that is an alpha/beta barrel (yellow), even though the other domains in both proteins are very different from one another. There is no sequence similarity or functional relationship between the similar domains.

The number of protein folds is large but limited

As more protein structures are determined experimentally, it is increasingly found that new structures look like old structures. Sometimes an entire "new" structure will resemble that of another protein whose structure is already known. In most cases, however, the overall polypeptide fold of the protein will be "new", but the structure will be divisible into a number of domains, at least one of which resembles the tertiary structure previously observed in another protein (Figure 1-45).

It appears that the number of different protein folds in nature is limited. They are used repeatedly in different combinations to create the diversity of proteins found in living organisms. Building new proteins, it would seem, is like assembling a four-course dinner from a set of *á la carte* choices—the possible **domain folds**. Although the size of the menu is not yet known, it is much smaller than the total number of gene products—perhaps as small as a few thousands—and almost all of the tertiary structure folds that have been discovered so far are known to appear in many different proteins. Thus, a complete protein can be described by specifying which folds each domain has and how they interact with each other. This approach to describing protein structure is appealing both for its logical form and because it reflects our prejudice that proteins fold up domain by domain.

Protein structures are modular and proteins can be grouped into families on the basis of the domains they contain

Although many proteins are composed of a single structural domain, most proteins are built up in a modular fashion from two or more domains fused together. In some cases, each domain has a characteristic biochemical function and the function of the entire protein is determined by the sum of the individual properties of the domains. Proteins involved in signal transduction and cell-cycle control are often constructed in this fashion (Figure 1-46). One example is the cancer-associated kinase Src-Lck, which has a catalytic kinase domain that phosphorylates proteins on tyrosine residues, an SH2 domain that binds phosphotyrosine residues, an SH3 domain that recognizes proline-rich sequences, and a phosphotyrosine region that can interact with its own or other SH2 domains. When the modules that form proteins of this type fold and function independently, the order in which they occur in the polypeptide is not necessarily always important. Thus module swapping and the recruitment of new functions by adding modules is often simple, either through the course of evolution or artificially.

Src/Lck	—	SH3	SH2	KINASE	P					
Btk/Tec	PH	SH3	SH2	KINASE	-					
Zap-70	SH2	—	SH2	—	KINASE	-				
SHP	SH2	-	SH2	—	PTPase	P				
PLC-γ	PH	PLC	P	SH2	-	SH2	-	SH3	H	PLC
p120 GAP	SH2	SH3	SH2	-	PH	—	GAP			
p85 P13K	SH3	Pro	—	SH2	—	SH2	-			
Grb2/Drk/Sem-5	—	SH3	SH2	SH3	-					
c-Crk	—	SH2	SH3	P	SH3					
Shc	PTB	—	P	SH2	-					
Stat	—	SH3	-	SH2	P	-				
p47phox	—	SH3	SH3	Pro	·					
p67phox	SH3	—	SH3							
dig	SH3	G-KINASE								
P130cas	SH3	P P P P P P P P	—							
b-ARK	—	KINASE	PH	—						
Sos	—	PH	Cdc25	—	Pro					
Dynamin	GTPase	PH	Pro							
IRS-1	—	PH	PTB	P P P P P P P P						
GAP1m	—	GAP	PH							

Figure 1-46 Schematic diagram of the domain arrangement of a number of signal transduction proteins The different modules have different functions; Pro = proline-rich regions that bind SH3 domains; P = phospho-tyrosine-containing regions that bind SH2 domains; PH = pleckstrin homology domains that bind to membranes; PTPase = phospha-tase domain; kinase = protein kinase domain; G-kinase = guanylate kinase domain; GAP = G-protein activation domain; PLC = phospho-lipase C catalytic domain. The function of the individual modules is sometimes, but not always, independent of the order in which they appear in the protein.

Definitions

domain fold: the particular topographical arrange-ment of secondary structural elements that character-izes a single domain. Examples are an antiparallel arrangement of four helices in a four-helix bundle, or an open twisted beta sandwich with a particular sequence that binds nucleotides.

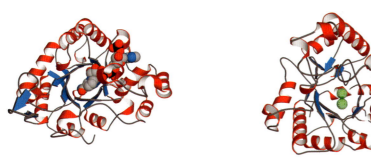

Figure 1-47 Structures of aldose reductase (left) and phosphotriesterase (right) Although the overall folds are very similar, aldose reductase uses NADPH to reduce sugars, while phosphotriesterase hydrolyzes phosphate groups using a bimetallic cofactor. (PDB 1ads and 1dpm)

Because sequence determines structure, which in turn determines function, it is tempting to classify proteins whose function cannot be recognized from sequence similarity alone into families based on the structures of the domains they contain. Often this approach is successful: proteins with a kinase domain are nearly always kinases; proteins with an alpha/beta hydrolase domain nearly always hydrolyze small-molecule substrates, and so forth. But often it is not the case that structural families share a common function. There are hundreds of proteins that contain a particular eight-stranded parallel beta barrel with surrounding alpha helices called a TIM barrel, but even two very similar single-domain TIM-barrel proteins can have completely different biochemical functions (Figure 1-47). Nor is it always the case that all proteins that perform the same biochemical function will have the same domains: amino-acid transamination, for example, can be catalyzed by two completely different folds (Figure 1-48). The coupling between overall structure and function can be quite loose. Nevertheless, grouping proteins into families on the basis of their domain architecture is, at a minimum, very useful for studying the way new protein functions may have evolved.

The modular nature of protein structure allows for sequence insertions and deletions

Deletions and insertions of amino acids can obscure evolutionary relationships, but how is it that long stretches of amino acids, sometimes an entire domain, can be inserted in or deleted from a protein sequence (see Figure 1-42) without disrupting the basic structure of a domain? The answer lies in the nature of domain folds. Domains are made up of secondary structure elements that are packed together to form tertiary structure. The loops that join the helices and sheets in most proteins are usually located on the surface, and often make few contacts with the rest of the domain. Within a given protein family, insertions and deletions nearly always occur in these surface loops, where variation in length has little effect on the packing of helices and sheets. Indeed, a rough rule of domains, and ultimately of the structural evolution of proteins, is that the framework tends to remain fairly constant in both sequence and structure while the loops change a great deal over evolutionary time. In the case of immunoglobulin (see section 1-10), the loops form the antigen-binding site and variation due to somatic recombination and mutation of immunoglobulin genes accounts for the diversity of antibody molecules.

Many models for protein evolution propose the shuffling of exon-coded segments to produce new protein molecules. Insertion of a new exon into an existing domain could change its properties dramatically, but of course the new molecule would still have to fold stably. Stable folding would be more likely if the new exon were inserted into a surface loop. Examination of intron/exon junctions in proteins whose three-dimensional structures are known shows that many exon boundaries do indeed occur in sequence positions corresponding to loops in the structure. Important exceptions include the immunoglobulins.

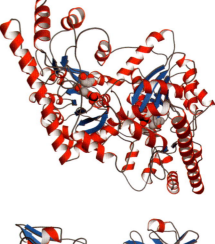

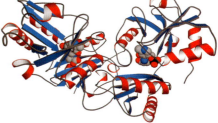

Figure 1-48 Structures of aspartate aminotransferase (top) and D-amino acid aminotransferase (bottom) Both enzymes catalyze the same reaction; but they have no structural similarity to each other at either the sequence or the tertiary level. Only the active sites, shown by the presence of the cofactor in space-filling representation, are very similar. (PDB 1yaa and 3daa)

References

Branden, C. and Tooze, J.: *Introduction to Protein Structure* 2nd ed. (Garland, New York, 1999).

Patthy, L.: **Genome evolution and the evolution of exon-shuffling.** *Gene* 1999, **238**:103–114.

Richardson, J.S.: **Introduction: protein motifs.** *FASEB J.* 1994, **8**:1237–1239.

Richardson, J.S. *et al.*: **Looking at proteins: representation, folding, packing and design.** *Biophys. J.* 1992,

63:1185–1209.

Richardson, J.S. and Richardson, D.C.: **Principles and patterns of protein conformation** in *Prediction of Protein Structure and the Principles of Protein Conformation* Fasman, G.D. ed. (Plenum Press, New York, 1990), 1–98.

Salem, G.M. *et al.*: **Correlation of observed fold frequency with the occurrence of local structural motifs.** *J. Mol. Biol.* 1999, **287**:969–981.

Thornton, J.M. *et al.*: **Protein folds, functions and

evolution.** *J. Mol. Biol.* 1999, **293**:333–342.

Internet resources on protein structure comparison and classification:

http://www.ebi.ac.uk/dali/

http://scop.mrc-lmb.cam.ac.uk/scop/

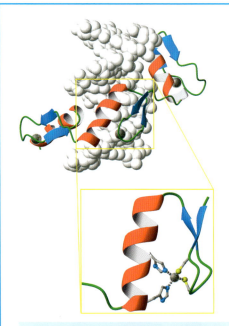

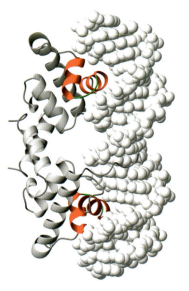

Figure 1-49 Zinc finger motif A fragment derived from a mouse gene regulatory protein is shown, with three zinc fingers bound spirally in the major groove of a DNA molecule. The inset shows the coordination of a zinc atom by characteristically spaced cysteine and histidine residues in a single zinc finger motif. The image is of Zif268. (PDB 1aay)

Protein motifs may be defined by their primary sequence or by the arrangement of secondary structure elements

The term **motif** is used in two different ways in structural biology. The first refers to a particular amino-acid sequence that is characteristic of a specific biochemical function. An example is the so-called zinc finger motif, CXX(XX)CXXXXXXXXXXXXHXXXH, which is found in a widely varying family of DNA-binding proteins (Figure 1-49). The conserved cysteine and histidine residues in this **sequence motif** form ligands to a zinc ion whose coordination is essential to stabilize the tertiary structure. Conservation is sometimes of a class of residues rather than a specific residue: for example, in the 12-residue loop between the zinc ligands, one position is preferentially hydrophobic, specifically leucine or phenylalanine. Sequence motifs can often be recognized by simple inspection of the amino-acid sequence of a protein, and when detected provide strong evidence for biochemical function. The protease from the human immunodeficiency virus was first identified as an aspartyl protease because a characteristic sequence motif for such proteases was recognized in its primary structure.

The second, equally common, use of the term motif refers to a set of contiguous secondary structure elements that either have a particular functional significance or define a portion of an independently folded domain. Along with the functional sequence motifs, the former are known generally as **functional motifs**. An example is the helix-turn-helix motif found in many DNA-binding proteins (Figure 1-50). This simple **structural motif** will not exist as a stably folded domain if expressed separately from the rest of its protein context, but when it can be detected in a protein that is already thought to bind nucleic acids, it is a likely candidate for the recognition element. Examples of structural motifs that represent a large part of a stably folded domain include the four-helix bundle (Figure 1-51), a set of four mutually antiparallel alpha helices that is found in many hormones as well as other types of proteins; the Rossmann fold, an alpha/beta twist arrangement that usually binds NAD cofactors; and the *Greek-key motif*, an all-beta-sheet arrangement found in many different proteins and which topologically resembles the design found on ancient vases. As these examples indicate, these structural motifs sometimes are suggestive of function, but more often are not: the only case here with clear functional implications is the Rossmann fold.

Identifying motifs from sequence is not straightforward

Because motifs of the first kind—sequence motifs—always have functional implications, much of the effort in bioinformatics is directed at identifying these motifs in the sequences of newly discovered genes. In practice, this is more difficult than it might seem. The zinc finger motif is always uninterrupted, and so is easy to recognize. But many other sequence motifs are discontinuous, and the spacing between their elements can vary considerably. In such cases, the term sequence motif is almost a misnomer, since not only the spacing between the residues but also the order in which they occur may be completely different. These are really functional motifs whose presence is detected from the structure rather than the sequence. For example, the "catalytic triad" of the serine proteases (Figure 1-52), which consists of an aspartic acid, a histidine and a serine, all interacting with one another, comprises residues aspartic acid 102, histidine 57

Figure 1-50 Helix-turn-helix The DNA-binding domain of the bacterial gene regulatory protein lambda repressor, with the two helix-turn-helix motifs shown in color. The two helices closest to the DNA are the reading or recognition helices, which bind in the major groove and recognize specific gene regulatory sequences in the DNA. (PDB 1lmb)

Definitions

convergent evolution: evolution of structures not related by ancestry to a common function that is reflected in a common **functional motif**.

functional motif: sequence or structural **motif** that is always associated with a particular biochemical function.

motif: characteristic sequence or structure that in the case of a **structural motif** may comprise a whole domain or protein but usually consists of a small local arrangement of secondary structure elements which then coalesce to form domains. **Sequence motifs**, which are recognizable amino-acid sequences found in different proteins, usually indicate biochemical function. Structural motifs are less commonly associated with specific biochemical functions.

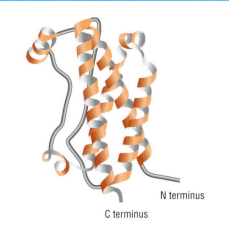

N terminus

C terminus

and serine 195 in one family of serine proteases. However, in another, unrelated family of serine proteases, the same triad is made up by aspartic acid 32, histidine 64, and serine 221 (see Figure 4-35). This is a case in which both the spacing between the residues that define the motif and the order in which they occur in the primary sequence are different. Nevertheless, these residues form a catalytic unit that has exactly the same geometry in the two proteases, and that carries out an identical chemical function. This is an example of **convergent evolution** to a common biochemical solution to the problem of peptide-bond hydrolysis. One of the major tasks for functional genomics is to catalog such sequence-based motifs, and develop methods for identifying them in proteins whose overall folds may be quite unrelated.

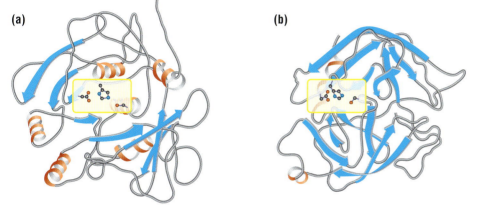

(a)

(b)

Figure 1-52 Catalytic triad The catalytic triad of aspartic acid, histidine and serine in **(a)** subtilisin, a bacterial serine protease, and **(b)** chymotrypsin, a mammalian serine protease. The two protein structures are quite different, and the elements of the catalytic triad are in different positions in the primary sequence, but the active-site arrangement of the aspartic acid, histidine and serine is similar.

Identifying structural motifs from sequence information alone presents very different challenges. First, as we have seen, many different amino-acid sequences are compatible with the same secondary structure; so there may be literally hundreds of different unrelated sequences that code for four-helix bundles. Sequence similarity alone, therefore, cannot be used for absolute identification of structural motifs. Hence, such motifs must be identified by first locating the secondary structure elements of the sequence. However, secondary structure prediction methods are not completely accurate, as pointed out earlier. Second, a number of structural motifs are so robust that large segments of additional polypeptide chain, even specifying entire different domains, can sometimes be inserted into the motif without disrupting it structurally. A common example is the so-called TIM-barrel domain, which consists of a strand of beta sheet followed by an alpha helix, repeated eight times. Protein domains are known that consist of nothing but this set of secondary structure elements; others are known in which an additional structural motif is inserted; and yet others are found in which one or more additional entire domains interrupt the pattern, but without disrupting the barrel structure (Figure 1-53).

References

Aitken, A.: **Protein consensus sequence motifs.** *Mol. Biotechnol.* 1999, **12**:241–253.

de la Cruz, X. and Thornton, J.M.: **Factors limiting the performance of prediction-based fold recognition methods.** *Protein Sci.* 1999, **8**:750–759.

Ponting, C.P. *et al.*: **Evolution of domain families.** *Adv. Protein Chem.* 2000, **54**:185–244.

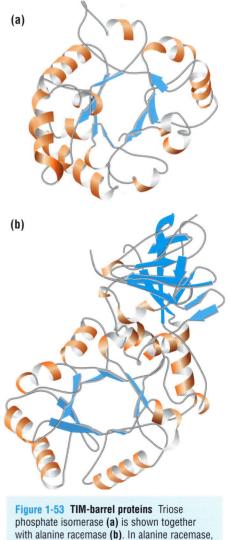

(a)

(b)

Figure 1-53 TIM-barrel proteins Triose phosphate isomerase **(a)** is shown together with alanine racemase **(b)**. In alanine racemase, the TIM-barrel domain is interrupted by an inserted domain.

Figure 1-54 Myohemerythrin A protein composed of a single four-helical bundle domain. (PDB 2mhr)

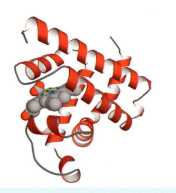

Figure 1-55 Myoglobin A protein composed of a single globin fold domain. (PDB 1a6k)

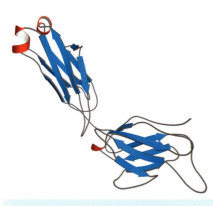

Figure 1-56 Immunoglobulin A protein composed of several beta domains (light chain only shown). (PDB 1a3l)

Protein domains can be classified according to their secondary structural elements

It is useful to group domain folds into five broad classes, based on the predominant secondary structure elements contained within them. **Alpha domains** are comprised entirely of alpha helices. **Beta domains** contain only beta sheet. **Alpha/beta domains** contain beta strands with connecting helical segments. **Alpha+beta domains** contain separate beta sheet and helical regions. And **cross-linked domains** have little, if any, secondary structure but are stabilized by several disulfide bridges or metal ions. Within each class, many different arrangements of these elements are possible; each distinct arrangement is a structural motif.

Two common motifs for alpha domains are the four-helix bundle and the globin fold

The preference for certain helix-crossing angles (see section 1-10) leads to two common motifs for interacting helices. One of them is a bundle of four antiparallel alpha helices, each crossing the next at an angle of about –20°, so that the entire motif has a left-handed twist. This **four-helix bundle** has been found in a wide variety of alpha domains, where it serves such diverse functions as oxygen transport, nucleic acid binding, and electron transport. Examples of four-helix bundle proteins include myohemerythrin, an oxygen-storage protein in marine worms (Figure 1-54), and human growth hormone, which helps promote normal body growth.

Another common alpha-domain motif, the **globin fold**, consists of a bag of about eight alpha helices arranged at +90° and +50° angles with respect to each other. This motif leads to the formation of a hydrophobic pocket in the domain interior in which large, hydrophobic organic and organometallic groups can bind (Figure 1-55). This fold gets its name from the protein myoglobin, a single-domain oxygen-storage molecule in which eight helices wrap around a heme group. It reappears in somewhat different form in the electron transport proteins called cytochromes, which also have bound heme groups. Interestingly, at least one heme-binding protein, cytochrome b562, is a four-helix bundle instead of a globin fold.

Beta domains contain strands connected in two distinct ways

Domains that contain only beta sheet, tight turns and irregular loop structures are called beta domains. Proteins made up of beta domains include immunoglobulins (Figure 1-56), several enzymes such as superoxide dismutase, and proteins that bind to sugars on the surfaces of cells. Because there are no helices to make long connections between adjacent strands of the beta sheet, all-beta domains contain essentially nothing but antiparallel beta structure, the strands of which are connected with beta turns and larger loops.

The patterns of connections between strands give rise to beta sheets with two distinct topologies. The directionality of the polypeptide chain dictates that a strand in an antiparallel beta sheet can only be linked to a strand an odd number of strands away. The most common connections are to an immediately adjacent strand or to one three strands away. If all the connections link adjacent strands, the beta sheet has an **up-and-down structural motif** (Figure 1-57). A particularly striking example is found in the enzyme neuraminidase from the influenza virus, which consists of a repeating structural motif of four antiparallel strands. Each up-and-down motif forms the blade of a so-called beta-propeller domain.

Definitions

alpha domain: a protein domain composed entirely of alpha helices.

alpha/beta domain: a protein domain composed of beta strands connected by alpha helices.

alpha+beta domain: a protein domain containing separate alpha-helical and beta-sheet regions.

beta domain: a protein domain containing only beta sheet.

beta sandwich: a structure formed of two antiparallel beta sheets packed face to face.

cross-linked domain: a small protein domain with little or no secondary structure and stabilized by disulfide bridges or metal ions.

four-helix bundle: a structure of four antiparallel alpha helices. Parallel bundles are possible but rare.

globin fold: a predominantly alpha-helical arrangement observed in certain heme-containing proteins.

Greek-key motif: an arrangement of antiparallel beta strands in which the first three strands are adjacent but the fourth strand is adjacent to the first, with a long connecting loop.

jelly roll fold: a beta sandwich built from two sheets with topologies resembling a Greek key design. The sheets pack almost at right-angles to each other.

up-and-down structural motif: a simple fold in which beta strands in an antiparallel sheet are all adjacent in sequence and connectivity.

Connection to the third strand leads to a motif called a **Greek key**, so named because it resembles the Greek-key design on ancient vases (Figure 1-58). An example of this motif is provided by pre-albumin, which contains two Greek-key motifs. The characteristic fold of the immunoglobulins, which is also found in a number of proteins that interact with other proteins on the cell surface, is a central Greek-key motif flanked on both sides by additional antiparallel strands.

Antiparallel beta sheets can form barrels and sandwiches

Antiparallel sheets in beta domains tend to be oriented with one face on the surface of the protein, exposed to the aqueous surroundings, and the other face oriented toward the hydrophobic core. This internal face is packed against another section of beta sheet with the inward-facing side chains of both packing together to form a hydrophobic core. Thus, in beta domains, the sheet tends to be amphipathic, with one face predominantly hydrophilic while the other is almost entirely composed of hydrophobic amino acids. This characteristic may make it possible to recognize such domains from the distribution of polar and nonpolar residues in the amino-acid sequence if secondary structure prediction methods become more accurate.

There are two ways to form structures in which antiparallel beta sheets can pack against each other. These give rise to beta barrels and **beta sandwiches**. In a beta-barrel motif, a single beta sheet forms a closed cylindrical structure in which all strands are hydrogen bonded to one another; the last strand in the sheet is hydrogen bonded to the first. Both types of beta-sheet connectivity are compatible with a beta barrel: pre-albumin is an example of a beta barrel constructed using the Greek-key motif (Figure 1-58), and human plasma retinol-binding protein, which carries vitamin A (retinol) in the serum, is an example of a beta barrel that is formed from an up-and-down motif (see Figure 1-19).

In a beta sandwich two separate beta sheets pack together face-to-face like two slices of bread. This arrangement differs from a barrel because the end strands of each sheet segment are not hydrogen bonded to one another. Their hydrogen-bonding potentials are satisfied chiefly by interactions with side chains or with water molecules. The two sheets in a beta sandwich are often at right angles to one another. Once again, both types of antiparallel sheet connectivity can be accommodated in this arrangement. The immunoglobulin fold (see Figure 1-56) is an example of a beta sandwich built with two Greek-key motifs. A variation of this theme is the **jelly roll fold** that comprises the major domain of the coat proteins of many spherical viruses. Bacteriochlorophyll A protein contains an antiparallel beta sandwich with up-and-down topology (Figure 1-59). The sandwich/barrel distinction is useful but not absolute: in a number of immunoglobulin domains the first or seventh strand switches sheets, forming a partial barrel.

The fibrous protein silk provides a particularly striking example of a beta sandwich. Silk is a beta-sheet protein composed largely of glycine, alanine, and serine, and every other residue in its sequence is a glycine. Because of the up-down alternation of residues in beta sheets, all glycines are on one side of the sheet, and the alanines and serines are on the other. The alanine and serine side chains of one sheet pack nicely between the alanine and serine side chains of another, producing a two-sheet structure. The tensile strength of silk derives from this interaction plus the hydrogen-bonded stability of the individual beta sheets themselves (see Figure 1-1).

Figure 1-59 Bacteriochlorophyll A protein This protein contains a domain with an up-and-down beta sandwich, a motif known as a jelly roll. (PDB 1ksa)

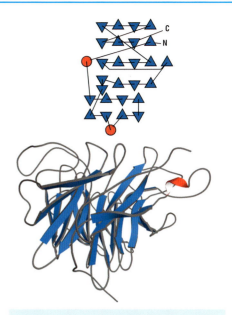

Figure 1-57 Neuraminidase beta-propeller domain A subunit of the four-subunit neuraminidase protein composed of repeating up-and-down beta motifs. (PDB 1a4q)

Greek key

Figure 1-58 Pre-albumin An example of a beta domain made up of Greek-key motifs. (PDB 1tta) Only one subunit of the two-subunit structure is shown.

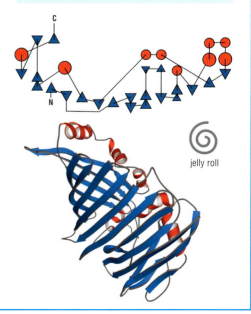

jelly roll

References

Bork, P. *et al.*: **The immunoglobulin fold. Structural classification, sequence patterns and common core.** *J. Mol. Biol.* 1994, **242**:309–320.

Branden, C. and Tooze, J.: *Introduction to Protein Structure* 2nd ed. (Garland, New York, 1999).

Richardson, J.S. and Richardson, D.C.: **Principles and patterns of protein conformation** in *Prediction of Protein Structure and the Principles of Protein Conformation* 2nd ed. Fasman, G.D. ed. (Plenum Press, New York, 1990), 1–98.

Weber, P.C. and Salemme, F.R.: **Structural and functional diversity in 4 alpha-helical proteins.** *Nature* 1980, **287**:82–84.

For a comprehensive analysis of domain folds, an all-against-all structure comparison can be found at:

http://www.ebi.ac.uk/dali/

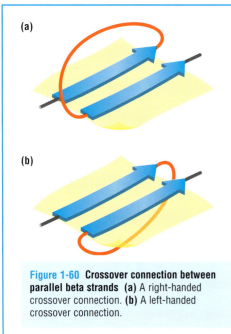

(a)

(b)

Figure 1-60 Crossover connection between parallel beta strands (a) A right-handed crossover connection. **(b)** A left-handed crossover connection.

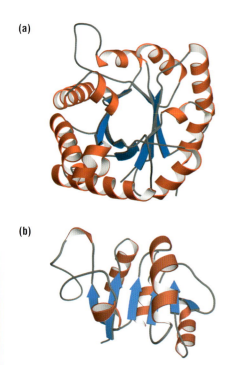

(a)

(b)

In alpha/beta domains each strand of parallel beta sheet is usually connected to the next by an alpha helix

In alpha/beta domains the beta sheet is composed of parallel or mixed strands; the parallel strands must be joined by long connections because the linking segment has to traverse the length of the sheet, and these connections are usually made by alpha helices connecting parallel adjacent strands, giving rise to beta-alpha-beta-alpha units. As illustrated in Figure 1-60, the crossover connection between the two parallel beta strands can be either right-handed or left-handed. The right-handed twist of the beta strand (see section 1-7), however, produces an enormous bias toward the right-handed crossover topology: it is observed in more than 95% of alpha/beta structures. This crossover rule is obeyed even when the connected strands are not adjacent or when the connecting segment is a loop, not a helix.

There are two major families of alpha/beta domains: barrels and twists

Just as two motifs predominate in antiparallel barrels and sandwiches, two motifs also account for nearly all alpha/beta domains. One of these is a closed structure called an **alpha/beta barrel** (Figure 1-61a). The other is an open twisted beta structure that looks somewhat like a saddle; we will call it an **alpha/beta twist** (Figure 1-61b).

The most regular form of alpha/beta structure is the alpha/beta barrel, in which the beta-alpha-beta-alpha motif is repeated four or more times. In this motif, the strand order is consecutive, and the combination of the twist of the beta sheet itself and the adjacent laying down of strands produces a closed barrel. This fold is particularly stable when there are eight strands in the barrel (Figure 1-61a). It is often called a **TIM barrel** because it was first discovered in the three-dimensional structure of the enzyme triosephosphate isomerase, which is abbreviated TIM.

The core of the alpha/beta barrel motif is its parallel beta sheet, which is surrounded by alpha helices that shield it from solvent. The helices are amphipathic and their nonpolar sides pack against the hydrophobic face of one side of the sheet. The center of a beta barrel is usually filled with hydrophobic side chains from the other face of the beta sheet; thus in alpha/beta barrels, the sheet is almost entirely hydrophobic. The TIM-barrel structure is one of the few domain folds that is relatively easy to recognize from the amino-acid sequence. Because it is the most common domain fold yet observed, occuring in 10% of all enzyme structures, it is a good bet that any sequence predicted to have a relatively nonpolar beta strand followed by an amphipathic alpha helix, repeated eight times, will form a TIM barrel.

The parallel beta strands in alpha/beta twists form an open sheet that is twisted into a saddle-shaped structure. The strand order in the sheet is not consecutive because the sheet is built in two halves. The first beta strand in the primary sequence forms a strand in the middle of the sheet. Additional strands are laid down consecutively outward to one edge, whereupon the chain returns to the middle of the sheet (the so-called "switch point") and forms the strand that hydrogen bonds to the outside of the first strand (Figure 1-61b). From there the chain continues out to the other edge. This mode of winding places the helices on one side for half of the sheet, and on the opposite side for the other half of the sheet. Again, the helices tend to

Figure 1-61 Alpha/beta domains (a) Alpha/beta barrel: the TIM barrel. (PDB 1tim) **(b)** Alpha/beta twist: aspartate semi-aldehyde dehydrogenase. (PDB 1brm) The connecting segments are usually alpha helices.

Definitions

alpha/beta barrel: a parallel beta barrel formed usually of eight strands, each connected to the next by an alpha-helical segment. Also known as a **TIM barrel**.

alpha/beta twist: a twisted parallel beta sheet with a saddle shape. Helices are found on one side of the sheet for the first half and the other side for the second half.

nucleotide-binding fold: an open parallel beta sheet with connecting alpha helices that is usually used to bind NADH or NADPH. It contains a characteristic

sequence motif that is involved in binding the cofactor. Also known as the Rossmann fold.

TIM barrel: another name for the alpha/beta barrel fold.

zinc finger: a small, irregular domain stabilized by binding of a zinc ion. Zinc fingers usually are found in eukaryotic DNA-binding proteins. They contain signature metal-ion binding sequence motifs.

be amphipathic whereas the sheet is predominantly hydrophobic. In its classic form, the alpha/beta twist motif has six parallel beta strands and five connecting helices, as shown in Figure 1-61b. Whenever this fold occurs in an enzyme, the switch region is always part of the catalytic site of the protein. Another name for this structure is the **nucleotide-binding fold**, which is indicative of the function it performs in many proteins.

In contrast to the antiparallel beta sheet, which always has one face in contact with water, most parallel beta structures are shielded from direct interaction with water by their coating of alpha helices. In the alpha/beta barrel motif, the interhelical packing angle is always +50°, and the same value is common for the helices that coat the surfaces of the alpha/beta twist motifs as well. The preference for this angle over the –20° and +90° alternatives reflects the need to nest the helices in the grooves on the surface of the twisted beta structure.

Alpha+beta domains have independent helical motifs packed against a beta sheet

Alpha+beta domains contain both beta sheets and alpha helices, but they are segregated. No special organizing principles can be stated for this class, but their individual secondary structure regions follow all of the principles we have described for alpha helices and beta sheets separately. The helical motifs in alpha+beta domains are usually just clusters of interacting helices, while the beta sheets tend to be antiparallel or mixed. One example is a saddle-shaped, antiparallel sheet with a layer of alpha helices covering one face (Figure 1-62). This arrangement leaves the other face of the sheet exposed to the solvent, which is a preference of antiparallel beta structures that we have already noted. Sometimes the layer of helices is used to form a recognition site, such as the peptide-binding groove in the major histocompatibility proteins.

Metal ions and disulfide bridges form cross-links in irregular domains

The final class of domain structure, the cross-linked irregular domain, is found in small single-domain intra- and extracellular proteins. There are two subclasses, which represent distinct solutions to the problem of structural stability in a domain that is too small to have an extensive hydrophobic core or a large number of secondary structural interactions. Both solutions involve cross-linking different parts of the domain via covalent interactions. In small irregular extracellular domains this cross-linking derives from disulfide bond formation, usually involving a number of cysteine pairs. In small irregular intracellular domains, metal ions (usually zinc but sometimes iron) form the cross-links, connecting different parts of the domain through ligation by nucleophilic side chains.

Disulfide-linked extracellular small proteins are often toxins that inhibit essential cellular proteins and prevent them from functioning. Most of these proteins are unusually stable to proteolytic digestion and heat denaturation. This class includes cobra venom neurotoxin, scorpion toxin (Figure 1-63), the ragweed pollen allergy factor Ra5, several secreted protease inhibitors, and toxic proteins from marine snails.

Metal ion cross-linked domains are found, for example, in **zinc finger** transcription factors (Figure 1-64) and iron–sulfur proteins called ferredoxins. A number of other metal-stabilized domains have been found. Although their structures are not as well characterized as that of the zinc finger, they too can be recognized at the sequence level because of characteristic sequence patterns in the vicinity of the residues that contribute metal ligands.

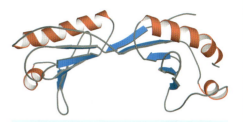

Figure 1-62 Alpha+beta saddle The structure of the TATA-binding protein that binds to DNA at the so-called TATA box that specifies the site at which gene transcription is initiated in eukaryotes. The beta sheet that forms the seat of the saddle binds in the minor groove of the DNA, bending it significantly. (PDB 1tgh)

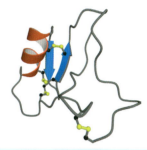

Figure 1-63 Disulfide-linked protein Scorpion toxin: a small irregular extracellular protein with no large hydrophobic core and minimal secondary structure. It is stabilized by four disulfide bridges. (PDB 1b7d)

Figure 1-64 Zinc finger A domain from a larger transcription factor, that is stabilized by the coordination of two histidines and two cysteines to a zinc ion. In the absence of the metal ion, this domain is unfolded, presumably because it is too small to have a hydrophobic core. This domain is the most abundant one in the human genome. (PDB 1aay)

References

Bellamacina, C.R.: **The nicotinamide dinucleotide binding motif: a comparison of nucleotide binding proteins.** *FASEB J.* 1996, **10**:1257–1269.

Branden, C. and Tooze, J.: *Introduction to Protein Structure*, 2nd ed. (Garland, New York, 1999).

Chothia, C.: **Asymmetry in protein structure.** *Ciba Foundation Symp.* 1991, **162**:36–49.

Leon, O. and Roth, M.: **Zinc fingers: DNA binding and**

protein-protein interactions. *Biol. Res.* 2000, **33**:21–30.

Reardon, D. and Farber, G.K.: **The structure and evolution of alpha/beta barrel proteins.** *FASEB J.* 1995, **9**:497–503.

Richardson, J.S. and Richardson, D.C.: **Principles and patterns of protein conformation** in *Prediction of Protein Structure and the Principles of Protein Conformation* 2nd ed. Fasman, G.D. ed. (Plenum Press, New York, 1990), 1–98.

Many proteins are composed of more than one polypeptide chain

Many proteins self-associate into assemblies composed of anything from two to six or more polypeptide chains. They may also associate with other, unrelated proteins to give mixed species of the form (ab), (a2b2), and so on (Figure 1-65a-c). The acetylcholine receptor, a membrane protein of vital importance for neuromuscular communication, is a five-chain molecule of the form (a2bcd) (Figure 1-65d). Proteins also assemble with other kinds of macromolecules.

Protein assemblies composed of more than one polypeptide chain are called **oligomers** and the individual chains of which they are made are termed **monomers** or subunits. Oligomers containing two, three, four, five, six or even more subunits are known as **dimers**, **trimers**, **tetramers**, **pentamers**, **hexamers**, and so on. Much the commonest of these are dimers. Some oligomers, as we have mentioned, contain only one kind of monomer, while others are made up of two or more different chains. Oligomers composed of only one type of monomer are sometimes prefixed homo-: for example, keratin, which is made up of three alpha-helical polypeptides coiled around one another, is composed of three identical chains and is thus a **homotrimer**. Oligomers composed of monomers encoded by different genes are prefixed hetero-: for example, hemoglobin, which contains two alpha and two beta chains, is a **heterotetramer** built from two homodimers. The number and kinds of subunits in an assembly, together with their relative positions in the structure, constitute the quaternary structure of an assembly.

It is very common for the subunits of hetero-oligomers to resemble one another structurally, despite being encoded by different genes and in some cases having little or no sequence similarity. This is true, for example, for hemoglobin, where the alpha and beta chains have nearly identical folds, and for the acetylcholine receptor, where the four different gene products that make up the pentamer are closely related structurally. One can speculate that this pattern reflects the origin of many hetero-oligomers in the duplication of a gene that coded for the single subunit of an ancestral homo-oligomeric protein.

Macromolecular assemblies form spontaneously when the right amounts of the appropriate components are present. The interactions between subunits are tight and specific, and they exclude "wrong" molecules from interfering with self assembly.

All specific intermolecular interactions depend on complementarity

Protein surfaces are irregular. This is what enables proteins to bind specific ligands and to associate specifically with other proteins, and it underlies the formation of quaternary structure. The "fit" between one protein surface and another depends on much more than shape. It extends to the weak bonds that hold complexes together; hydrogen-bond donors are opposite acceptors, nonpolar groups are opposite other nonpolar groups, and positive charges are opposite negative charges (Figure 1-66). This property of complementarity is observed in all binding interactions, whether between a protein and a small molecule or between a protein and another kind of macromolecule.

Complementarity is necessary because an intermolecular interface is composed of many weak interactions. Any single hydrogen bond or van der Waals interaction will break quite often at body temperature (see section 1-12). For a complex to be stable long enough to function, the strength of binding must be greater than about 15–20 kJ/mole. As free energies are additive, tight binding can be achieved if there is a large number of weak interactions, and the number

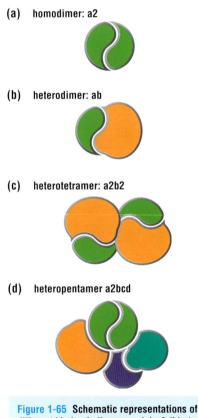

(a)   homodimer: a2

(b)   heterodimer: ab

(c)   heterotetramer: a2b2

(d)   heteropentamer a2bcd

Figure 1-65 Schematic representations of different kinds of oligomers (a) a2 **(b)** ab **(c)** a2b2 **(d)** a2bcd. Many other arrangements are possible and are observed (see Figure 1-74).

Definitions

coiled coil: a protein or a region of a protein formed by a dimerization interaction between two alpha helices in which hydrophobic side chains on one face of each helix interdigitate with those on the other.

dimer: an assembly of two identical (homo-) or different (hetero-) subunits. In a protein, the subunits are individual folded polypeptide chains.

heptad repeat: a sequence in which hydrophobic residues occur every seven amino acids, a pattern that is reliably indicative of a **coiled-coil** interaction between two alpha helices in which the hydrophobic side chains of each helix interdigitate with those of the other.

heterotetramer: an assembly of four subunits of more than one kind of polypeptide chain.

hexamer: an assembly of six identical or different subunits. In a protein the subunits are individual folded polypeptide chains.

homotrimer: an assembly of three identical subunits: in a protein, these are individual folded polypeptide chains.

monomer: a single subunit: in a protein, this is a folded polypeptide chain.

oligomer: an assembly of more than one subunit: in a protein, the subunits are individual folded polypeptide chains.

pentamer: an assembly of five identical or different subunits: in a protein, these are individual folded polypeptide chains.

quaternary structure: the subunit structure of a protein.

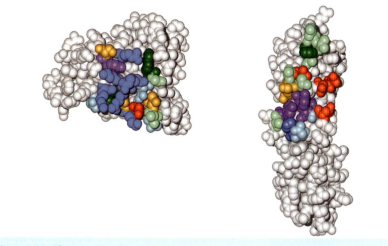

Figure 1-66 "Open-book" view of the complementary structural surfaces that form the interface between interleukin-4 (left) and its receptor (right) The contact residues are colored as follows: red, negatively charged; dark blue, positively charged; light blue, histidine; cyan, glutamine and asparagine; purple, tyrosine; yellow, serine/threonine; green, hydrophobic. Note that this interface contains a mixture of interaction types. Graphic kindly provided by Walter Sebald and Peter Reineme.

Figure 1-67 Coiled-coil alpha-helical interactions (a) Two interacting alpha helices of tropomyosin shown in a chain representation; (b) a space-filling representation of the separate alpha helices of tropomyosin with the hydrophobic side chains shown as dark protrusions; (c) the tropomyosin dimer showing how the hydrophobic side chains interdigitate in the coiled coil in a knobs in holes arrangement. (Taken from Cohen, C. and Parry, D.A.: **Alpha-helical coiled coils and bundles: how to design an alpha-helical protein.** *Proteins* 1990, **7**:1–15.)

and strength of weak interactions is maximized if contact surfaces fit closely together. Complementarity ensures that all possible van der Waals contacts are made, and that hydrogen-bond donors and acceptors at the interface between the two molecules pair with each other instead of making hydrogen bonds to water.

A particularly well characterized example of complementarity between interacting surfaces occurs in the case of coiled-coil structures (Figure 1-67). **Coiled coils** are dimers of alpha helices formed through the ridges and grooves arrangement we have already mentioned as the basis for tertiary structural interactions between alpha helices (see section 1-10). In such interacting helices, hydrophobic side chains, often those of leucines, are repeated at intervals of seven amino acids in the chain, forming the "ridge" of hydrophobic side chains that fit into spaces on the interacting helix. This pattern is known as the **heptad repeat**, and is characteristic of all dimeric structures formed through interacting alpha helices. It is one of the few cases in which structure can reliably be predicted from sequence.

Although all intermolecular interactions depend on surface complementarity, not all of them occur between preexisting complementary surfaces: one of the surfaces involved, or both, may be an unfolded region of the peptide in the absence of its partner. In coiled-coil proteins, for example, the two subunits are frequently unfolded as monomers and assume their folded structure only on dimerization. This is the case for the so-called leucine zipper family of transcriptional regulators which bind DNA on dimerization through a leucine-rich heptad repeat (Figure 1-68).

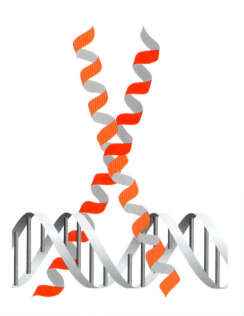

Figure 1-68 Peptide–peptide interactions in the coiled coil of the leucine zipper family of DNA-binding proteins The monomers of the leucine zipper are disordered in solution but fold on dimerization through hydrophobic coiled-coil interactions in their carboxy-terminal regions and on contact with DNA through their basic amino-terminal regions.

tetramer: an assembly of four identical or different subunits.

trimer: an assembly of three identical or different subunits.

References

Anston, A.A. *et al.*: **Circular assemblies.** *Curr. Opin. Struct. Biol.* 1996, **6**:142–150.

Bosshard, H.R. *et al.*: **Energetics of coiled coil folding:** **the nature of the transition state.** *Biochemistry* 2001, **40**:3544–3552.

Creighton, T.E.: *Proteins: Structure and Molecular Properties* 2nd ed. (Freeman, New York, 1993), 233–236.

Gonzalez, L. Jr *et al.*: **Buried polar residues and structural specificity in the GCN4 leucine zipper.** *Nat. Struct. Biol.* 1996, **3**:1011–1018.

Jones, S. and Thornton, J.M.: **Principles of protein-protein interactions.** *Proc. Natl Acad. Sci. USA* 1996, **93**:13–20.

Myers, J.K. and Oas, T.G.: **Reinterpretation of GCN4-p1 folding kinetics: partial helix formation precedes dimerization in coiled coil folding.** *J. Mol. Biol.* 1999, **289**:205–209.

Perham, R.N.: **Self-assembly of biological macromolecules.** *Philos. Trans. R. Soc. Lond. B.* 1975, **272**:123–136.

Zielenkiewicz, P. and Rabczenko, A.: **Methods of molecular modelling of protein-protein interactions.** *Biophys. Chem.* 1988, **29**:219–224.

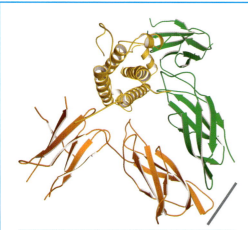

Protein assemblies built of identical subunits are usually symmetric

Protein complexes are built up through interactions across complementary binding surfaces. If one subunit has binding region A, the subunit it binds to must have the complementary region A'. If the interacting subunits are not identical, then nothing definite can be said about the spatial relationship of the monomers in the complex and the complex is said to be asymmetric. The human growth hormone–receptor complex is an example of an asymmetric complex (Figure 1-73).

If the subunits are identical, however, interactions across complementary surfaces nearly always produce symmetric complexes, in which the subunits are related to one another with one of a few kinds of geometry (Figure 1-74). Identical subunits form symmetric complexes because, in order to interact, each subunit must possess binding region A and its complement A'. (This is in contrast to non-identical subunits, each of which has only one or the other.) Depending on the location of A and A' on the surface, subunits can associate to form closed structures (Figure 1-75a and b), with dimers and trimers being most common, or, much more rarely, open-ended chains, with helical arrangements being most common.

The repeating unit from which a symmetric complex is built can be either a monomer or an association of unlike polypeptide chains. For example, hemoglobin, which is constructed from four polypeptide chains, (a2b2), is a symmetric dimer of two (ab) units. The asymmetric unit from which a symmetric complex is built is referred to as the **protomer**.

If the subunit has a second set of complementary binding regions, B and B', in addition to A and A', it can associate to form more elaborate complexes (Figure 1-75c). A second binding region can allow symmetric rings to pair, with pairs of dimers that form tetramers and pairs of trimers that form hexamers being the most common. Insulin is an example of a hexameric protein that is built in this way (Figure 1-74f). In a similar way, open-ended chains can associate side-by-side to form multistranded helices. This is what happens in sickle-cell hemoglobin, when an additional binding site is created by mutation (see Figure 1-71). Subunits with two sets of complementary binding regions can also associate into more complex structures, usually described by reference to geometric figures—tetrahedra, octahedra, icosahedra—with the same symmetry. Type II 3-dehydroquinate dehydratase, for example, crystallizes as a dodecamer in which a tetramer of trimers forms a tetrahedron (Figure 1-74j); and the rhinovirus that causes the common cold is a large multisubunit icosahedron (Figure 1-74k).

So powerful is the tendency of subunits to form symmetric arrangements that this even influences the structure of oligomers made up of non-identical polypeptide chains. Many of these proteins are **pseudosymmetric**, as we have already seen for hemoglobin. In this protein the alpha and beta subunits are similar in sequence and hence nearly identical in structure, so it is a nearly symmetrical tetramer of four monomers. The giant multisubunit proteolytic complex called the proteasome is another example of a pseudosymmetric structure (Figure 1-74l).

Figure 1-73 The human growth hormone–receptor complex Structure of the human growth hormone (yellow) complexed with two identical molecules of its receptor (orange and green). The receptor is a membrane protein, but only the extracellular hormone-binding portion is shown. The plane of the membrane is indicated by the slanted line. A molecule of the monomeric hormone binds to two identical receptor molecules. Similar regions of the two receptor molecules are used to bind two distinct regions of the hormone; the conformational flexibility of these regions allows for this versatility. (PDB 3hhr)

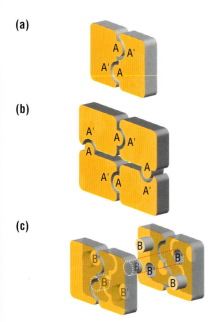

(a)

(b)

(c)

Figure 1-74 Examples of quaternary arrangements observed for oligomeric proteins The structures shown in **a-k** are homo-oligomers. The proteasome (**l**) is a pseudo-symmetric structure, in which the subunits are not identical. (**a**) D-amino acid aminotransferase (PDB 3daa); (**b**) KDGP aldolase (PDB 1fq0); (**c**) neuraminidase (PDB 1a4q); (**d**) lactate dehydrogenase (PDB 1ldn); (**e**) cholera toxin (PDB 1chp); (**f**) insulin (PDB 4ins); (**g**) molybdenum cofactor biosynthesis protein C (PDB 1ekr); (**h**) GroES co-chaperonin (PDB 1g31); (**i**) galactonate dehydratase; (**j**) 3-dehydroquinate dehydratase (PDB 2dhq); (**k**) rhinovirus (PDB 1aym): this multisubunit protein has the same geometry as a soccer ball; (**l**) proteasome (PDB 1g65).

Figure 1-75 Interactions underlying different geometric arrangements of subunits Subunits with a pair of complementary binding sites A and A' may form symmetric dimers (**a**) or tetramers (**b**) depending on the positions of the two binding sites. More complex assemblies may be formed by subunits with a second pair of complementary binding sites B and B' that could for example allow the formation (**c**) of a tetrameric complex of two dimers.

Definitions

protomer: the asymmetric repeating unit (or units) from which an oligomeric protein is built up.

pseudosymmetric: having approximate but not exact symmetry. A protein with two non-identical subunits of very similar three-dimensional structure is a pseudosymmetric dimer.

References

Goodsell, D.S. and Olson, A.J.: **Structural symmetry and protein function.** Annu. Rev. Biophys. Biomol.

Struct. 2000, **29**:105–153.

Matthews, B.W. and Bernhard, S.A.: **Structure and symmetry of oligomeric enzymes.** Annu. Rev. Biophys. Bioeng. 1973, **2**:257–317.

Milner-White, E.J.: **Description of the quaternary structure of tetrameric proteins. Forms that show either right-handed and left-handed symmetry at the subunit.** Biochem. J. 1980, **187**:297–302.

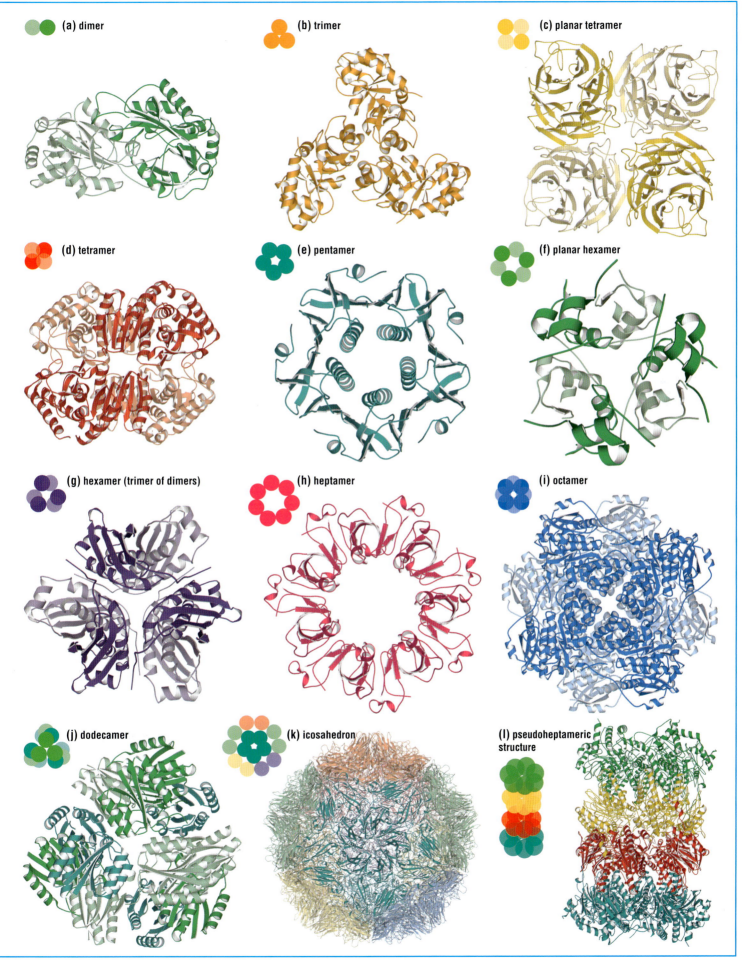

(a) dimer

(b) trimer

(c) planar tetramer

(d) tetramer

(e) pentamer

(f) planar hexamer

(g) hexamer (trimer of dimers)

(h) heptamer

(i) octamer

(j) dodecamer

(k) icosahedron

(l) pseudoheptameric structure

Types of Motion Found in Proteins (all values approximate)			
Motion	Spatial displacement (Å)	Characteristic time (s)	Energy source
Fluctuations (e.g., atomic vibrations)	0.01 to1	10^{-15} to 10^{-11}	k_bT
Collective motions (A) fast, infrequent (e.g., Tyr, Phe ring flips) (B) slow (e.g., domain movement; hinge-bending)	0.01 to > 5	10^{-12} to 10^{-3}	k_bT
Triggered conformational changes	0.5 to > 10	10^{-9} to 10^{3}	Binding interactions

Figure 1-76 Table of protein motions

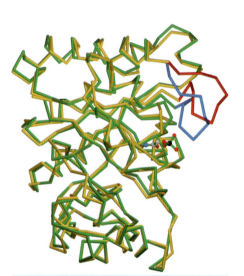

Figure 1-77 Triosephosphate isomerase
Binding of substrate or inhibitor to the active site of the enzyme triosephosphate isomerase induces a 10 Å rigid-body movement in an eight-residue loop (red; open) which closes down over the active site (blue; closed) and shields the substrate from solvent. The inhibitor can be seen just below the loop.

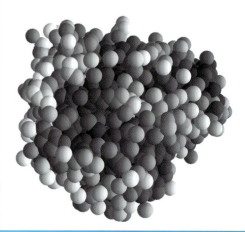

Proteins are flexible molecules

The pictures of protein structures that emerge from X-ray crystallography and NMR seem rigid and static; in reality, proteins are highly flexible. Because the forces that maintain the secondary and tertiary folds are weak, there is enough energy available at body temperature to break any particular interaction. When existing weak interactions are broken, the groups that are released can make new interactions of comparable energy. These rearrangements can occur on a time scale that is faster than the time required to determine the structure by tools such as X-ray crystallography. Thus, the three-dimensional structures of proteins determined by physical techniques are average structures.

Protein motions can be classified in terms of their relationship to the average structure (Figure 1-76). The fastest motions are atomic fluctuations such as interatomic vibrations and the rotations of methyl groups. Next come collective motions of bonded and non-bonded neighboring groups of atoms, such as the wig-wag motions of long side chains or the flip-flopping of short peptide loops. The slowest motions are large-scale, ligand-induced conformational changes of whole domains.

Conformational fluctuations in domain structure tend to be local

It is almost as important to understand what types of conformational change are not observed in proteins as it is to realize that they are flexible in the first place. Whole folded domains never undergo large, thermally driven distortions at ordinary temperatures. Transitions from one type of folding motif to another are rarely seen except in pathological cases; an all alpha-helical protein will not normally refold to an all beta-sheet protein, except, for example, in the cases of amyloid and prion diseases. Smaller-scale refolding does occur in some proteins, however. Ligand binding may induce disordered polypeptide segments to become ordered. Ligands can also induce the disordering of a previously ordered strand, although this is less common. Association and dissociation of subunits can also be triggered by ligand binding, and the ligand can be as small as a proton if it changes the charge of a crucial residue.

Perhaps the most common ligand-induced conformational change is the lid-like movement of a polypeptide segment to cover a ligand-binding site (Figure 1-77). When the lid is open, there is free access to the ligand-binding site. Once the site is occupied, the loop interacts with the ligand to stabilize the closed conformation, and closure isolates the bound ligand from the surrounding solvent. Most loop closures involve rigid-body movement of the loop on two hinges. The internal conformation of the loop does not change appreciably because its side chains are packed closely together, making it function like a solid lid for the ligand-binding site. Mobile loops both act as gates for ligand binding and can make interactions that stabilize the complex. They play an important part in many enzymes.

Protein motions involve groups of non-bonded as well as covalently bonded atoms

At body temperature, the atoms in most protein molecules fluctuate around their average positions by up to an Ångstrom or occasionally even more, depending on their position in the protein (Figure 1-78). In the tightly packed interior, atomic motions are restricted to less than an Ångstrom. The closer to the surface of the molecule, the greater the increase in mobility until, for surface groups that are not surrounded by other atoms, the mean fluctuation may be several Ångstroms. Proteins have been called "semi-liquid" because the movements of their atoms are larger than those found in solids such as NaCl, but smaller than those observed in a liquid like water.

In a protein, the covalent structure of the polymer sets limits on the motions of atoms and groups of atoms. Chemical groups such as methyl groups or aromatic side chains display collective motions. Methyl groups rotate on a picosecond time-scale; aromatic rings, even those in the interior of the protein, flip at average rates of several thousand per second. The actual ring flip takes only about a picosecond, but it happens only about once every 10^9 picoseconds.

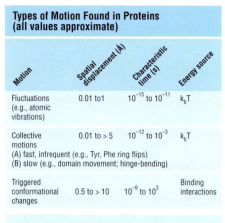

Figure 1-78 Protein shaded according to flexibility Space-filling model of sperm whale myoglobin in which each atom is shaded according to its average motion as determined by X-ray crystallography. The darker the atom, the more rigid it is. Note that the surface is not uniform in its flexibility. (PDB 1a6k)

Flipping an aromatic ring inside a protein, where the packing density is high, requires that surrounding atoms move out of the way. The probability that they will all move in the right direction at the same time is very low: hence the relatively long interval between flips. In the interior of proteins, close atomic packing couples the motions of non-bonded neighboring atoms. If a methyl group in the center of a protein is next to another methyl group, the motions of both will be correlated by virtue of their tendency to collide. Thus both the extent of motion of every group and its preferred directions depend on non-bonded as well as bonded contacts. Only for surface side chains and protruding loops are non-bonded interactions of little importance, and residues in such unrestrained positions are always the most flexible parts of a protein structure.

At biological temperatures, some proteins alternate between well-defined, distinct conformations (Figure 1-79). In order for two conformational states to be distinct, there must be a free-energy barrier separating them. The motions involved to get from one state to the other are usually much more complex than the oscillation of atoms and groups about their average positions. It is often the case that only one of the alternative conformations of a protein is biologically active.

Triggered conformational changes can cause large movements of side chains, loops, or domains

Of most importance for protein function are those motions that occur in response to the binding of another molecule. Ligand-induced conformational changes can be as modest as the rearrangement of a single side chain, or as complex as the movement of an entire domain. In all cases, the driving force is provided by ligand–protein interactions.

Often, the motion enables some part of the structure to make contact with a ligand. For example, the binding of aspartate to a large domain of the enzyme aspartate aminotransferase causes a smaller domain to rotate by 10°. This rotation moves the small domain by more than 5 Å, bringing it into closer contact with the rest of the protein (Figure 1-80). When the binding of a specific ligand causes a protein to change from an inactive to an active conformation, the process is described as **induced fit**. The driving force for induced fit in aspartate aminotransferase appears to be the formation of a salt bridge between an arginine residue in the mobile domain and the alpha-carboxylate of the bound aspartate. Mutant enzymes that are unable to carry out this triggered conformational change are inactive. We discuss the use of conformational changes to regulate enzymes in more detail in Chapter 2.

Ligand-induced conformational changes can also change the quaternary structure of proteins. This usually involves repacking of the interfaces between subunits so that the relative positions of the monomers are altered; this happens when oxygen binds to the tetramer hemoglobin. Sometimes, however, the stoichiometry of the oligomer changes on ligand binding. One example is the polymerization of actin monomers, driven by the binding of ATP, into linear helical polymers called thin filaments or microfilaments. Regulated polymerization of actin is essential for the formation and disassembly of cytoskeletal components needed for cell movement. Changing the oligomeric state of actin is a mechanism for controlling what it does.

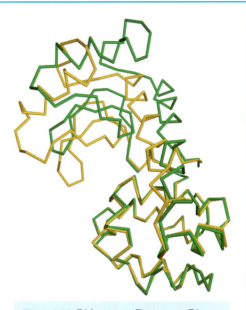

Figure 1-79 **T4 lysozyme** The enzyme T4 lysozyme contains two domains connected by a hinge. In different crystal forms of the protein, an open and closed state have been observed, related to each other by a hinge-bending motion. It is presumed that the protein in solution can exist in an equilibrium between both states at physiological temperature. (PDB 1l96 and 1l97)

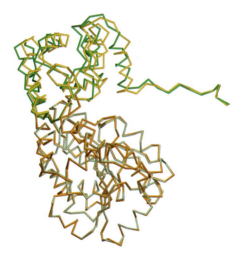

Figure 1-80 **Aspartate aminotransferase, open and closed forms** The enzyme L-aspartate aminotransferase contains two domains with the active site lying between them. Substrate binding induces a movement of the small domain (green) to a new position (yellow) in which the active site is more enclosed. This movement is essential to position some of the residues important for catalysis, and only the specific substrates of the enzyme induce it. (PDB 1ars and 1art)

Definitions

induced fit: a change in the conformation of a protein induced by the binding of a ligand. In the case of an enzyme, this may result in catalytic activation.

References

Arrondo, J.L. and Goni, F.M.: **Structure and dynamics of membrane proteins as studied by infrared spectroscopy.** *Prog. Biophys. Mol. Biol.* 1999, **72**:367–405.

Daggett, V.: **Long timescale simulations.** *Curr. Opin. Struct. Biol.* 2000, **10**:160–164.

Ishima, R. and Torchia, D.A.: **Protein dynamics from NMR.** *Nat. Struct. Biol.* 2000, **7**:740–743.

Karplus, M. and Petsko, G.A.: **Molecular dynamics simulations in biology.** *Nature* 1990, **347**:631–639.

Petsko, G.A. and Ringe, D.: **Fluctuations in protein structure from X-ray diffraction.** *Annu. Rev. Biophys. Bioeng.* 1984, **13**:331–371.

Ringe, D. and Petsko, G.A.: **Mapping protein dynamics by X-ray diffraction.** *Prog. Biophys. Mol. Biol.* 1985, **45**:197–235.

Wall, M.E. *et al.*: **Large-scale shape changes in proteins and macromolecular complexes.** *Annu. Rev. Phys. Chem.* 2000, **51**:355–380.

2

From Structure to Function

There are many levels of protein function, ranging from atomic reorganizations to changes in the development of an organism, but all of them involve binding to other molecules, large and small. Sometimes this specific molecular recognition is the sole biochemical function of a protein, but in other cases the protein also promotes a chemical transformation in the molecule that it binds. This chapter looks first at how the structural features described in Chapter 1 dictate the ability of proteins to recognize specifically and bind a wide variety of ligands. The second part of the chapter looks at how these structural features dictate the ability of proteins to catalyze the wide variety of chemical transformations on which life depends.

There are many levels of protein function

It is a fundamental axiom of biology that the three-dimensional structure of a protein determines its function. Understanding function through structure is a primary goal of structural biology. But this is not always simple, partly because a biologically useful definition of the function of a protein requires a description at several different levels. To the biochemist, function means the biochemical role of an individual protein: if it is an enzyme, function refers to the reaction catalyzed; if it is a signaling protein or a transport protein, function refers to the interactions of the protein with other molecules in the signaling or transport pathway. To the geneticist or cell biologist, function includes these roles but will also encompass the cellular roles of the protein, as judged by the phenotype of its deletion, for example, or the pathway in which it operates. A physiologist or developmental biologist may have an even broader view of function.

We can take as an example tubulin, which not only has several cellular functions, but also has more than one biochemical function: it is an enzyme that hydrolyzes GTP, and also a structural protein that polymerizes to form stiff hollow tubes. In the cell, it forms a network of microtubules growing out of the centrosome (Figure 2-1), creating a system of tracks along which proteins, vesicles and organelles can be moved from one part of the cell to another. In a dividing cell it forms the mitotic spindle that segregates the chromosomes equally into the two daughter cells. In certain motile eukaryotic cells it forms the cilia and flagella that provide propulsion or sweep fluid over the cell surface. Tubulin is a protein whose functions cannot be condensed into a single sentence. This is likely to be the case for most gene products, especially in higher organisms.

Not surprisingly, biochemical function is generally the easiest to deduce from sequence and structure, although in some cases it is possible to go further. In the age of genomics, function will be derived in a partly empirical way from many different techniques employed together, augmented by comparative sequence analysis across genomes and the recognition of functional motifs in both the primary and tertiary structure. We shall illustrate in Chapter 4 how this operates in some selected cases. In this chapter and the next, we outline the general principles that have been experimentally established about the relationship of structure to the biochemical function of proteins.

There are four fundamental biochemical functions of proteins

We illustrated in Chapter 1 four biochemical functions of proteins: binding, catalysis, switching, and as structural elements (see Figure 1-1). The most fundamental of these is binding, which underlies all the other biochemical functions of proteins. Enzymes must bind *substrates*, as well as cofactors that contribute to catalysis and regulatory molecules that either activate or inhibit them. Structural proteins are, at their simplest, assemblages of a single type of protein molecule bound together for strength or toughness; in more complex cases they bind to other types of molecules to form specialized structures such as the actin-based intestinal microvilli or the spectrin-based mesh that underlies the red blood cell membrane and helps maintain its integrity as the cells are swept round the body. Protein switches such as the *GTPases* (see Figure 2-1) depend on both binding and catalytic functions of proteins: their switching properties rely fundamentally on the binding and the hydrolysis of GTP, which they catalyze. They must also bind the molecules with which they interact when GTP is hydrolyzed plus the regulatory molecules that activate GTP hydrolysis and that exchange GDP for GTP to enable the cycle to start again.

References

Alberts, B. *et al.*: *Molecular Biology of the Cell* 4th ed. Chapter 3 (Garland Publishing, New York, 2002).

Desai, A. and Mitchison, T.J.: **Microtubule polymerization dynamics.** *Annu. Rev. Cell Dev. Biol.* 1997, **13**:83–117.

Nogales, E.: **Structural insights into microtubule function.** *Annu. Rev. Biochem.* 2000, **69**:277–302.

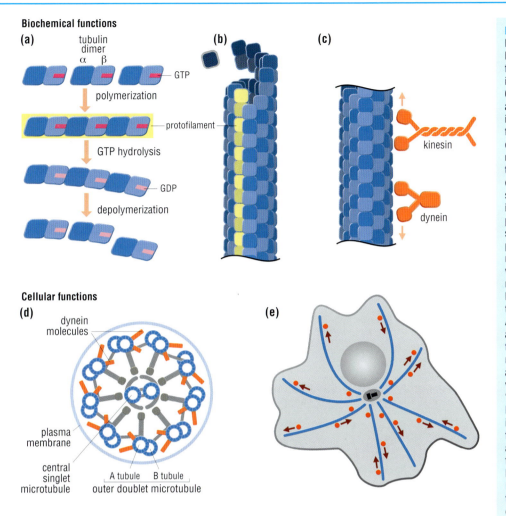

Biochemical functions

(a)

tubulin dimer
α   β

GTP

polymerization

protofilament

GTP hydrolysis

GDP

depolymerization

(b)

(c)

kinesin

dynein

Cellular functions

(d)

dynein molecules

plasma membrane

central singlet microtubule

A tubule   B tubule

outer doublet microtubule

(e)

Figure 2-1 The functions of tubulin (a) The biochemical functions of tubulin include binding of tubulin monomers to each other to form a polymeric protofilament, a process that is reversed by the hydrolysis of bound GTP to GDP. Tubulin-catalyzed hydrolysis of GTP acts as a switching mechanism, in that protofilaments in the GDP form rapidly depolymerize unless the concentration of free tubulin is very high or other proteins stabilize them. **(b)** This nucleotide-dependent mechanism is used by the cell to control the assembly and disassembly of the protofilaments and the more complex structures built from them. These structures include microtubules, which consist of 13 protofilaments arranged as a hollow tube (here shown in growing phase). **(c)** Binding to motor proteins such as kinesin or dynein allows the microtubules to form molecular machines in which these motor proteins "walk" along microtubules in a particular direction, powered by ATP hydrolysis. **(d, e)** The functions of these machines are defined at the cellular level. Assemblies of microtubules, motor proteins and other microtubule-associated proteins form the flagella that propel sperm, for example (d); microtubules and associated motor proteins also form a network of "tracks" on which vesicles are moved around in cells (e). The "role" of tubulin thus encompasses both biochemical and cellular functions. The individual functions of proteins work in concert to produce the exquisite machinery that allows a cell, and ultimately a multicellular organism, to grow and survive. The anti-cancer drug taxol blocks one essential cellular function of tubulin. It binds to the polymerized protein, preventing the disassembly of microtubules that must occur during cell division.

We start this chapter, therefore, with the surface properties of proteins that determine where and how they bind to the other molecules with which they interact, whether these are small-molecule *ligands* or other macromolecules, and whether they bind stably, as oligomeric complexes, or dynamically, as in intracellular signaling pathways. And we shall see how they create the specialized microenvironments that promote the specific interactions required for catalysis.

We then explore in detail the structural basis for enzyme catalysis, and how the special chemical properties of some of the charged amino acids described in Chapter 1 (see section 1-1) make a critical contribution to enzyme action. We shall see that protein flexibility plays an essential part in the binding interactions of proteins and their catalytic actions.

Protein switches, and the more dramatic conformational changes that proteins can undergo in response to ligand binding, are described in Chapter 3, where we discuss the ways in which the activities of proteins are regulated.

Protein functions such as molecular recognition and catalysis depend on complementarity

The functions of all proteins, whether signaling or transport or catalysis, depend on the ability to bind other molecules, or **ligands**. The ligand that is bound may be a small molecule or a macromolecule, and binding is usually very specific. Ligand binding involves the formation of noncovalent interactions between ligand and protein surface; these are the same types of bonds that are involved in stabilizing folded proteins (see section 1-4) and in interactions between protein subunits (see section 1-19). Specificity arises from the complementarity of shape and charge distribution between the ligand and its binding site on the protein surface (Figure 2-2), and from the distribution of donors and acceptors of hydrogen bonds. Changes in the conformation of a protein may accompany binding or be necessary for binding to occur. Alternatively, even a small change in the structure of a ligand or protein can abolish binding.

Molecular recognition depends on specialized microenvironments that result from protein tertiary structure

Specific binding occurs at sites on the protein that provide the complementarity for the ligand. These are called **ligand-binding sites** if their sole function is molecular recognition (the ligand may be as small as a proton or as large as another macromolecule) or **active sites** if they promote chemical catalysis. Such sites are formed as a consequence of the three-dimensional structure of the protein. When a polypeptide sequence folds into a compact three-dimensional structure, it creates internal cavities where the side-chain packing is not perfect, and also pockets or clefts of various sizes on its surface. These regions can have a microscopic environment that is quite different from that provided by the bulk solution around the protein. If the residues that line the cavity or pocket are hydrophobic, for example, the environment inside can resemble a nonpolar organic solvent more than it does water, enabling the protein to bind highly hydrophobic ligands such as lipids. If the residues all have, say, a negative charge, the cavity or pocket can have a very strong local electrostatic field, which would enable it to bind highly charged ligands like a calcium ion, as occurs in ion-transport proteins. Such arrangements can occur even if at first sight they seem to be unfavorable energetically—for instance, crowding a number of like charges together—because the enormous number of other, favorable interactions throughout the rest of the structure can more than make up for a small number of unfavorable ones in one place.

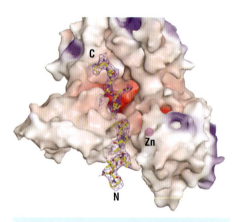

Figure 2-2 Substrate binding to anthrax toxin lethal factor Lethal factor (LF) is a component of anthrax toxin that acts as a protease to cut mitogen-activated protein kinase kinase (MAPKK-2), thereby blocking the cell cycle. This figure shows part of the surface of LF colored by charge (red, negative; blue, positive), with the model of the MAPKK-2 amino-terminal peptide shown in ball-and-stick representation. Where the model or map would be hidden by the protein surface, the surface is rendered as translucent. The active-site cleft of LF is complementary in shape and charge distribution to the substrate. Taken from Pannifer, A.D. *et al.: Nature* 2001, **414**:229–233. Graphic kindly provided by Robert Liddington.

Specialized microenvironments at binding sites contribute to catalysis

Most enzymes operate, in part, through *general acid-base catalysis*, in which protons are transferred between donating or accepting atoms on the substrate and key basic and acidic side chains in the enzyme active site (we discuss this in detail later, in section 2-12). Proton transfer can be promoted in two ways by specialized microenvironments on the surface of proteins.

In one case, a strong electrostatic field is produced in which acids and bases are close to each other but cannot react with each other. For instance, the close juxtaposition of an acidic side chain such as glutamic acid and a basic one such as lysine is possible. In aqueous solution at neutral pH a carboxylic acid would give up its proton to the more basic amino group, forming a carboxylate–ammonium ion charge pair. Indeed, such interactions are used to stabilize protein structure (see Figure 1-10, salt bridge). But a folded protein can also position these two

---

Definitions

active site: asymmetric pocket on or near the surface of a macromolecule that promotes chemical catalysis when the appropriate **ligand** (substrate) binds.

ligand: small molecule or macromolecule that recognizes and binds to a specific site on a macromolecule.

ligand-binding site: site on the surface of a protein at which another molecule binds.

residues in reasonable proximity, yet provide microenvironments around each one that make it unfavorable for them to exchange a proton. This occurs in many enzyme active sites; for example, in aspartate aminotransferase, a lysine and an aspartic acid are both involved in binding the pyridoxal phosphate cofactor, but in the environment of the active site their proton affinities are adjusted so that they do not transfer a proton between them. Thus, sites can be created that have both reasonably strong acids and reasonably strong bases in them, which is very difficult to achieve in free solution but very useful for general acid-base catalysis.

The other type of environment is one in which the affinity of a functional group for protons has been altered dramatically. Placing two lysine side chains close to one another will lower the proton affinity of both of them, producing a stronger acid. At physiological pH the two side chains will exist as a mixture of protonated and unprotonated states, enabling one of them to function as an acid-base catalyst (Figure 2-3). Analogously, burying a single lysine side chain in a hydrophobic pocket or cavity will lower its affinity for protons so that it will tend to be unprotonated at neutral pH, a state in which it is more reactive to certain chemical processes. One example of this effect is the enzyme pyruvate decarboxylase, where the active site is hydrophobic. Model compounds that mimic the substrates of this enzyme react thousands of times faster in non-polar solvents than they do in water. The enzyme favors the reaction by providing a nonpolar environment that destabilizes the substrate, which is negatively charged, and stabilizes the product, which is uncharged.

Once again, it is important to note that both of these situations are energetically unfavorable, but the energy cost is offset by favorable interactions elsewhere in the protein. In addition, long-range electrostatic effects in proteins can create a microenvironment in which the affinity of an isolated side chain for a proton can be perturbed. The significance of this effect is to allow a particular ionization state to predominate so that it can be exploited in the chemical reaction.

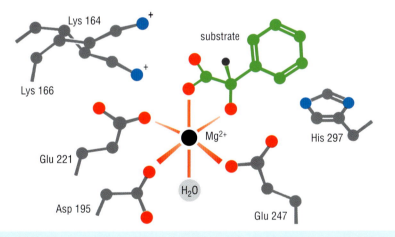

Figure 2-3 Schematic of the active site of mandelate racemase showing substrate bound Lysine 164 is located very close to the catalytically important residue lysine 166, shown at the top left of the figure. The proximity of these two positive charges lowers the proton affinity of both of them, making lysine 166 a better proton shuttle for the metal-bound substrate. Landro, J.A. et al.: Biochemistry 1994, **33**:635–643.

References

Badger, J. et al.: **Structural analysis of antiviral agents that interact with the capsid of human rhinoviruses.** Proteins 1989, **6**:1–19.

Highbarger, L.A. et al.: **Mechanism of the reaction catalyzed by acetoacetate decarboxylase. Importance of lysine 116 in determining the pK_a of active-site lysine.** Biochemistry 1996, **35**:41–46.

Landro, J.A. et al.: **The role of lysine 166 in the mechanism of mandelate racemase from Pseudomonas**

putida: **mechanistic and crystallographic evidence for stereospecific alkylation by (R)-alpha-phenylglycidate.** Biochemistry 1994, **33**:635–643.

Pannifer, A.D. et al.: **Crystal structure of the anthrax lethal factor.** Nature 2001, **414**:229-233.

Ringe, D.: **What makes a binding site a binding site?** Curr. Opin. Struct. Biol. 1995, **5**:825–829.

2-2 Flexibility and Protein Function

The flexibility of tertiary structure allows proteins to adapt to their ligands

In order to bind a ligand specifically a protein must have, or be able to form, a binding site whose stereochemistry, charge configuration, and potential hydrogen-bond-forming groups are complementary to those of the ligand. The classic analogy is that of a key fitting into a lock. This analogy holds for many proteins and ligands: more than one key will sometimes fit into a lock but most other keys will not, and usually only the right key will open the lock. Similarly, more than one ligand will sometimes fit into a binding site on the surface of a protein, but most other ligands will not, and usually only the right ligand will produce the "right" biological function. However, the lock-and-key analogy implies rigidity of the protein (the lock) and of the ligand (the key). In reality, both proteins and the ligands that bind to them are naturally flexible, so the classic view has been augmented by a model of **induced fit**: during binding, each can adjust its structure to the presence of the other and the protein can be said to "snuggle" around the ligand, optimizing interactions between them (Figure 2-4). More fundamentally, this give and take on the part of protein and ligand is essential for the biochemical activities of proteins.

Such conformational changes are allowed because of the inherent flexibility of proteins. In undergoing conformational changes, the protein is responding to changes in the balance between the forces that hold the tertiary structure together and the new interaction forces provided by association with a ligand. It is because both the ligand and the molecule to which it binds are flexible that many drugs that do not obviously resemble the biological ligand are nonetheless able to bind tightly to a ligand-binding site (Figure 2-5). Both the drug and the site can adjust somewhat to accommodate the difference in shape of the small molecule, provided that enough favorable interactions can be made to overcome any energetic cost of the adjustments.

Protein flexibility is essential for biochemical function

Although some proteins serve only to bind ligand, in most cases binding is followed by some action. That action may be the chemical transformation of the ligand (catalysis), a conformational change in the protein, translocation of the protein to another part of the cell, transport

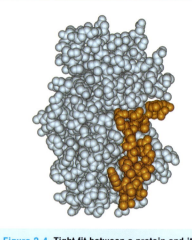

Figure 2-4 Tight fit between a protein and its ligand A space-filling representation of the catalytic domain of protein kinase A (blue) bound to a peptide analog (orange) of its natural substrate shows the snug fit between protein and ligand, achieved by mutual adjustments made by the two molecules. (PDB 1atp)

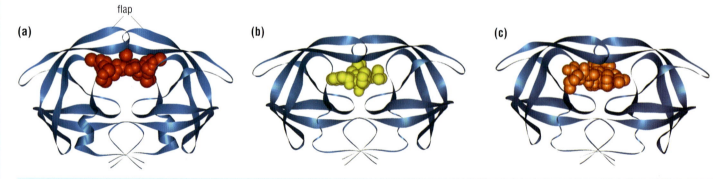

Figure 2-5 HIV protease, an enzyme from the virus that causes AIDS, bound to three different inhibitors The anti-viral action of some drugs used in AIDS therapy is based on their ability to bind to the active site of viral protease and inhibit the enzyme. The protease inhibitors haloperidol **(a)** and crixivan **(b)** are shown, with a peptide analog **(c)** of the natural substrate also shown bound to the enzyme. Each inhibitor clearly has a quite different structure and two of them (a, b) are not peptides, yet all bind tightly to the active site and induce closure of a flap that covers it, a conformational change that also occurs with the natural substrate. (PDB 1aid, 1hsg, 1a8k)

Definitions

induced fit: originally, the change in the structure of an enzyme, induced by binding of the substrate, that brings the catalytic groups into proper alignment. Now generalized to the idea that specific ligands can induce the protein conformation that results in optimal binding interactions.

References

Ding, X., et al.: **Direct structural observation of an acyl-enzyme intermediate in the hydrolysis of an ester substrate by elastase.** Biochemistry 1994, **33**:9285–9293.

Hammes, G.G.: **Multiple conformational changes in enzyme catalysis.** Biochemistry 2002, **41**:8221–8228.

Koshland, D.E. Jr. et al.: **Comparison of experimental binding data and theoretical models in proteins containing subunits.** Biochemistry 1966, **5**:365–385.

Rasmussen, B.F. et al.: **Crystalline ribonuclease A loses function below the dynamical transition at 220 K.** Nature 1992, **357**:423–424.

Rutenber, E. et al.: **Structure of a non-peptide inhibitor complexed with HIV-1 protease. Developing a cycle of structure-based drug design.** J. Biol. Chem. 1993, **268**:15343–15436.

Tsou, C.L.: **Active site flexibility in enzyme catalysis.** Ann. NY Acad. Sci. 1998, **864**:1–8.

Wrba, A. et al.: **Extremely thermostable D-glyceralde-**

Figure 2-6 Differences in the temperature dependence of the specific activity of D-glyceraldehyde-3-phosphate dehydrogenase (GAPDH) from two organisms The red line shows the increase in enzyme activity as the temperature rises for the enzyme from yeast, a mesophile (with optimal growth rate at about 35 °C). This is as expected because most reaction rates increase with increasing temperature. Above 35 °C, however, the activity drops precipitously as the protein becomes thermally denatured. The blue line shows the behavior of GAPDH from an extreme thermophile, *Thermotoga maritima* (whose optimal growth rate is at about 85 °C). At room temperature the activity of the thermophilic enzyme is less than that of the mesophilic enzyme. One reason for this could be that the enzyme is more rigid at this temperature than the mesophilic enzyme. At temperatures where the mesophilic enzyme begins to denature, the thermophilic enzyme is still active and continues to be so beyond 70 °C. The denaturation temperature for this enzyme is above 85 °C. Wrba, A. *et al.*: *Biochemistry* 1990, **29**:7584–7592.

of the ligand, alteration of the properties of the ligand if it is a macromolecule, or some combination of these. In all cases, the structure of the protein must be flexible enough that the net free energy released by binding and/or chemical transformation of the ligand can drive the required changes in the protein's structure or properties. Binding can range from weak (dissociation constant $K_d \sim 10^{-3}$ M) to extremely strong ($K_d \sim 10^{-12}$ M or even tighter). In the case of an enzyme, the protein must undergo a series of adjustments in order to allow binding, response to changes in ligand structure and subsequent release of ligand. Yet, if the protein is too flexible, neither specific recognition nor specific action can occur. For example, adding small quantities of a denaturant such as urea or the detergent sodium dodecyl sulfate (SDS) to a protein will often cause it to become less specific for the ligands it binds; adding a bit more will greatly reduce or abolish biochemical activity, and these effects can occur long before the structure completely unfolds.

Conversely, observations on mutant enzymes that are stable at higher temperatures than normal and have been shown to be more rigid, have shown that, at least for some such proteins, rigidifying the structure abolishes function. Comparisons of enzymes from organisms that live at normal temperatures and from extreme thermophiles also show that the thermophilic enzymes are less active than the normal ones at lower temperatures (Figure 2-6). The exact mechanism of these effects is still under investigation, but it is possible that stability at higher temperatures has been achieved at the expense of flexibility at lower ones, and that the proper balance between flexibility and rigidity is necessary for many, if not most, proteins.

Protein flexibility is a natural consequence of the weak forces that hold the tertiary structure together; because the free energies of these interactions are close to the kinetic energy available at ordinary temperatures (this energy is sometimes denoted as kT), such interactions are frequently breaking and reforming under physiological conditions. The marginally stable nature of proteins also helps make them easier to degrade, and we discuss in Chapter 3 how degradation enables the protein complement of a cell to be regulated.

The degree of flexibility varies in proteins with different functions

Not all proteins are equally flexible. Although some very small rearrangements in atomic positions always occur on ligand binding, a number of proteins behave as though they were relatively rigid. Many of these are extracellular proteins, and their rigidity may help them survive in the more hostile environment outside the cell. Other proteins undergo very large shape changes when the correct ligand binds (Figure 2-7): we discuss these large ligand-induced conformational changes in Chapter 3, in the context of ligands that regulate protein function.

hyde-3-phosphate dehydrogenase from the eubacterium *Thermotoga maritima*. *Biochemistry* 1990, **29**:7584–7592.

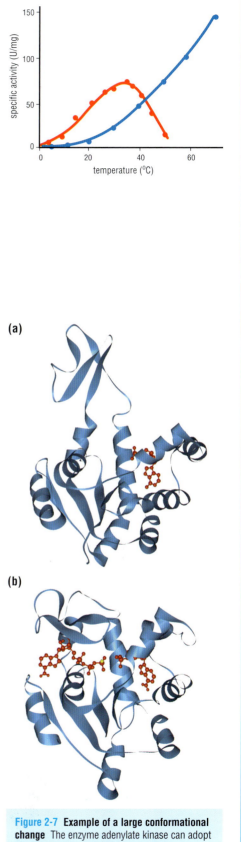

(a)

(b)

Figure 2-7 Example of a large conformational change The enzyme adenylate kinase can adopt either an open or closed conformation depending on which substrates are bound. **(a)** In the presence of AMP alone, no conformational change occurs. **(b)** On binding of the *cosubstrate* ATP, here in the form of the analog AMPPNP, a large rearrangement occurs that closes much of the active site. (PDB 2ak3, 1ank)

2-3 Location of Binding Sites

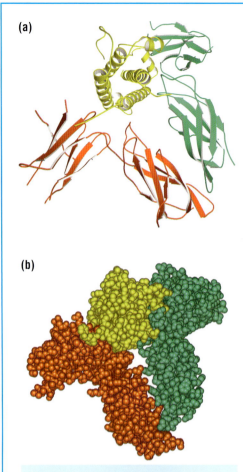

(a)

(b)

Figure 2-8 **The complex between human growth hormone and two molecules of its receptor** **(a)** Ribbon diagram of the complex. Two different protein–protein interfaces can be made by one molecule of growth hormone (yellow) with two identical receptor molecules (orange and green). **(b)** A space-filling model of the complex shows the tight fit at both interfaces. Kossiakoff, A.A. and De Vos, A.M.: *Adv. Protein Chem.* 1998, **52**:67–108.

Binding sites for macromolecules on a protein's surface can be concave, convex, or flat

The specific recognition of a macromolecule by a protein usually involves interactions over a large contiguous surface area (hundreds of square Ångstroms) or over several discrete binding regions (Figure 2-8). A macromolecule will make many points of contact with the protein's surface; these add up to provide a great deal of binding energy, so a binding site for a macromolecule can occur, in theory, anywhere on a protein's surface. Consequently, accurate prediction of a macromolecule-binding site on the surface of an experimentally determined or modeled protein structure is difficult. The most frequently observed sites are protruding loops or large cavities because these provide specific shape complementarity, but relatively flat binding sites are also found. Many binding sites for RNA or DNA on proteins are protruding loops or alpha helices that fit into the major and minor grooves of the nucleic acid (Figures 2-9a, b). These protrusions do not have any obvious common features that make them identifiable from a simple examination of the protein's structure, and most do not even have the same structure in the unbound protein and when ligand is bound.

Binding sites for small ligands are clefts, pockets or cavities

Many important biological ligands are small molecules: examples are the substrates for enzyme catalysis; cofactors that bind to the active sites of enzymes and contribute to catalysis; and *allosteric effectors*, which bind at sites remote from the active site yet modulate enzyme activity. (These will be discussed later, in sections 2-6 to 2-16.) Such small-molecule ligands bind at depressions on the protein surface, except in certain cases when they are buried within the protein's interior. Deep binding pockets allow the protein to envelop the ligand and thus use complementarity of shape to provide specificity (see Figure 2-2). They hinder access of water

(a)   **(b)**

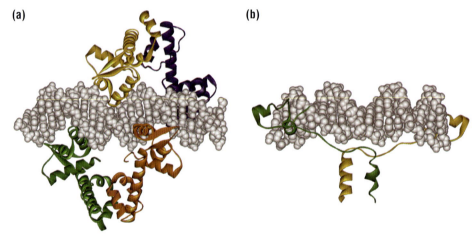

Figure 2-9 **Two protein–DNA complexes** **(a)** The complex between the bacterial diphtheria toxin gene repressor protein and the *tox* operator DNA sequence to which it binds. The repressor is a natural homodimer and two dimers bind to the pseudo-symmetrical operator sequence. The DNA sequence is recognized in the major groove by a helix in a helix-turn-helix motif, a feature often seen in DNA-binding proteins. **(b)** Structure of the complex between the eukaryotic Gal4 transcription factor and DNA. The interaction occurs in the major groove again, but this time the recognition unit contains a loop of chain that is stabilized by a cluster of zinc ions (not shown). Marmorstein, R. *et al.*: *Nature* 1992, **356**:408–414.

Definitions

cavity: a completely enclosed hole in the interior of a protein. Cavities may contain one or more disordered water molecules but some are believed to be completely empty.

structural domain: a compact part of the overall structure of a protein that is sufficiently independent of the rest of the molecule to suggest that it could fold stably on its own.

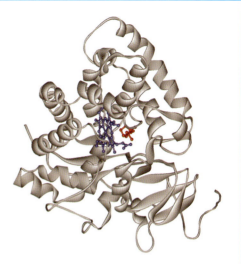

Figure 2-10 **Structure of bacterial cytochrome P450 with its substrate camphor bound** The active site of this enzyme contains a catalytic heme group (purple) most of which is completely buried inside the protein. There is no obvious route from the exterior to the active-site pocket in the average structure. Fluctuations in the structure must open a transient path or paths for the substrate camphor (red) to bind. Poulos, T.L. *et al.*: *J. Mol. Biol.* 1987, **195**:687–700.

to the bound ligand, which can be important for many enzyme reactions. Clefts or cavities can easily provide unusual microenvironments. And they enable even a small molecule to have enough contact points to bind strongly if that is needed. This characteristic of ligand-binding sites means that they can often be identified even in the structure of an unliganded protein: one looks for a large cleft or pocket on the protein surface or, if none is obvious, an internal **cavity** large enough to accommodate the ligand.

Binding in an interior cavity requires that the ligand diffuses through the protein structure within a reasonable time frame. Protein structural flexibility can allow such penetration, even for large substrates (Figure 2-10). It is not known for certain whether there are multiple pathways by which a ligand can "worm" its way through to a buried cavity, but computer simulations suggest that there can be. Another possibility is that the enzyme could exist in an open and closed state, with substrate binding to the open form triggering a conformational change (domain or flap closure) that sequesters the active site.

Catalytic sites often occur at domain and subunit interfaces

Enzyme active sites and most receptor binding sites for small ligands are found at generally predictable locations on protein surfaces. If a protein has more than one **structural domain**, then the catalytic site will nearly always be found at the interface between two of them, or all of them. If the protein is composed of more than one subunit, then the active site will often be found at an intersubunit interface. And if both conditions apply, then the active site will usually be located at a site corresponding to both an interdomain and an intersubunit interface (Figure 2-11).

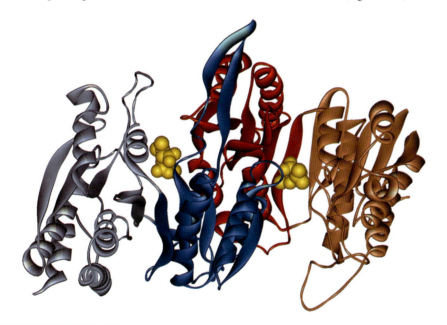

Figure 2-11 **Structure of the dimeric bacterial enzyme 3-isopropylmalate dehydrogenase** The two active sites are indicated by the presence of the bound cofactor NADPH (yellow). Each site occurs at an interface between the two subunits (blue/light-blue and red/brown) of the enzyme and also at an interface between the two domains of each subunit (blue and light-blue; red and brown). Imada, K. *et al.*: *J. Mol. Biol.* 1991, **222**:725–738.

References

Imada, K. *et al.*: **Three-dimensional structure of a highly thermostable enzyme, 3-isopropylmalate dehydrogenase of *Thermus thermophilus* at 2.2 Å resolution.** *J. Mol. Biol.* 1991, **222**:725–738.

Kossiakoff, A.A. and De Vos, A.M.: **Structural basis for cytokine hormone-receptor recognition and receptor activation.** *Adv. Protein Chem.* 1998, **52**:67–108.

Marmorstein, R. *et al.*: **DNA recognition by GAL4: structure of a protein-DNA complex.** *Nature* 1992,

356:408–414.

Poulos, T.L. *et al.*: **High-resolution crystal structure of cytochrome p450 cam.** *J. Mol. Biol.* 1987, **195**:687–700.

Ringe, D.: **What makes a binding site a binding site?** *Curr. Opin. Struct. Biol.* 1995, **5**:825–829.

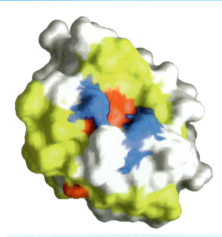

Figure 2-12 Surface view of the heme-binding pocket of cytochrome c6, with hydrophobic residues indicated in yellow The area around the heme (red) is very nonpolar because this protein must bind to another protein via this site to form an electron-transport complex involving the heme. The blue area indicates the presence of two positively charged residues important for heme binding. Graphic kindly provided by P. Roesch. Beissinger, M. *et al.*: *EMBO J.* 1998, **2**:27–36.

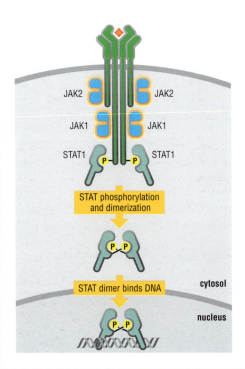

Binding sites generally have a higher than average amount of exposed hydrophobic surface

Protein surfaces are never completely polar as there are always isolated nonpolar groups in contact with the solvent. Ligand-binding sites, however, are generally distinguished by a much higher than average amount of exposed hydrophobic surface area: they are sites where nonpolar groups tend to be clustered on the protein surface, and this physical-chemical characteristic can sometimes be used to recognize them (Figure 2-12).

Binding sites for small molecules are usually concave and partly hydrophobic

Large hydrophobic areas on the surface of a protein lead to self-association and indeed they are the basis for oligomerization (see section 1-20). Hydrophobic ligand-binding sites do not, however, lead to oligomerization of the protein because they are usually too small and too concave to allow the protein to self-associate. But they will readily associate with a small-molecule ligand that they can interact with more favorably than with the water that covers them. Thus the combination of a high degree of concavity plus considerable exposed hydrophobic surface is reasonably diagnostic of a binding site for a small molecule on the surface of the protein.

Weak interactions can lead to an easy exchange of partners

As well as the large hydrophobic areas that lead to stable oligomerization, many proteins have smaller hydrophobic patches that are important in more transient protein–protein interactions. These occur, for example, in associations between the components of signal transduction pathways which must form and dissociate according to need. These hydrophobic patches are generally not only smaller but also less hydrophobic than those involved in oligomerization, reflecting the need for the two partners to exist independently in the aqueous environment of the cell. Because the favorable interactions at the interface between such proteins are often relatively few, and each individual interaction at the interface is weak, some of them, perhaps many, may be broken at any given time. Thus, given an existing protein–protein complex, and other proteins that can make weak interactions of a similar total energy with the partners in the complex, the dissociated components of the complex can combine with the other proteins to form new associations. This process is termed **partner swapping** and is the basis for many of the dynamic protein associations needed for signal transduction pathways.

In signal transduction, proteins such as kinases, phosphatases, and G-protein effectors are targeted to other proteins by specialized modules such as the SH2 domain, which recognizes phosphotyrosine-containing peptides on other proteins. For example, many hormones, growth factors, and immune regulators signal to the cell nucleus through STAT (signal transducer and activator of transcription) molecules, which bind to phosphotyrosine on the receptor tails via their SH2 domains. The STAT molecules are then themselves phosphorylated, and their SH2 domains dissociate from the receptor and bind to the phosphotyrosine on the other STAT

Figure 2-13 Partner swapping in a signaling pathway Binding of a signaling molecule to its receptor induces tyrosine phosphorylation by JAK kinase molecules of the cytoplasmic domains of the receptor, which then bind STAT molecules. This leads to tyrosine phosphorylation of the STAT molecules which dissociate from the receptor and instead bind through the same domains to the phosphotyrosine on the other STAT molecule, thereby forming dimers which migrate to the nucleus and bind to DNA, activating genes with various cellular effects.

Definitions

affinity: the tightness of a protein–ligand complex.

anisotropic: behaving differently in different directions; dependent on geometry and direction.

domain swapping: the replacement of a structural element of one subunit of an oligomer by the same structural element of the other subunit, and vice versa. The structural element may be a secondary structure element or a whole domain.

partner swapping: exchange of one protein for another in multiprotein complexes.

References

Beissinger, M. *et al.*: **Solution structure of cytochrome c6 from the thermophilic cyanobacterium *Synechococcus elongatus*.** *EMBO J.* 1998, **2**:27–36.

Bennett, M.J. *et al.*: **Domain swapping: entangling alliances between proteins.** *Proc. Natl Acad. Sci. USA* 1994, **91**:3127–3131.

Bourne, Y. *et al.*: **Crystal structure of the cell cycle-regulatory protein suc1 reveals a beta-hinge conformational switch.** *Proc. Natl Acad. Sci. USA* 1995, molecule to form an active signaling dimer (Figure 2-13). Such partner swapping is easy because the interactions between the STAT and the receptor are weak and each possible complex has about the same energy. However, because the STAT molecules become bound to one another by two SH2–phosphotyrosine interactions instead of one, as in the initial STAT–receptor interaction, the STAT dimer once formed is relatively stable and this association is unlikely to be reversed.

Domain swapping has been found in viral coat proteins (Figure 2-14) and signal transduction proteins. The inactive form of the enzyme PAK1 protein kinase is a domain-swapped dimer in which a regulatory domain from each monomer inhibits the active site of the other monomer. When an activator protein, such as Cdc42 in its GTP-bound form, binds to this regulatory domain, it relieves the domain swapping and frees the active site.

Displacement of water also drives binding events

A protein in solution is completely surrounded by water (see Figure 1-25). Some of these water molecules will interact more or less tightly with the protein surface. In fact, it is generally accepted that at least a single layer of bound water molecules should be considered an integral part of a protein structure. However, in order for a ligand which is itself surrounded by water molecules to bind to a solvated protein, both water layers must be disrupted and, at least partially, displaced. Thus, the protein and the ligand would exchange a layer of waters for favorable interactions with each other, and the enthalpic cost of releasing the surface waters can be balanced by the favorable enthalpy of the new interactions as well as by the hydrogen bonds the water can make with other solvent molecules.

The relationship between these energetic contributions is not simple. Although it might seem that the free energy of a water hydrogen bond to a protein group would be comparable to that of a hydrogen bond with another water molecule, the difference in enthalpy could be either positive or negative, depending on the microenvironment on the protein surface. For example, in an environment of reduced polarity, a hydrogen bond between a water molecule and a serine side chain could be stronger than in aqueous solution because the electrostatic attraction is greater in a medium of low polarity. Further, it is likely that some protein-bound waters gain entropy when they are displaced. Thus, the free energy of ligand binding will depend on the tightness of the water interactions with the protein surface. Although there are many potential binding sites on the irregular, largely polar protein surface, the sites where ligands actually bind will be those where favorable interactions can occur and where bound solvent can also be displaced.

Contributions to binding affinity can sometimes be distinguished from contributions to binding specificity

Although the displacement of water may provide some of the energetic driving force for ligand binding, from a practical point of view one often needs to distinguish between contributions to the **affinity**, or strength, of such binding and contributions to ligand specificity. It is generally accepted that the affinity between a protein and its ligand is chiefly due to hydrophobic interactions, which are non-directional, whereas specificity of binding is chiefly due to **anisotropic**, or directional, forces such as hydrogen bonding (Figure 2-15). The evidence for this view comes largely from measurements of ligand affinities between mutant proteins and altered ligands. However, the relative contributions of the different types of interactions to specificity and affinity will clearly vary from case to case, so conclusions drawn from mutational analysis may be misleading in the absence of a thorough investigation of the structure of the mutant with bound ligand and a thermodynamic analysis of the changes in binding energy.

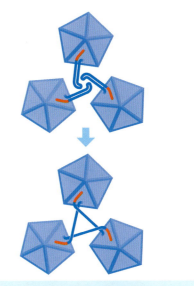

Figure 2-14 Domain swapping in the papilloma virus capsid protein Schematic representation of the conformational rearrangement of the carboxy-terminal arm of the papillomavirus capsid protein required for the stable assembly of the virus coat. On the structure of isolated capsid protein (top), the carboxy-terminal region of the polypeptide (red) is folded back and interacts with the rest of the protein. On the full-size virus particle, the arm "invades" the adjacent subunit on the surface of the virus, and makes a similar interaction with the other molecule. The domain swapping stabilizes trimers of the coat protein.

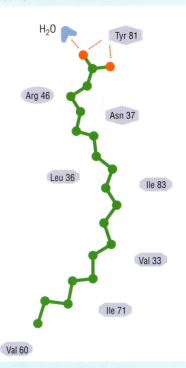

Figure 2-15 Ligand binding involving hydrophobic and hydrogen-bond interactions The binding of the lipid oleate (green) to the maize lipid-transport protein nsLTP involves mainly hydrophobic interactions of uncharged polar or nonpolar residues with the lipid tail, and hydrogen bonds (red dotted lines) to the charged head group.

92:10232–10236.

Darnell, J.E. Jr.: **STATs and gene regulation.** *Science* 1997, **277**:1630–1635.

Han, G.W. *et al.*: **Structural basis of non-specific lipid binding in maize lipid-transfer protein complexes revealed by high-resolution X-ray crystallography.** *J. Mol. Biol.* 2001, **308**:263–278.

Jones, S. and Thornton, J.M.: **Principles of protein-protein interactions.** *Proc. Natl Acad. Sci. USA* 1996, **93**:13–20.

Liu, Y. *et al.*: **The crystal structure of a 3D domain-swapped dimer of RNase A at a 2.1-Å resolution.** *Proc. Natl Acad. Sci. USA* 1998, **95**:3437–3442.

Modis, Y. *et al.*: **Atomic model of the papillomavirus capsid.** *EMBO J.* 2002, **21**:4754–4762.

Ringe, D.: **What makes a binding site a binding site?** *Curr. Opin. Struct. Biol.* 1995, **5**:825–829.

Szwajkajzer, D. and Carey, J.: **Molecular constraints on ligand-binding affinity and specificity.** *Biopolymers* 1997, **44**:181–198.

2-5 Functional Properties of Structural Proteins

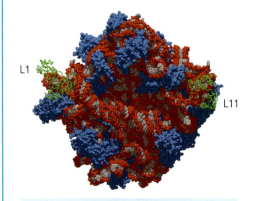

Figure 2-16 Structure of the 50S (large) subunit of the bacterial ribosome The ribosomal RNA is shown in red and grey, while most of the structural proteins are shown as blue space-filling models. Two of the proteins (L1 and L11) are shown as backbone diagrams in green because their positions are not known and have been arrived at by modeling. Graphic kindly provided by Poul Nissen and Thomas Steitz.

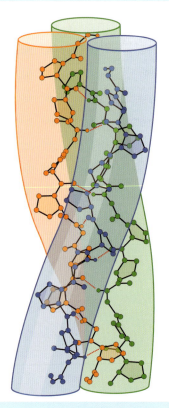

Figure 2-17 Structure of collagen The collagen triple helix is a coiled-coil structure in which each of the three protein chains is made up of repeating GlyXY sequences, where X is often proline (in the example shown here, Y is also proline). The hydrophobic nature of this repeat results in a set of regularly spaced hydrophobic sites along each chain; these complementary sites plus interchain hydrogen bonds (red dotted lines) hold the triple helix together. In collagen fibers, multiple triple helices are aligned end-to-end and side-by-side in a regular fashion, producing the light and dark bands observed when collagen fibers are imaged in an electron microscope.

Proteins as frameworks, connectors and scaffolds

Cells are not just structureless bags filled with freely floating molecules. All cells are surrounded by a protein-reinforced membrane; some have a cell wall that is primarily protein and carbohydrate. Internal structures within the cell also are made up of particular structural proteins that confer shape, strength and flexibility on these cellular structures. In some cases, structural proteins are assisted by DNA, RNA, lipid and carbohydrate molecules; in other cases the structure is built up from a large number of different proteins. The ribosome, for example, has over a hundred different protein components that stabilize the folded form of the ribosomal RNA, which provides the catalytic function (Figure 2-16).

There is a dynamic character to many of these subcellular structures. In some, for example muscle, the structure itself can change shape in response to external stimuli; in other cases, the structural proteins provide a framework for dynamic processes to occur, driven by other types of proteins. The actin filaments that provide the tracks along which many proteins and protein complexes run within the cell are an example of such a framework. Some structural proteins form temporary structures that are then destroyed when no longer needed. Fibrinogen, the primary component of blood clots, is such a protein: it polymerizes to form a dense fibrous mass but dissolves when the wound is healed. The inappropriate formation of such structures can underlie serious human diseases. Unregulated clotting can lead to fatal thromboses; aggregation of beta-amyloid protein, which is not normally a structural protein but becomes one after proteolytic cleavage, produces the amyloid fibrils associated with Alzheimer's disease.

Some structural proteins only form stable assemblies

Many of the structural components of cells and organisms are designed to be permanent: they are neither altered nor destroyed during the lifetime of the organism. Such assemblies can be constructed from proteins alone, as in the case of silk, collagen, elastin or keratin, or the coat proteins of a virus (see Figure 1-74k), or from protein plus some other component, as in the case of cartilage, which is composed of protein plus carbohydrate.

There are two ways in which these structures can be stabilized. One is by protein–protein interactions alone. Although such interactions are noncovalent and thus relatively weak in energetic terms, there can be enormous numbers of them in any given structural assembly, which produces an overall stabilization equal to many covalent bonds. To make the assembly take on a particular shape, the interactions also need to be specific. One way to achieve a large number of specific weak interactions is to place the complementary surfaces on simple repeating secondary structure elements such as alpha helices and beta strands. Stable assemblies are thus often coiled coils of long helices or stacks of beta sheets, and to achieve regularity the sequences of the component proteins are often made up of simple repeating motifs, which can be recognized easily in sequence analysis. Collagen, the fibrous component of tendons, is one example of such a structure: the basic component of collagen is a triple helix of three protein chains made up of repeating GlyXY sequences (Figure 2-17). Variability at the third position in the repeat can impart special local properties. Silk is an example of a stack of beta sheets (see Figure 1-1). The other way of stabilizing structural assemblies is by covalent cross-linking of their protein components. For example, the structure of collagen is stabilized further by covalent cross-linking. This is initiated by the enzyme lysyl oxidase which converts the terminal amino groups in the side chains of lysine residues to peptidyl aldehydes. These then undergo a number of uncatalyzed reactions that cross-link the chains. Elastin, the protein that gives lung tissues their elastic properties, is stabilized by the same cross-linking reaction: the cross-links hold the

Definitions

scaffold protein: a protein that serves as a platform onto which other proteins assemble to form functional complexes.

individual polypeptide chains in a rubber-like network. Deficiency in lysyl oxidase activity is associated with the genetic diseases Type IX Ehlers-Danlos syndrome and Menkes syndrome, which are characterized by loss of stability and elasticity in the connective tissues.

Some catalytic proteins can also have a structural role

The regular assemblies formed by structural proteins very often have cellular functions that require time-dependent changes in shape or conformation or some other property of the assembly. These changes are often brought about by changes in the structure of a single component of the multicomponent assembly, which in turn derive from the energy released by a chemical process. This may be merely the binding of another protein or small molecule—as small as a proton if the change is pH-driven as in the conformational changes required for fusion of viruses to cell membranes—but is usually a protein-catalyzed chemical transformation such as the hydrolysis of ATP.

Muscle is an example of such a multicomponent assembly, in which the dynamic component is the motor protein myosin II. Muscle is composed chiefly of interdigitating actin filaments and myosin filaments, held together by other structural proteins. Myosin II is a homodimer of two main subunits, each with a long helical "tail" and a head domain that contains both a binding site for an actin filament and a catalytic site. The two tails form an extended coiled coil which associates with those of other myosin molecules to form the thick myosin filament, from which the heads protrude at regular intervals. Hydrolysis of ATP at the catalytic site produces a conformational change in the head domain that results in a change in the position of the head on the actin filament; this motion causes the myosin and actin filaments to slide against each other, thus causing the muscle fiber to contract. Myosin II is thus a structural protein (it forms filaments), a catalytic protein (an ATPase), and a motor protein involved in cell motility.

Some structural proteins serve as scaffolds

When signals external to the cell are transmitted inside the cell, the kinases, phosphatases and transcription factors that make up the intracellular signal transduction pathways must find each other in order to carry out the sequence of reactions that transmit the signal. Diffusion of a protein across a typical eukaryotic cell can occur in a few milliseconds, but for two randomly distributed proteins the time to interact by diffusion will be much longer, implying that specific recruitment is needed. Sometimes this recruitment occurs by localizing one or more of the components to the cell membrane, but in other cases specific structural proteins serve as **scaffold proteins** onto which the other members of the pathway assemble, forming a signaling complex.

Mitogen-activated protein kinase (MAPK) signal transduction cascades form such molecular assemblies within cells. The spatial organization for these is provided by scaffold proteins. Yeast Ste5p was the first MAPK cascade scaffold protein to be described. Ste5p selectively tethers the MAP kinases MAPKKK, MAPKK and MAPK, which act sequentially in the pheromone-stimulated yeast mating pathway. Recent work indicates that Ste5p is not a passive scaffold but plays a direct part in the activation of the MAPKKK by interacting with a heterotrimeric G protein and another kinase. This activation event requires the formation of an active Ste5p oligomer and proper recruitment of Ste5p to the cell cortex, forming a "signalosome" linked to a G protein (Figure 2-18). Many other such scaffold proteins have been identified. They can often be detected in sequence analysis because they frequently contain repeats of known protein–protein interaction domains such as RING-H2, WW and WD40 domains. These will be discussed in detail in Chapter 3.

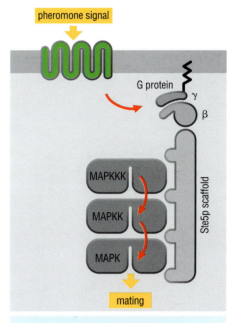

Figure 2-18 The Ste5p scaffold Ste5p is essential for the yeast response to mating pheromone and binds the sequential kinase components of a *mitogen-activated protein kinase* (MAPK) cascade: these are the MAPKKK, the MAPKK and the MAPK. Pheromone stimulation releases the *G-protein* components Gβγ, which recruit Ste5p and a cytoplasmic kinase (not shown) to the plasma membrane, activating the MAPK cascade. Pathway activation is thought to be coordinated with the conversion of a less active closed form of Ste5p containing a protected RING-H2 domain into an active Ste5p dimer that can bind to Gβγ and form a multimeric scaffold lattice upon which the kinase components of the MAPK cascade can assemble.

References

Ban, N. *et al.*: **The complete atomic structure of the large ribosomal subunit at 2.4 Å resolution.** *Science* 2000, **289**:905–920.

Berisio, R., *et al.*: **Recent progress on collagen triple helix structure, stability and assembly.** *Curr. Pharm. Des.* 2002, **9**:107–116.

Elion, E.A.: **The Ste5p scaffold.** *J. Cell Sci.* 2001, **114**:3967–3978.

Houdusse, A. *et al.*: **Three conformational states of scallop myosin S1.** *Proc. Natl Acad. Sci. USA* 2000, **97**:11238–11243.

Rochet, J.C. and Lansbury, P.T. Jr.: **Amyloid fibrillogenesis: themes and variations.** *Curr. Opin. Struct. Biol.* 2000 **10**:60–68.

Sette, C. *et al.*: **Mutational analysis suggests that activation of the yeast pheromone response mitogen-activated protein kinase pathway involves conformational changes in the Ste5 scaffold protein.** *Mol. Biol. Cell* 2000, **11**:4033–4049.

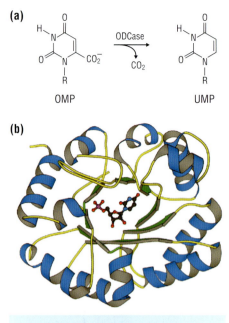

(a)

OMP   UMP

(b)

Figure 2-19 The enzyme orotidine 5′-monophosphate decarboxylase catalyzes the transformation of orotidine 5′-monophosphate to uridine 5′-monophosphate (a) The reaction catalyzed by orotidine 5′-monophosphate decarboxylase (ODCase). This reaction does not occur readily at room temperature in the absence of a catalyst. ODCase accelerates the rate of the reaction 10^{17}-fold. R = the ribose phosphate group of the nucleotide. (b) The structure of ODCase. The product, UMP, is shown bound in the active site.

Catalysts accelerate the rate of a chemical reaction without changing its overall equilibrium

In the ligand-binding events we have discussed so far, the ligand is unchanged on binding the protein. We now turn to the actions of enzymes—biological catalysts whose ligands undergo chemical transformation during metabolism, biosynthesis and in many of the signaling and motor activities of cells. A **catalyst** is any substance that accelerates the rate of a chemical reaction without itself becoming permanently altered in the process. The word permanently is important, because, as we shall see, enzymes often undergo transient alteration in both their conformation and covalent structure while catalyzing reactions; but by the end of the overall reaction they are always restored to their original condition. Although most enzymes are proteins, a number of RNA molecules also function as catalysts (termed *ribozymes*), including the ribosomal RNA on the large subunit of the ribosome, which has been shown to be the catalyst of peptide-bond formation during protein synthesis.

Enzymes can be extraordinarily efficient catalysts. The transformation of orotidine 5′-monophosphate (OMP) to uridine 5′-monophosphate (UMP) (Figure 2-19a) is a key decarboxylation reaction in purine biosynthesis. In free solution, the uncatalyzed reaction is estimated to take approximately 78 million years to go halfway to completion (obviously, this is an estimated, not measured, rate). But when the same reaction is catalyzed by the enzyme orotidine 5′-monophosphate decarboxylase (OMP decarboxylase or ODCase; Figure 2-19b), it is completed in less than a second—an increase in reaction rate of 10^{17}-fold. OMP decarboxylase is not the only enzyme that is known to accelerate a reaction by more than a billion fold; catalytic accelerations of this magnitude or more are common in biology (Figure 2-20), and are the reason that living organisms can exist at moderate temperatures. In the absence of an efficient catalyst, a reaction such as OMP decarboxylation would require very high temperatures to proceed at a measurable rate. Since primordial enzymes are unlikely to have been very efficient—the assumption being that catalytic proficiency requires billions of years of evolution by trial-and-error through random mutations—it is thought that the earliest living organisms were probably extreme thermophiles, unicellular microorganisms that live at high temperatures. Such organisms can still be found today, in hot springs such as those in America's Yellowstone National Park.

All chemical transformations, even those that are not obviously reversible, actually represent equilibria that involve forward and backward reactions. Each has a characteristic **equilibrium constant**, which is the ratio of the concentrations of the products (multiplied together) to the concentrations of the reactants (multiplied together) that are present when **equilibrium** is attained. The equilibrium constant can also be determined by the relative rates of the forward and reverse processes. A reaction whose forward rate is faster than its reverse will have more product than reactant at equilibrium, whereas one where the back reaction is faster will have a preponderance of reactant. As the chemical transformations in the back reaction just follow the same path but in the opposite direction, any catalyst, including an enzyme, will speed up the forward and the back reaction by the same amount, leaving the relative concentrations of product(s) and reactant(s) at equilibrium unchanged. Thus, catalysis accelerates the rate at which equilibrium is reached, but not the final equilibrium concentrations (or the equilibrium constant).

Catalysis usually requires more than one factor

For over a century biochemists have been fascinated by the enormous catalytic power of enzymes. Rate accelerations of up to 10^{17}, as with ODCase, are so far beyond those achievable in ordinary laboratory reactions that it was long thought that there must be some special, undreamed-of catalytic principle that enzymes use. But after decades of intense research, including the determination of the atomic structure of many enzyme–substrate complexes, it has become apparent that there is no unique secret to enzymatic catalysis. Enzymes have at their disposal a variety of simple contributory factors to help them attain huge rate accelerations; each enzyme uses a combination of several of these, but the particular combination, and especially the relative importance of each one, vary from enzyme to enzyme.

Some of these factors are physical in nature: they depend on the structure and physical properties of the enzyme and on the ability to orient the ligand very precisely relative to catalytic residues in the active site. Others are chemical: they involve the chemical properties of the amino acids (and cofactors, if any) that make up the enzyme, including their ability to stabilize unstable chemical species by weak interactions, their ability to polarize bonds, and their ability to form covalent adducts. There are some enzymes in which a single catalytic factor predominates, but these are the exception. Similar enzymes from different organisms may also have subtle differences in the balance of contributions of the different factors. It is the net contribution of many simple effects, not any one special feature unique to living organisms, that accounts for the extraordinary power of enzymes to speed up reactions, thereby providing the chemistry necessary to sustain life at ordinary temperatures and pressures.

Catalysis is reducing the activation-energy barrier to a reaction

For a chemical reaction to occur, the reactants (which in enzymology are generally called **substrates**) must undergo rearrangements in stereochemistry, charge configuration, and covalent structure. If the reactant is a stable compound, there will be a free-energy barrier to such transformations, even if the product of the reaction is more stable. The higher this barrier, the slower the reaction and the more difficult the chemical step is to achieve. The energy required to overcome this barrier is known as the **activation energy**, and the barrier is called the **activation-energy barrier**, because, in order to react, the substrate must attain a higher free-energy state and in this state is said to be activated. We shall see examples of how activation is achieved in later sections. The **transition state** is the highest point in free energy on the reaction pathway from substrate to product; it is the top of the activation-energy barrier (see TS_u in Figure 2-21). Chemically, it is a species that exists for about the time required for a single atomic vibration to occur (about 10^{-15} s). In the transition state, the making or breaking of chemical bonds in the reaction is not yet complete: the atoms are "in flight". The stereochemistry and charge configuration of the transition state is thus likely to be quite different from that of either the substrate or the product, although it may resemble one more than it does the other.

The activation free-energy barrier must be overcome for the reaction to proceed. One way to speed up the reaction is therefore to lower the barrier, either by raising the free energy, or **ground state**, of the reactant—or of the product in the reverse direction (all relative to some standard state). An alternative way of lowering the barrier is to lower the free energy of the transition state: this would correspond to stabilizing it, as molecules of lower energy are more stable. Still another is to cause the reaction to take a different path, one in which there may well be more free-energy "hills" than for the uncatalyzed reaction, but along which every hill is smaller. In such cases there will be local "valleys" of free energy between the hills; these metastable molecules, usually of relatively high free energy, are called the **intermediates** in the reaction. This third option is shown in Figure 2-21. Enzymes, as we shall see, employ all these catalytic tricks.

References

Bruice, T.C. and Benkovic, S.J.: **Chemical basis for enzyme catalysis.** Biochemistry 2000, **39**:6267-6274.

Hammes, G.G.: **Multiple conformational changes in enzyme catalysis.** Biochemistry 2002, **41**:8221-8228.

Jencks, W.P.: Catalysis and Enzymology (Dover Publications, New York, 1987).

Miller, B.G. and Wolfenden, R.: **Catalytic proficiency: the unusual case of OMP decarboxylase.** Annu. Rev.

Biochem. 2002, **71**:847-885.

Radzicka, A. and Wolfenden, R.: **A proficient enzyme.** Science 1995, **267**:90–93.

Silverman, R.B.: The Organic Chemistry of Enzyme-Catalyzed Reactions (Academic Press, New York, 2000).

Walsh, C.: Enzymatic Reaction Mechanisms (Freeman, San Francisco, 1979).

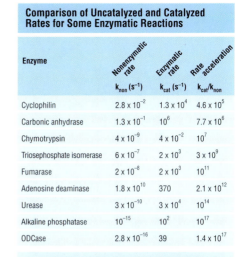

Comparison of Uncatalyzed and Catalyzed Rates for Some Enzymatic Reactions

Enzyme	Nonenzymatic rate k_{non} (s^{-1})	Enzymatic rate k_{cat} (s^{-1})	Rate acceleration k_{cat}/k_{non}
Cyclophilin	2.8×10^{-2}	1.3×10^4	4.6×10^5
Carbonic anhydrase	1.3×10^{-1}	10^6	7.7×10^6
Chymotrypsin	4×10^{-9}	4×10^{-2}	10^7
Triosephosphate isomerase	6×10^{-7}	2×10^3	3×10^9
Fumarase	2×10^{-8}	2×10^3	10^{11}
Adenosine deaminase	1.8×10^{10}	370	2.1×10^{12}
Urease	3×10^{-10}	3×10^4	10^{14}
Alkaline phosphatase	10^{-15}	10^2	10^{17}
ODCase	2.8×10^{-16}	39	1.4×10^{17}

Figure 2-20 Table of the uncatalyzed and catalyzed rates for some representative enzymatic reactions k is the rate constant for the reaction. Adapted from Radzicka, A. and Wolfenden, R.: *Science* 1995, **267**:90–93.

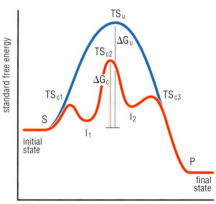

Figure 2-21 Energetics of catalysis The energy course of a hypothetical reaction from substrate S to product P can be described in terms of transition states and intermediates. For the uncatalyzed reaction (blue curve) a single transition-state barrier determines the rate at which product is formed. In the presence of a catalyst (red curve), which in this case is acting by changing the pathway of the reaction and introducing additional smaller activation-energy barriers, intermediate I_1, formed by crossing transition-state barrier TS_{c1}, leads to transition-state barrier TS_{c2}. Its free energy (ΔGc), although the highest point in the reaction, is considerably lower than the free energy (ΔGu) of the uncatalyzed transition state, TS_u. After formation of a second intermediate, I_2, a third transition state, TS_{c3}, leads to product. Because TS_{c2} is the highest transition state in the catalyzed reaction, the rate at which the reactants pass over this barrier determines the overall rate and thus it is said to be the rate-determining transition state of the catalyzed reaction. The rate-determining step of this reaction is thus the conversion of I_1 to I_2.

2-7 Active-Site Geometry

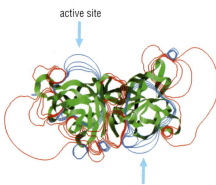

active site

active site

Figure 2-22 The electrostatic potential around the enzyme Cu,Zn-superoxide dismutase Red contour lines indicate net negative electrostatic potential; blue lines net positive potential. The enzyme is shown as a homodimer (green ribbons) and two active sites (one in each subunit) can be seen at the top left and bottom right of the figure where a significant concentration of positive electrostatic potential is indicated by the blue contour lines curving away from the protein surface. The negative potential elsewhere on the protein will repel the negatively charged superoxide substrate ($O_2^{-\cdot}$) and prevent non-productive binding, while the positive potential in the active site will attract it. Graphic kindly provided by Barry Honig and Emil Alexov.

Reactive groups in enzyme active sites are optimally positioned to interact with the substrate

In any enzyme-catalyzed reaction, the first step is the formation of an enzyme–substrate complex in which the substrate or substrates bind to the active site, usually noncovalently. Specificity of binding comes from the close fit of the substrate within the active-site pocket, which is due primarily to van der Waals interactions between the substrate and nonpolar groups on the enzyme, combined with complementary arrangements of polar and charged groups around the bound molecule. This fit is often so specific that even a small change in the chemical composition of the substrate will abolish binding. Enzyme–substrate dissociation constants range from about 10^{-3} M to 10^{-9} M; the lower the value the more tightly the substrate is bound. It is important that an enzyme does not hold onto its substrates or products too tightly because that would reduce its efficiency as a catalyst: the product must dissociate to allow the enzyme to bind to another substrate molecule for a new catalytic cycle.

Formation of a specific complex between the catalyst and its substrates does more than just account for the specificity of most enzymatic transformations: it also increases the probability of productive collisions between two reacting molecules. All chemical reactions face the same problem: the reacting molecules must collide in the correct orientation so that the requisite atomic orbitals can overlap to allow the appropriate bonds to be formed and broken. If we consider, in general terms, that a molecule can have a "reactive side" where the chemical changes take place and an "unreactive side" where they do not (at least immediately; the

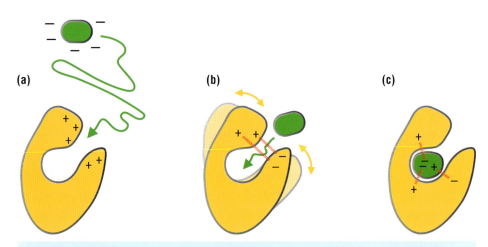

(a)   (b)   (c)

Figure 2-23 Schematic diagram showing some of the ways in which electrostatic interactions can influence the binding of a ligand to a protein **(a)** Electrostatic forces and torques can steer the ligand (green) into its binding site on the protein (shown in yellow). **(b)** Some binding sites are normally shielded from the solvent and can be kept "closed" by salt links between groups on the protein surface. If the correct substrate disrupts these salt links it can gain access to the binding site. This is known as "**gated**" **binding**. Alternatively, the dynamics of the protein may open and close such a site transiently (as indicated by the yellow arrows). **(c)** Electrostatic interactions, particularly salt links and hydrogen bonds, between ligand and protein can contribute to the affinity and specificity of binding and to the orientation of the ligand in the binding site and the structure of the complex formed. All three of these ways of exploiting electrostatic interactions can be used by a single enzyme. Adapted from Wade, R.C. et al.: *Proc. Natl Acad. Sci. USA* 1998, **95**:5942–5949.

Definitions

gated binding: binding that is controlled by the opening and closing of a physical obstacle to substrate or inhibitor access in the protein.

reaction sub-site: that part of the active site where chemistry occurs.

specificity sub-site: that part of the active site where recognition of the ligand takes place.

molecule may rearrange during the reaction), then in a simple reaction in which two molecules combine, both of them must collide reactive side-to-reactive side. Any other orientation and the collision will be non-productive. Thus, if both molecules first bind to an enzyme active site, and do so in such a way that their reactive portions are juxtaposed, the probability of a reaction is optimized. In solution, when two molecules collide but do not react they bounce off each other more or less randomly. On the enzyme, however, once the first reactive molecule has bound, it will stay there for some time, waiting for the second to come along. If that molecule does not bind productively, the first one may still remain associated with the enzyme (depending on the affinity constant) long enough for many other collisions to be tried.

In addition, enzyme active sites may even have evolved to attract their substrates so that finding the active site is not a random process. Most biological molecules are charged (so they can be retained within a cell by their insolubility in the hydrophobic membrane that surrounds it). Although active sites have exposed hydrophobic patches, the overall electrostatic field produced by the protein with all its polar and charged groups can yield an electrostatic potential with a net charge in the active-site region (Figure 2-22). It is possible that this net potential may "draw" the substrate into the oppositely charged active site, increasing the probability of productive binding (Figure 2-23a). Other ways in which electrostatics can aid in binding are illustrated in Figure 2-23b and c.

In most cases, there is a further critical factor in the facilitation of reactions by enzyme active sites. Usually, the enzyme itself supplies one or more of the chemical groups that participate in catalysis. Groups that are already part of the active-site structure in the folded protein before the substrate binds are already oriented properly for catalysis, or become so as the enzyme binds its substrate. The folding energy of the protein has already paid most, if not all, of the cost of positioning these groups, so there will be no unproductive collisions because they are in the wrong orientation. Every substrate molecule that forms the enzyme–substrate complex will therefore be exposed to an environment in which the catalytic groups are positioned correctly, relative to the substrate, for the desired reaction to take place.

In fact, the reactive portion of the substrate need not be the part that is used to hold it at the enzyme surface. An enzyme can recognize and interact with the remote parts of the substrate molecule, parts not involved in the chemistry, and use these interactions to hold and orient the substrate. This is a general principle: enzyme active sites consist of a **specificity sub-site** and a **reaction sub-site**, and in these protein groups are positioned around different parts of the substrate. In the specificity sub-site the enzyme uses polar and nonpolar groups to make weak interactions with the substrate; in the reaction sub-site other groups on the enzyme carry out the chemistry (Figure 2-24). In some cases the same amino-acid residue may participate in both specific substrate binding and catalysis. This design feature makes excellent sense. During the catalytic reaction, portions of the substrate molecule will undergo changes in geometry, charge and covalent bonding. If an enzyme had binding interactions with parts of the substrate that had to undergo rearrangement, those interactions would have to be broken before the substrate could change its structure, which could slow the enzyme down. The dual nature of enzyme active sites is being exploited in medicine and industry to design new catalysts. Amino-acid changes can often be made in the specificity sub-site of an enzyme without affecting its catalytic sub-site. So an enzyme that originally catalyzed a reaction involving positively charged substrates, for example, can sometimes be altered to perform exactly the same chemistry on new substrates that are negatively charged.

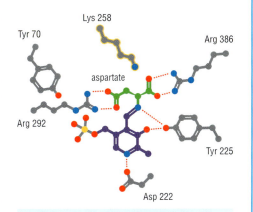

Figure 2-24 Schematic diagram of the active site of *E. coli* aspartate aminotransferase The enzyme uses a pyridoxal phosphate (PLP) cofactor (purple) and lysine (yellow outline) to carry out chemistry. The substrate amino acid (green) reacts with the cofactor to form an adduct (as shown in this model) which then rearranges to give product. Substrate specificity for the negatively charged aspartic acid substrate is determined by the positively charged guanidino groups of arginine 386 and arginine 292, which have no catalytic role. Mutation of arginine 292 to aspartic acid produces an enzyme that prefers arginine to aspartate as a substrate. Adapted from Cronin, C.N. and Kirsch, J.F.: *Biochemistry* 1988, **27**:4572–4579; Almo, S.C. *et al.*: *Prot. Eng.* 1994, **7**:405–412.

References

Almo, S.C. *et al.*: **The structural basis for the altered substrate specificity of the R292D active site mutant of aspartate aminotransferase from *E. coli*.** *Prot. Eng.* 1994, **7**:405–412.

Cronin, C.N. and Kirsch, J.F.: **Role of arginine-292 in the substrate specificity of aspartate aminotransferase as examined by site-directed mutagenesis.** *Biochemistry* 1988, **27**:4572–4579.

Wade, R.C. *et al.*: **Electrostatic steering and ionic tethering in enzyme-ligand binding: insights from simulations.** *Proc. Natl Acad. Sci. USA* 1998, **95**:5942–5949.

Some active sites chiefly promote proximity

Binding of substrates in the correct orientation probably makes a significant contribution to the catalytic efficiency of all enzymes, but in some cases it accounts for virtually all their effectiveness. If the substrate molecules are intrinsically reactive, simply holding them close to each other in the proper orientation may be all that is needed to facilitate the appropriate chemistry. This is often referred to as the **proximity factor,** or sometimes, the **propinquity factor**.

A striking example of this factor in action is provided by the metabolic enzyme aspartate transcarbamoylase (ATCase), which promotes the condensation of carbamoyl phosphate and aspartic acid to yield carbamoyl aspartate (Figure 2-25a). The crystal structure of ATCase with an inhibitor bound to the active site has revealed the potential interactions the enzyme makes with these substrates. The inhibitor, PALA (N-phosphonoacetyl-L-aspartate), is a bisubstrate analog: it has features that resemble both substrates and it occupies both binding sites simultaneously (Figure 2-25b). Examination of the structure of the inhibitor–enzyme

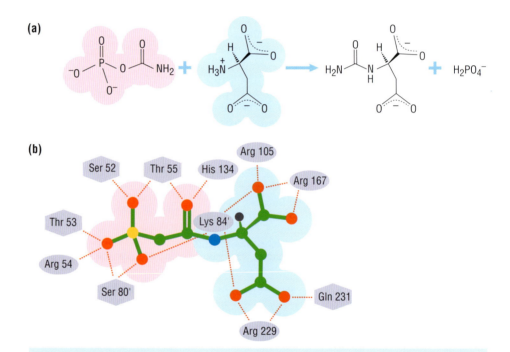

Figure 2-25 Catalysis of the reaction of carbamoyl phosphate and aspartate by the enzyme aspartate transcarbamoylase depends on holding the substrates in close proximity and correct orientation in the active site (a) The reaction catalyzed by aspartate transcarbamoylase (ATCase). Carbamoyl phosphate (shaded in pink) and aspartate (shaded in blue) undergo a condensation reaction to form N-carbamoyl aspartate. This is an essential step in pyrimidine biosynthesis. **(b)** Schematic diagram of the active site of ATCase with the inhibitor PALA (N-phosphonoacetyl-L-aspartate) (green) bound. The amino acids forming the active site and binding PALA by noncovalent bonds (red dotted lines) are represented by the purple shapes. (Ser 80' and Lys 84' come from an adjacent subunit to the other residues shown.) PALA resembles both carbamoyl phosphate and aspartate, as can be seen by comparison with (a), and binds to both the binding sites for these substrates in the active site. Note the absence of catalytic amino acids in a position to interact with the H_3N^+ group of the normal substrate aspartate, the group that reacts to form the bond with carbamoyl phosphate.

Definitions

ground-state destabilization: raising the free energy (relative to some reference state), of the ground state, usually referring to the bound substrate in the active site before any chemical change has occurred. Geometric or electronic strain are two ways of destabilizing the ground state.

propinquity factor: another term for **proximity factor**.

proximity factor: the concept that a reaction will be facilitated if the reacting species are brought close

together in an orientation appropriate for chemistry to occur.

complex shows the surprising fact that there are no chemically reactive side chains positioned anywhere near those parts of the substrates where the bond between them would be formed. The active site grips other parts of both substrates, but any chemistry comes chiefly from the reactivities of the molecules themselves. They condense because they are held in close proximity while being oriented so that the atomic orbitals that must form the new chemical bond are positioned to overlap.

Another example of a reaction that is aided primarily by proximity is provided by the enzyme chorismate mutase, which catalyzes a reaction in the pathway for the biosynthesis of aromatic amino acids. The reaction is an internal rearrangement of chorismate, of a type known as a pericyclic rearrangement. Crystal structures of the enzyme with substrate or substrate-analog inhibitors bound show that the enzyme promotes the reaction primarily by binding the substrate in an unusual "chair" conformation, which in solution would be energetically unfavorable. This conformation positions the group to be moved in an orientation that facilitates the internal rearrangement (Figure 2-26). Compounds designed to mimic this conformation are particularly effective inhibitors of the enzyme because the active site has evolved to be complementary to it.

Some active sites destabilize ground states

The bound conformation of chorismate (or of prephenate, the product of the chorismate mutase reaction) in the active site of chorismate mutase is not only a conformation that aids in catalysis by positioning the reacting atoms near one another; it is also a conformation that is much higher in free energy (less stable) than the conformation of substrate or product that normally exists in free solution. By binding the substrate in a less stable conformation (relative to some standard state used for comparison), the enzyme has started it out farther up the free-energy hill towards the transition state (see section 2-6), thereby increasing the reaction rate through **ground-state destabilization**.

The enzyme ODCase, which we encountered in section 2-6, may provide a striking example of the power of ground-state destabilization; it is possible that this factor is the major contributor to the 10^{17}-fold rate acceleration achieved by this enzyme. Crystal structures of ODCase with substrate analogs bound show that the negatively charged carboxylate group of the substrate is not interacting with any positively charged groups in the active site. Instead, it is located in a hydrophobic pocket. This unfavorable environment for an ionized group should greatly increase the free energy of the bound substrate and facilitate the transformation of the carboxylate to the neutral product, CO_2, which will have a much lower free energy in such a site.

Figure 2-26 The pericyclic rearrangement of chorismate to prephenate via the proposed "chair-like" transition state When chorismate is bound to the active site of the enzyme chorismate mutase, the atoms involved in the internal rearrangement become arranged in a chair configuration. In the active site, the substrate (and, indeed, the product) are held in a conformation that resembles the single transition state for the chemical transformation (highlighted in yellow). The dotted lines indicate partial bond formation that occurs in the presumed transition state. The part of the molecule that is rearranged is indicated in blue.

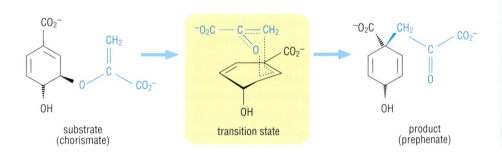

substrate
(chorismate)

transition state

product
(prephenate)

References

Chook, Y.M. *et al*.: **Crystal structures of the monofunctional chorismate mutase from *Bacillus subtilis* and its complex with a transition state analog.** *Proc. Natl Acad. Sci. USA* 1993, **90**:8600–8603.

Jencks, W.P.: **Binding energy, specificity, and enzymic catalysis: the Circe effect.** *Adv. Enzymol. Relat. Areas Mol. Biol.* 1975, **43**:219–410.

Krause, K.L. *et al*.: **2.5 Å structure of aspartate carbamoyltransferase complexed with the bisubstrate analog *N*-(phosphonoacetyl)-L-aspartate.** *J. Mol. Biol.* 1987, **193**:527–553.

Lau, E.Y. *et al*.: **The importance of reactant positioning in enzyme catalysis: a hybrid quantum mechanics/molecular mechanics study of a haloalkane dehalogenase.** *Proc. Natl Acad. Sci. USA* 2000, **97**:9937–9942.

Mesecar, A.D. *et al*.: **Orbital steering in the catalytic power of enzymes: small structural changes with large catalytic consequences.** *Science* 1997, **277**:202–206.

Radzicka, A. and Wolfenden, R.: **A proficient enzyme.** *Science* 1995, **267**:90–93.

Strater, N. *et al*.: **Mechanisms of catalysis and allosteric regulation of yeast chorismate mutase from crystal structures.** *Structure* 1997, **5**:1437–1452.

Wu, N. *et al*.: **Electrostatic stress in catalysis: structure and mechanism of the enzyme orotidine monophosphate decarboxylase.** *Proc. Natl Acad. Sci. USA* 2000, **97**:2017–2022.

Some active sites primarily stabilize transition states

Many enzyme active sites are complementary to the transition states of the reactions they catalyze, both in stereochemistry and charge configuration, or become that way during the reaction. The differences in structure between the transition state and the ground state for a given reaction are important because it is the differential binding of the enzyme for these two states that leads to catalysis. If an enzyme binds both substrate and transition state with equal affinity, the reaction will not be facilitated. But if the transition state can be bound more tightly than the substrate, then the free-energy difference between substrate and transition state will be reduced. The more that difference is reduced, the lower the barrier to reaction, and the more likely reaction becomes (Figure 2-27). This is the definition of catalysis.

The sources of the binding energy required to reduce that free-energy difference are the weak interactions that occur in the active site between enzyme and substrate. Transition-state complementarity often involves the placement in an active site of charged groups at positions where charges of opposite sign will develop when transition states appear. If a transition state has new hydrogen-bond donors or acceptors, or positions existing ones in new places, the enzyme can also stabilize the transition state by having appropriately placed acceptors and donors. Another complementarity that can occur is purely steric. Because bonds are forming and breaking in the transition state, its geometry differs from that of the substrate, and active sites fit transition states sterically better than they do their substrates. Stabilization of the transition state facilitates the electron shifts that are part of the reaction mechanism. Citrate synthase is an example of an enzyme that has certain catalytic groups prearranged so as to stabilize the transition state, which differs considerably in its geometry from the substrate (Figure 2-28).

If an enzyme active site does not start out perfectly complementary to the transition state before substrate binds, the enzyme may undergo conformational changes that increase that complementarity. Phosphoglycerate kinase (PKA) is an example of an enzyme in which a conformational change that occurs when both substrates are bound generates new interactions that stabilize the transition state (Figure 2-29). The change in conformation enables certain amino acids of the active site, together with the enzyme's bound magnesium ion, to stabilize the oxygens of the transferring phosphoryl group to a greater extent than in either the substrate or product. By stabilizing this transition state, PKA catalyzes phosphoryl transfer.

Many active sites must protect their substrates from water, but must be accessible at the same time

Water participates in many chemical reactions, but there are other reactions that cannot proceed rapidly, or even at all, in an aqueous environment. The ability of some enzymes to shield their substrates from aqueous solvent by taking advantage of conformational changes that close off the active site from contact with bulk solvent is important in enabling them to accelerate the rates of the reactions they catalyze. There is a problem connected with these conformational shifts, however. When an active site is in its closed conformation, anything already bound is protected from water, but substrates cannot enter it, and products cannot leave it. Similarly, when it is in its open conformation, such an enzyme is not an effective catalyst. This problem is solved by having the resting enzyme exist in an open state to which substrates can bind readily, and then having substrate binding trigger the conformational changes to the closed form. The rate at which these conformational shifts occur could potentially limit the rate at which an enzyme that depends on them could operate.

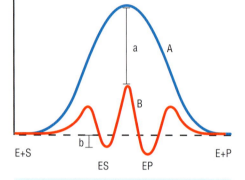

Figure 2-27 Effect of binding energy on enzyme catalysis A hypothetical case in which substrate and product are of equal energy, with a large free-energy barrier between them (A). Binding of the substrate and transition state to the enzyme can lower the free energies of these complexes by the amounts b and a respectively. If the transition state is bound more tightly than the substrate (B; a > b), which may occur, in part, through strain (destabilization) in the enzyme–substrate (ES) complex, or through the development of new interactions with the transition state, or both, the net effect will be catalysis. If, however, the enzyme binds both the substrate and the transition state with equal affinity (a = b), the barrier between the ES complex and the transition state will have the same height as the barrier for the uncatalyzed reaction, and no catalysis will occur. This is still true if the substrate and product are of unequal energies, which is the usual case. As always in discussing stabilization of ground or transition states by enzymes, external reference states must be defined.

Definitions

hydride ion: a hydrogen atom with an extra electron.

References

Bernstein, B.E. and Hol, W.G.: **Crystal structures of substrate and products bound to the phopshoglycerate kinase active site reveal the catalytic mechanism.** *Biochemistry* 1995, **37**:4429–4436.

Bernstein, B.E. *et al.*: **Synergistic effects of substrate-induced conformational changes in phosphoglycerate kinase activation.** *Nature* 1997, **385**:275–278.

Karpusas, M. *et al.*: **Proposed mechanism for the condensation reaction of citrate synthase: 1.9Å structure of the ternary complex with oxaloacetate and carboxymethyl coenzyme A.** *Biochemistry* 1990, **29**:2213–2219.

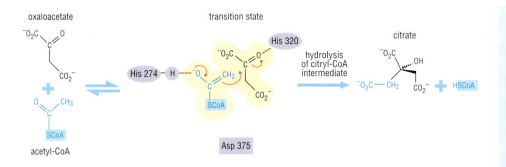

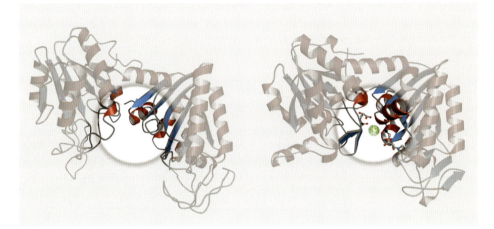

Figure 2-28 The active site of citrate synthase stabilizes a transition state with a different geometry from that of the substrate Citrate is synthesized by an aldol addition reaction between oxaloacetate and the acetyl group of acetyl-CoA. In the active site, histidine 274 and aspartic acid 375 catalyze the formation of the enol of acetyl-CoA, the rate-limiting step in the reaction. The enol is stabilized by a hydrogen bond to histidine 274. The acetyl-CoA enol then attacks the carbonyl carbon of oxaloacetate, resulting in the addition of the elements of the acetyl group at this position. Red arrows show the movement of electron pairs in making and breaking bonds. The resulting intermediate, citryl-CoA, is unstable and is hydrolyzed in the active site. Adapted from Karpusas, M. *et al.*: *Biochemistry* 1990, **29**:2213–2219.

NAD-dependent dehydrogenases transfer a **hydride ion** (H⁻) during their catalytic reactions. Hydride ions are unstable in water, so the active site must be shielded from bulk solvent during hydride transfer. These enzymes usually have a flexible loop that closes down over the active site when the cosubstrate NADH or NAD⁺ binds. A set of noncovalent interactions makes the closed-lid conformation more favorable energetically than the open-lid conformation when substrate is present (Figure 2-30). For many enzymes that use such a mechanism, the opening and/or closing of the lid is the rate-determining step in the reaction.

Figure 2-29 Phosphoglycerate kinase (PGK) undergoes a conformational change in its active site after substrate binds PGK catalyzes a reversible reaction in which a phosphoryl group is transferred between 1,3-bisphosphoglycerate and ADP to form 3-phosphoglycerate and ATP. Structures of unliganded PGK (left, from horse) and the ternary complex of PGK and its substrates 1,3-bisphosphoglycerate and ADP (right, from *Trypanosoma brucei*), show the change in conformation that occurs on substrate binding: the two domains shift position to close the active site and a metal-ion-binding site is created by the ligand and close proximity of previously distant residues. (The appearance of elements of secondary structure in the liganded enzyme that are absent from the unliganded enzyme may be an incidental consequence of the crystallization conditions rather than an effect of ligand binding.) Substrates are in ball-and-stick representation. The green sphere represents a magnesium ion. The active-site region is highlighted by the circle. The change in conformation increases the complementarity of the active site for the transition state of the reaction. (PDB 2pgk and 13pk)

There are three further questions one might ask about this system: what drives the motions that open and close the lid, what makes the lid swing open once a bound substrate molecule has been reduced by hydride transfer, and how fast does the lid move? The answer to the first question is straightforward: the movements of the lid are driven by the thermal motions that apply to all molecules in solution. The answer to the second question is not known for any enzyme with a similar moving part. The interactions between enzyme and substrate are not altered in any way that would obviously favor the opening of the lid. The loop of polypeptide chain that forms the lid is only held in place by a few weak interactions with the substrate and the rest of the enzyme, and these interactions will tend to break periodically at ordinary temperatures. When they do, the loop can swing open, and product is released. Whatever the mechanism, the loop does not have to move very fast. A single molecule of the enzyme glyceraldehyde-3-phosphate dehydrogenase can carry out almost 1,000 hydride transfers in a second, so the loop must move at least 10 Å every 0.001 second. That corresponds to a velocity of only 10^{-6} km per hour (a car travels at 10^2 km/h). Peptide conformational transitions in aqueous solution are known to be at least 10^4 times faster than that.

Figure 2-30 NAD-dependent lactate dehydrogenase has a mechanism for excluding water from the active site once substrates are bound Lactate dehydrogenase (LDH) catalyzes the conversion of lactate (orange circle) to pyruvate (red circle), with the accompanying reduction of NAD⁺ (light-blue square) to NADH (dark-blue square). The mechanism of action of LDH involves a conformational change that closes a mobile "lid" down over the active site to exclude water when both substrates have bound. Residues in this loop make weak connections to substrate and other parts of the protein when the lid is closed. The products are released when the lid opens, probably spontaneously.

mobile loop

2-10 Redox Reactions

A relatively small number of chemical reactions account for most biological transformations

A mammalian cell produces over 10,000 different proteins, of which more than half are enzymes. These thousands of different enzymes would seem to catalyze thousands of different reactions with thousands of different substrates and products. But many of these reactions are actually similar to one another and the substrates and products are often similar as well. In fact, about three-quarters of the reactions involved in metabolism can be described by as few as four general types of chemical transformations: oxidation/reduction; addition/elimination; hydrolysis; and decarboxylation.

In catalyzing these reactions, enzymes face certain common chemical problems. Organic molecules are typically stable under physiological conditions and are replete with –C–H bonds, which are relatively unreactive. One problem, therefore, is how to activate these bonds so that chemistry can occur there under the mild conditions found in living organisms and organic molecules can be degraded to provide energy. Another is how to stabilize thermally or chemically unstable intermediates that form as these reactions proceed. Yet another is the prevention of side-reactions in a milieu containing 55 moles per liter of a reactive substance, namely water. All these tasks require highly specialized and sequestered microenvironments. We have already seen some of the physical properties and mechanisms that enzymes use to create such environments (see section 2-6). Because the chemistry of living cells can be described by a relatively small number of chemical reaction types, there are also certain chemical factors that all enzymes use to promote these transformations. We will discuss the most important of these in detail in the later sections on active-site chemistry and cofactors.

Oxidation/reduction reactions involve the transfer of electrons and often require specific cofactors

Oxidation/reduction reactions, or **redox reactions**, involve the transfer of electrons from one compound to another. **Oxidation** is the loss of electrons, usually from a carbon center; **reduction** is the reverse process, the gain of electrons. Oxidation and reduction always occur together, as whenever electrons are lost or gained by something, something else must either receive them or donate them. These reactions are central to metabolism; deriving energy from food is usually a process of oxidation in which the loss of electrons serves to break chemical bonds and release energy, whereas the biosynthesis of complex molecules usually involves reduction to form new chemical bonds. In general, these reactions require the participation of specialized organic or metal-ion containing groups called cofactors, which we discuss later in the chapter. Certain cofactors have evolved for use in certain sub-types of biological redox reactions: three of these reactions are illustrated in Figure 2-31. Oxidation and reduction at carbon–oxygen centers typically involves the nicotinamide-containing cofactors NAD (nicotinamide adenine dinucleotide) and NADP (nicotinamide adenine dinucleotide phosphate). The enzyme malate dehydrogenase in the tricarboxylic acid (TCA) cycle uses a molecule of NAD to oxidize a –C–OH group in malate to a –C=O group, producing oxaloacetate and reduced NAD (NADH) plus a proton (Figure 2-31a). Redox reactions at carbon–carbon bonds most frequently utilize the flavin-containing cofactors FAD (flavin adenine dinucleotide) or FMN (flavin mononucleotide). Another TCA-cycle enzyme, succinate dehydrogenase, converts succinate to fumarate by an oxidation that requires FAD (Figure 2-31b). Oxidation and reduction at –C–H centers and

Definitions

oxidation: the loss of electrons from an atom or molecule.

redox reactions: reactions in which oxidation and reduction occur.

reduction: the gain of electrons by an atom or molecule.

nitrogen-containing centers often involves cofactors that contain metal ions. One example is the enzyme cytochrome P450, which takes specific –C–H bonds in unreactive carbon compounds and inserts molecular oxygen to produce –C–OH groups, as in the biosynthetic pathway for steroid hormones such as pregnenolone (Figure 2-31c); this reaction, the biological equivalent of a propane torch, is carried out at room temperature with the aid of the iron-containing cofactor heme.

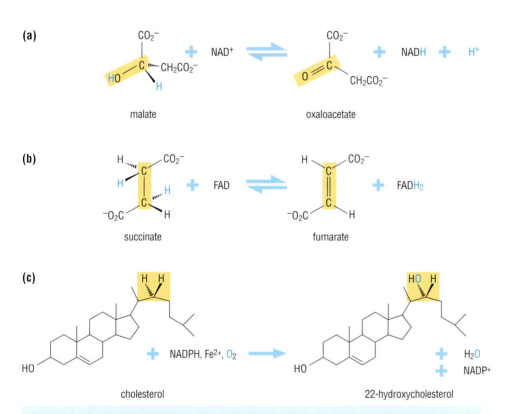

(a)

malate   oxaloacetate

(b)

succinate   fumarate

(c)

cholesterol   22-hydroxycholesterol

Figure 2-31 Examples of oxidation/reduction reactions (a) The oxidation of an alcohol to a ketone by NAD as illustrated by the reaction catalyzed by malate dehydrogenase. The reaction is driven by the abstraction by NAD of an electron and a hydrogen atom (a hydride ion, H⁻) from the carbon center of malate at which oxidation occurs, resulting in the subsequent transformation of the C–OH bond at this center to a C=O bond. In the course of the reaction the NAD is reduced to NADH. **(b)** Oxidation of a saturated carbon–carbon bond to an unsaturated carbon–carbon bond by FAD as in the reaction catalyzed by succinate dehydrogenase. **(c)** In the first step of the pathway for the conversion of cholesterol to pregnenolone—the formation of 22-hydroxycholesterol—the iron atom in the heme cofactor of cytochrome P450 inserts an atom of oxygen into a –C–H bond. Here, oxidation is the insertion of an oxygen atom derived from molecular oxygen. In all these reactions, the sites of bond rearrangement are shaded yellow and the transferred hydrogens or oxygens in blue.

References

Silverman, R.B.: *The Organic Chemistry of Enzyme-Catalyzed Reactions* (Academic Press, New York, 2000).

2-11 Addition/Elimination, Hydrolysis and Decarboxylation

Addition reactions add atoms or chemical groups to double bonds, while elimination reactions remove them to form double bonds

Addition reactions transfer atoms or chemical groups to the two ends of a double bond, forming a more highly substituted single bond. Elimination reactions reverse this process, forming a new double bond. In many addition/elimination reactions involving C=C bonds, the transferred species is a molecule of water. In the reaction catalyzed by the TCA-cycle enzyme fumarase a molecule of water is added to the C=C of fumarate (OH^- to one carbon and H^+ to the other) to produce a molecule of malate (Figure 2-32a). Another type of addition reaction, called an aldol condensation (the reverse elimination is called aldol cleavage) is used to make new carbon–carbon bonds. This reaction involves addition of an activated carbon center (for example, the acyl moiety of acyl-coenzyme A) to the C=O carbonyl carbon and is the most common way of making a new C–C bond in biology. In the TCA cycle, such a reaction is catalyzed by the enzyme citrate synthase (Figure 2-32b), which we encountered in section 2-9. Breaking a C–C bond is difficult and is often done by aldol cleavage. In glycolysis, the enzyme aldolase catalyzes an aldol cleavage reaction that splits a six-carbon sugar phosphate into two three-carbon fragments.

Esters, amides and acetals are cleaved by reaction with water; their formation requires removal of water

The breaking of C–N, C–O, S–O, P–O and P–N bonds is accomplished in biology by the reaction of compounds containing such bonds with water. Such reactions are referred to as hydrolysis; hydrolytic reactions of biochemical importance are the hydrolysis of amides (C–N), esters and complex carbohydrates such as acetals (C–O), sulfate esters (S–O), phosphoesters (P–O), and phosphonamides (P–N). Degradation of biopolymers such as proteins in digestion is almost entirely a hydrolytic process. All proteases, such as trypsin, which degrade the amide bonds in proteins use water in this way (Figure 2-33a) as do the nucleases that digest DNA and RNA by hydrolyzing P–O bonds between adjacent nucleotide residues (Figure 2-33b). A P–O phosphoanhydride bond is also hydrolyzed in the conversion of ATP to ADP and

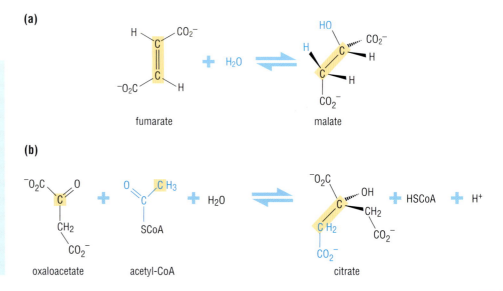

(a)

fumarate + H_2O ⇌ malate

(b)

oxaloacetate + acetyl-CoA + H_2O ⇌ citrate + HSCoA + H^+

Figure 2-32 Examples of addition/elimination reactions (a) Addition of the elements of water across the C=C of fumarate to create the HC–COH group of malate, a reaction catalyzed by fumarase. This reaction is readily reversible, but is driven in the direction of addition rather than elimination by the high concentration of water in the cellular milieu. **(b)** Addition of the elements of acetate to the carbonyl carbon of oxaloacetate in the aldol condensation reaction catalyzed by citrate synthase. Acetate is activated by coupling it to a cofactor called coenzyme A (CoA). The aldol condensation gives an activated *S*-citryl-CoA intermediate (not illustrated here) which is then hydrolyzed to give citrate, regenerated CoA (HSCoA) and H^+.

Definitions

decarboxylation: removal of carbon dioxide from a molecule.

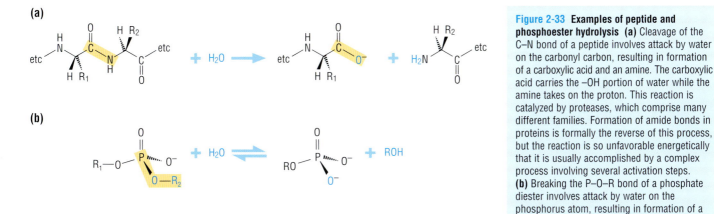

Figure 2-33 Examples of peptide and phosphoester hydrolysis (a) Cleavage of the C–N bond of a peptide involves attack by water on the carbonyl carbon, resulting in formation of a carboxylic acid and an amine. The carboxylic acid carries the –OH portion of water while the amine takes the proton. This reaction is catalyzed by proteases, which comprise many different families. Formation of amide bonds in proteins is formally the reverse of this process, but the reaction is so unfavorable energetically that it is usually accomplished by a complex process involving several activation steps. **(b)** Breaking the P–O–R bond of a phosphate diester involves attack by water on the phosphorus atom, resulting in formation of a phosphate monoester and an alcohol. When the phosphate diester is part of the backbone of DNA, R_1 is derived from the 3´-hydroxyl of the deoxysugar (ribose) of one nucleotide and R_2 is derived from the 5´-hydroxyl of the deoxysugar of another nucleotide. This reaction is catalyzed by endonucleases such as DNase. In the reverse direction, if the alcohol is on the sugar group of one deoxyribonucleotide, loss of water produces the phosphodiester linkage of the DNA backbone (both of these groups must be activated so the reaction is driven to completion). This biosynthetic reaction is catalyzed by DNA polymerase.

inorganic phosphate, for example, a reaction central to energy metabolism in all cells. Formation of these types of bonds involves the loss of a molecule of water in the reverse of hydrolysis, which is usually called condensation or sometimes dehydration. Reactions in which compounds are formed with the loss of water include the synthesis of proteins (C–N), acylglycerols and oligosaccharides (C–O), and polynucleotides (P–O) from their monomers, and the formation of ATP from ADP and inorganic phosphate during respiration. Since these types of molecules often have higher free energies than their monomers, a common strategy in biology is to first activate the components by, for example, formation of a phosphate ester, which can then be reacted with another activated species via the loss of water. DNA synthesis from activated nucleoside triphosphates is just a series of sequential phosphodiester bond formations in which water is lost, catalyzed by the enzyme DNA polymerase.

Loss of carbon dioxide is a common strategy for removing a single carbon atom from a molecule

Since C–C and C=O bonds are quite stable, shortening a molecule by one carbon atom is not an easy process chemically. In biology, it is usually accomplished by the loss of carbon dioxide, a process that is thermodynamically favored because CO_2 is a very stable molecule. Loss of CO_2 is termed **decarboxylation** and is usually assisted by cofactors. Several different cofactors can participate in decarboxylations. The most common are pyridoxal phosphate (PLP) and thiamine diphosphate (TDP; also referred to as TPP, thiamine pyrophosphate), but transition-metal ions such as manganese are sometimes used instead. Pyruvate decarboxylase, which converts pyruvate to acetaldehyde in the pathway for ethanol biosynthesis, uses TPP as a cofactor in its decarboxylation reaction (Figure 2-34).

Figure 2-34 Example of the decarboxylation of a carboxylic acid Shortening of the three-carbon unit of pyruvate to the two-carbon unit of acetaldehyde is accomplished by the loss of CO_2, catalyzed by the cofactor TPP bound at the active site of the enzyme pyruvate decarboxylase. The CO_2 molecule is derived from the carboxylate group of the acid. To help break the C–C bond, in this and similar reactions the carbon of the acid group is usually activated in some fashion, often by temporary chemical coupling to the enzyme's cofactor.

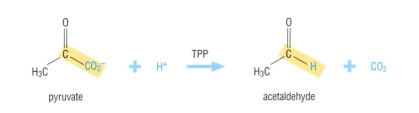

pyruvate   acetaldehyde

References

Silverman, R.B.: *The Organic Chemistry of Enzyme-Catalyzed Reactions* (Academic Press, New York, 2000).

Active sites promote acid-base catalysis

In essentially all biological reactions, regardless of the reaction type, there is at least one step that involves the transfer of a proton from one group to another. Groups donating protons are referred to as **acids**; the groups that accept them are called **bases**. Often the gain or loss of a proton will change the chemical reactivity of the group considerably. Catalysis in which a proton is transferred in going to or from the transition state is called **acid-base catalysis**.

The ease with which a proton can be transferred between an acid and a base depends on the relative proton affinities of the two groups. Proton affinity is measured by the **pKa value**, which over typical ranges of pKa can be thought of as the pH of an aqueous solution of the acid or base at which half of the molecules are protonated and the other half are deprotonated. Strong acids lose their protons readily to water, forming hydronium ions (H_3O^+). Strong acids have pKa values of 2 or lower. Strong bases tend to take protons from water, forming the

Table of Typical pKa Values

Acid (proton donor)		Conjugate base (proton acceptor)	pKa
HCOOH formic acid	⇌	HCOO⁻ formate ion	3.75
CH_3COOH acetic acid	⇌	CH_3COO^- acetate ion	4.76
OH | $CH_3CH — COOH$ lactic acid	⇌	OH | $CH_3CH — COO^-$ lactate ion	3.86
H_3PO_4 phosphoric acid	⇌	$H_2PO_4^-$ dihydrogen phosphate ion	2.14
$H_2PO_4^-$ dihydrogen phosphate ion	⇌	HPO_4^{2-} monohydrogen phosphate ion	6.86
HPO_4^{2-} monohydrogen phosphate ion	⇌	PO_4^{3-} phosphate ion	12.4
H_2CO_3 carbonic acid	⇌	HCO_3^- bicarbonate ion	6.37
HCO_3^- bicarbonate ion	⇌	CO_3^{2-} carbonate ion	10.25
C_6H_5OH phenol	⇌	$C_6H_5O^-$ phenolate ion	9.89
NH_4^+ ammonium ion	⇌	NH_3 ammonia	9.25
H_2O	⇌	OH^-	15.7

Figure 2-35 Table of pKa values for some common weak acids in biology Note that for compounds with more than one ionizable group (for example, phosphoric acid), the loss of the second (and third) proton is always more difficult than the loss of the first (and second). This is because, if each loss of a proton creates a negative charge, charge repulsion makes it harder to put more negative charge on the conjugate base.

Definitions

acid: a molecule or chemical group that donates a proton, either to water or to some other base.

acid-base catalysis: catalysis in which a proton is transferred in going to or from the transition state. When the acid or base that abstracts or donates the proton is derived directly from water (H^+ or OH^-) this is called specific acid-base catalysis. When the acid or base is not H^+ or OH^-, it is called general acid-base catalysis. Nearly all enzymatic acid-base catalysis is general acid-base catalysis.

base: a molecule or chemical group that accepts a proton, either from water or from some other acid.

pKa value: strictly defined as the negative logarithm of the equilibrium constant for the acid-base equation. For ranges of pKa between 0 and 14, it can be thought of as the pH of an aqueous solution at which a proton-donating group is half protonated and half deprotonated. pKa is a measure of the proton affinity of a group: the lower the pKa, the more weakly the proton is held.

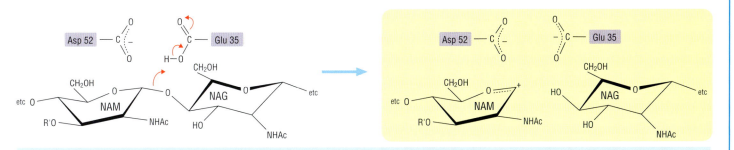

Figure 2-36 Active site of lysozyme The enzyme lysozyme hydrolyzes the acetal links between monomers in certain carbohydrate polymers. The substrate shown here is part of a polymer of N-acetyl glucosamine (NAG) and N-acetyl muramic acid (NAM). Two carboxylic acid side chains (aspartate and glutamate; purple) are found in the active site of lysozyme. In solution, these residues would be expected to have a pK_a around 4 or 5, just like acetic and lactic acids (see Figure 2-35). But in the microenvironment provided by the protein, their acidities differ considerably. Aspartic acid 52 has the expected pK_a, so at pH 7 it is ionized and can fulfill its function, which is to use its negative charge to stabilize the positive charge that builds up on the sugar during catalysis. Glutamic acid 35, however, is in a hydrophobic pocket, which raises its pK_a to around 7. In its protonated form it acts as a weak acid and donates a proton to the sugar –C–O–R group (where R is the next sugar in the chain), breaking the C–O bond. In its negatively charged form it helps stabilize the positive charge on the transition state. There is recent evidence that the mechanism may also involve a covalent intermediate between aspartic acid 52 and the substrate. The red arrows show the movement of electron pairs as bonds are made and broken.

hydroxide ion (OH^-). Strong bases have pK_a values greater than about 12. In this context, water is a very weak acid and a very weak base. Most biological acids and bases are weak; they only partially give up protons in aqueous solution at physiological pH and exist as an equilibrium between protonated and unprotonated species. If the pK_a of the group is between 4 and 7, it is a weak acid (the higher the pK_a the weaker the acid); if the pK_a is between 7 and 10, the group is a weak base (the lower the pK_a the weaker the base). Proton transfers occur efficiently from groups with low pK_a values to those of higher values. Figure 2-35 shows the pK_a values for some common biological acids and bases.

Missing from the list of pK_a values in Figure 2-35 is the weakest acid of importance in biology, the aliphatic carbon group, –C–H. Carbon has only a vanishingly small tendency to give up a proton in aqueous solution; the pK_a value of the –C–H groups in simple sugars is over 20. Yet the transfer of a proton to and from a carbon center is a common reaction in biology, occurring in almost half of the reactions of intermediary metabolism. That it can occur at all, and occur efficiently, is due to the ability of enzyme active sites to change effective pK_a values.

Enzymes can increase the efficiency of acid-base reactions by changing the intrinsic pK_a values of the groups involved. Thus, the alpha –C–H group in lactic acid can be made more acidic (that is, its pK_a can be lowered) by, for example, making a strong hydrogen bond to the –OH group attached to it. This hydrogen bond will tend to pull electrons away from the oxygen atom, which in turn will pull electrons away from the adjacent –C–H bond, weakening the affinity of the carbon for its hydrogen and thus lowering the pK_a. The pK_a of a weak acid such as the carboxylic acid side chain of lactic acid (pK_a ~ 3.9 in water) can be raised to 7 or higher by, for example, placing the group in a nonpolar environment. With no water molecules around to accept a proton, the carboxylic acid will tend to hang on to its hydrogen rather than lose it, thereby generating a negatively charged carboxylate anion in a hydrophobic region of the protein; thus, its pK_a will be raised and it will become an even weaker acid (and consequently a much stronger base). Figure 2-36 shows just this situation for the carboxylic acid side chain of glutamate in the active site of the enzyme lysozyme, where it is estimated that the pK_a of glutamic acid 35 is raised from about 4 to above 6, and it can donate a proton to catalyze the breaking of the C–O bond in the substrate.

References

Malcolm, B.A. *et al.*: **Site-directed mutagenesis of the catalytic residues Asp-52 and Glu-35 of chicken egg white lysozyme.** *Proc. Natl Acad. Sci. USA* 1989, **86**:133–137.

Vocadlo, D.J. *et al.*: **Catalysis by hen egg-white lysozyme proceeds via a covalent intermediate.** *Nature* 2001, **412**:835–838.

Voet, D. and Voet, J.: *Biochemistry* 2nd ed. Chapter 12 (Wiley, New York, 1995).

Many active sites use cofactors to assist catalysis

Not every biological reaction can be carried out efficiently using only the chemical properties of the 20 naturally occurring amino acids. Oxidation/reduction (redox) reactions, for example, in which electrons are transferred from one group to another, can be promoted to some extent by the cysteine sulfur atom, but most redox reactions need more help than this, as we saw in section 2-10. The creation of unpaired electrons (free radicals), which are useful in a number of chemical reactions, also requires chemical species that are not found in amino-acid side chains. To overcome such limitations, many enzyme active sites contain non-amino-acid **cofactors** that allow specialized chemical functions.

Cofactors can be as small as a metal ion or as large as a heterocyclic organometallic complex such as heme. They are tightly bound—sometimes covalently attached—to the proteins in which they function. Cofactors that are organic compounds and assist catalysis are often referred to as **coenzymes**. Cofactors may be imported into an organism from the food it eats (in humans most cofactors are derived from vitamins and minerals in the diet), or they can be synthesized from simple building blocks. Some common vitamin-derived cofactors are listed in Figure 2-37. In many organisms, specific sets of genes are devoted not only to the synthesis

Figure 2-37 Table of organic cofactors

Some Common Coenzymes

Coenzyme [vitamin from which it is derived]	Entity transferred	Representative enzymes that use coenzyme	Deficiency disease
thiamine pyrophosphate (TPP or TDP) [vitamin B_1, thiamin]	aldehydes	pyruvate dehydrogenase	beri beri
flavin adenine dinucleotide (FAD) [vitamin B_2, riboflavin]	hydrogen atoms	succinate dehydrogenase	(a)
nicotinamide adenine dinucleotide (NAD^+) [niacin]	hydride ion	lactate dehydrogenase	pellagra
nicotinamide adenine dinucleotide phosphate ($NADP^+$) [niacin]	hydride ion	isocitrate dehydrogenase	pellagra
pyridoxal phosphate (PLP) [vitamin B_6, pyridoxal]	amine groups	aspartate aminotransferase	(a)
coenzyme A (CoA) [pantothenic acid]	acyl groups	acetyl-CoA carboxylase	(a)
biotin (biocytin) [biotin]	CO_2	propionyl-CoA carboxylase	(a)
5'-deoxyadenosylcobalamin [vitamin B_{12}]	H atoms and alkyl groups	methylmalonyl-CoA mutase	pernicious anemia
tetrahydrofolate (THF) [folate]	one-carbon units	thymidylate synthase	megaloblastic anemia
lipoamide [lipoic acid]	two-carbon units; R-SH	pyruvate dehydrogenase	(a)
heme [no vitamin]	e^-, O_2, NO, CO_2	cytochrome oxidase	anemia, leukemia

(a) no specific name

Definitions

coenzyme: a cofactor that is an organic or organometallic molecule and that assists catalysis.

cofactor: a small, non-protein molecule or ion that is bound in the functional site of a protein and assists in ligand binding or catalysis or both. Some cofactors are bound covalently, others are not.

of some cofactors, but also to their insertion into the active sites of the particular proteins in which they are to function. It is likely that biochemical machinery ensuring that the right cofactor binds to the right protein at the right time is prevalent throughout the living world.

Metal-ion cofactors (Figure 2-38) are usually first-row transition metals and the most common are also among the most abundant in the Earth's crust, suggesting that many enzymes evolved to use whatever chemically reactive species were plentiful at the time. Enzymes are known that use molybdenum, nickel, cobalt, and manganese, but the majority of metalloenzymes use iron, copper, zinc, or magnesium. In some cases, more than one type of metal ion will work (that is, promote catalysis) in experiments carried out on a purified enzyme that has been stripped of its native metal, but it is thought that in the cell one type predominates or is used exclusively.

The structures of typical organic cofactors (coenzymes) reveal a striking pattern: most of them look like pieces of RNA; that is they are either ribonucleotides or derivatives of ribonucleotides. It has been suggested that this is a relic of an earlier "RNA world" that preceded the protein-based world we live in today. In this scenario, the earliest proteins would have looked for help from the most commonly available reactive chemical compounds, and these might have been RNA molecules or fragments thereof.

The function of organic cofactors is to transfer specific chemical species to a substrate. The groups transferred range from electrons and hydride ions (H^-) to small carbon-containing fragments up to about two carbons in length. These species are usually either extremely unstable on their own or would be damaging to the protein if they were not contained. The function of the coenzyme is to make them, transfer them, and/or sequester them. If the species transfer causes a change in the chemical structure of the cofactor, as in the case of NAD^+ (or its phosphorylated counterpart $NADP^+$, which does the same chemistry) it is necessary to recycle the cofactor back to its original reactive form before another enzyme turnover can take place. In many cases, specific reactions catalyzed by other enzymes perform this necessary function in a tightly coupled system.

Recently, it has been found that amino-acid side chains in a protein can be modified to produce an organic cofactor *in situ*. One striking example is found not in an enzyme but in the green fluorescent protein (GFP) from marine organisms, which is used in biology as an optical marker. Because the fluorescent chromophore is synthesized by the protein itself from the reaction of a tyrosine side chain with neighboring serine and glycine residues, GFP can be introduced genetically into any organism without the need for the organism to have other genes to make the cofactor. An example of this type of cofactor in enzymes is the unusual coenzyme lysine tyrosylquinone (LTQ) (Figure 2-39) in copper amine oxidase. LTQ is synthesized by the addition of a lysine side chain of the enzyme itself to the aromatic ring of an oxidized tyrosine residue elsewhere on the protein chain. The active enzyme is essential for the proper cross-linking of elastin and collagen in connective tissue because the cross-linking reaction is difficult and involves redox chemistry that the LTQ promotes.

Metal Ions and Some Enzymes Requiring Them	
Metal ion	**Enzyme**
Fe^{2+} or Fe^{3+}	cytochrome oxidase
	catalase
	peroxidase
Cu^{2+}	cytochrome oxidase
Zn^{2+}	DNA polymerase
	carbonic anhydrase
	alcohol dehydrogenase
Mg^{2+}	hexokinase
	glucose-6-phosphatase
	pyruvate kinase
Mn^{2+}	arginase
K^+	pyruvate kinase
Ni^{2+}	urease
Mo	nitrate reductase
Se	glutathione peroxidase

Figure 2-38 Table of metal-ion cofactors

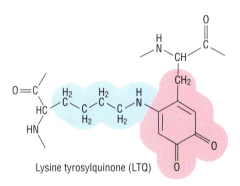

Lysine tyrosylquinone (LTQ)

Figure 2-39 The coenzyme lysine tyrosylquinone Lysine tyrosylquinone is formed on the enzyme by the addition of the lysine side chain (blue) to an oxidized form of tyrosine (pink). Formation of the cofactor is catalyzed by the copper ion (not shown) required as part of the overall reaction catalyzed by the enzyme. These types of cofactors are essential in certain enzymes that catalyze oxidation reactions involving a highly reactive radical intermediate. They stabilize the radical and make it available for the reaction.

References

Voet, D. and Voet, J. *Biochemistry* 2nd ed. (Wiley, New York, 1995).

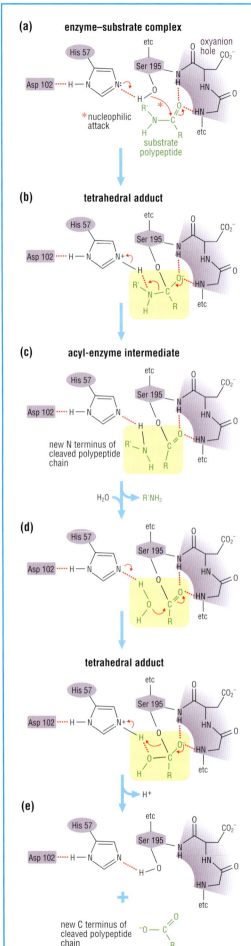

(a) enzyme–substrate complex

(b) tetrahedral adduct

(c) acyl-enzyme intermediate

new N terminus of cleaved polypeptide chain

(d) tetrahedral adduct

(e)

new C terminus of cleaved polypeptide chain

Some active sites employ multi-step mechanisms

The chemistry of most biological reactions is very difficult to carry out at ordinary temperatures and pressures in aqueous solution at neutral pH. To catalyze such reactions, enzymes must circumvent the high-energy transition state that prevents the reaction from occurring under moderate conditions (see Figure 2-21). One way of achieving this, as we have seen (section 2-9), is by stabilizing the transition state. Alternatively, enzymes may direct the reaction along a different route from substrates to products, one not accessible to the uncatalyzed reaction in aqueous solution. This often entails breaking the reaction up into a number of steps, each of which has a lower-energy transition state. The "product" of each individual step is a relatively unstable reaction intermediate (see section 2-6). In many instances, these intermediates can be trapped by physical (low temperature) or chemical (reduction) methods, and thus can be characterized completely. In other instances, the existence of these intermediates has only been inferred from kinetic or spectroscopic data, or from chemical logic.

Some intermediates are covalent: they involve the linkage of an amino-acid side chain or part of a cofactor to the substrate. The product, a covalent intermediate, then reacts further, forming other intermediates and/or the final product(s). The classic example of this strategy is the serine protease reaction, in which the peptide bonds between amino acids are broken to degrade proteins, as occurs in digestion. Hydrolysis of a peptide bond is a difficult chemical process: this is why proteins are normally stable in cells. In the uncatalyzed reaction, water attacks the carbonyl carbon of the amide bond. Because of the polarity of this bond, the carbon atom is somewhat electron-deficient and therefore susceptible to attack by a **nucleophile**, a group that is electron-rich. Water is a poor nucleophile. To overcome this problem, two strategies are possible: water can be activated to make it a better nucleophile, and/or the amide group can be activated to make it more susceptible to attack. Enzymes that catalyze the degradation of proteins by hydrolysis of the amide bonds of their backbones solve this problem by dividing the overall reaction into two steps; they activate both the attacking nucleophile (serine OH or H_2O) and the amide group in a reaction that has two steps with an intermediate (Figure 2-40).

In the first step (Figure 2-40a–c), instead of using the –OH group of water as the nucleophile, the enzyme uses the hydroxyl group of a serine side chain, which attacks the carbonyl carbon of the amide bond, producing an acyl-enzyme intermediate and releasing the amino-terminal portion of the peptide substrate. The –OH of a serine is also not a good nucleophile and requires some activation. It is able to break the amide bond in the first step of this reaction for two reasons. One is proximity and orientation: the serine is positioned adjacent to the carbonyl carbon of the substrate in the active site. The second is that the –OH of the serine is activated by hydrogen bonding with a histidine side chain which, in turn, is anchored by interaction with the carboxylate group of an aspartic acid (see Figure 2-40a). The aspartic acid ensures that the histidine is in the right charge state to pull a proton from the –OH of serine, moving the serine towards a serine alkoxide (–O⁻), which is a much better nucleophile. The Asp-His-Ser signature of the serine proteases is also called the **catalytic triad**. The energy barrier for the formation of the acyl-enzyme intermediate is further lowered by stabilization of the transition state (the tetrahedral adduct) leading to its formation, with concomitant activation of the carbonyl group of the peptide substrate for attack (Figure 2-40b). In the transition state, negative charge develops on the carbonyl oxygen. The protein is configured to stabilize this

Figure 2-40 The chemical steps in peptide hydrolysis catalyzed by the serine protease chymotrypsin (a) The substrate (green) binds to the active site such that the hydroxyl group of serine 195 is positioned to attack the peptide carbonyl carbon. Simultaneously, the carbonyl oxygen hydrogen bonds to two –NH groups of the protein and the histidine removes a proton from the serine –OH, thus activating it. The two –NH groups are part of the peptide backbone forming a loop called the oxyanion hole (shaded in purple). **(b)** In the transition state for formation of the acyl-enzyme intermediate, negative charge has developed on the carbonyl oxygen; this is also stabilized by hydrogen bonding in the oxyanion hole. The first product is beginning to be released in this step and the bond between the carbonyl carbon of the substrate and the serine oxygen is beginning to form. **(c)** The acyl-enzyme intermediate. In this intermediate, the serine side chain is covalently bound to the carbonyl carbon of the substrate. The histidine gives up the proton it acquired from the serine to the nitrogen of the substrate, making it easier for the first product to split off from the rest of the substrate. **(d)** In the transition state for hydrolysis of the acyl-enzyme intermediate, a water molecule attacks the carbonyl carbon of the acyl-enzyme ester as the histidine accepts a proton from it, analogous to the acylation step (b). The transient negative charge on the oxygen of the acyl enzyme is again stabilized by hydrogen bonding in the oxyanion hole. **(e)** Formation of product. The transition state collapses to release the other portion of the substrate and regenerate the serine –OH.

negative charge through the donation of two hydrogen bonds to the carbonyl oxygen from what has been called the **oxyanion hole** in the active site (Figure 2-40).

Step two of the reaction requires hydrolysis of the covalent acyl-enzyme intermediate (Figure 2-40d,e). When the intermediate is hydrolyzed, the carboxy-terminal portion of the peptide substrate is released and the serine on the enzyme is restored to its original state. Attack of water on the acyl enzyme is more facile than attack of water on the original amide substrate, for the hydrolysis of an ester is easier than hydrolysis of an amide. This is because the C–N bond of an amide has more double-bond character than the corresponding C–O bond of an ester. The transition state for hydrolysis of the acyl enzyme is similar to that already described for its formation, and it is stabilized in the same way, by interactions with the oxyanion hole (Figure 2-40d).

An obvious question raised by this example is why not just activate a molecule of water directly by pulling a proton off it, turning the weak nucleophile H_2O into the strong nucleophile OH^-, instead of activating a serine first. Interestingly, this is exactly what happens in another class of proteases, the metalloproteases. Instead of using a covalent enzyme intermediate, these enzymes use a metal-ion cofactor to activate water for direct attack on the substrate carbonyl, while at the same time activating the carbonyl. In contrast to the serine protease mechanism, metalloproteases such as carboxypeptidase and thermolysin do not proceed through a stable intermediate. They bind a molecule of water to a metal ion, usually Zn^{2+}. Binding H_2O to a transition metal ion lowers the pK_a of the bound water, making it a better acid. It is thus much easier to turn metal-bound H_2O into OH^-. The metal-bound OH^- can attack the amide bond of a peptide substrate directly, hydrolyzing it without the need for a covalent enzyme intermediate. The presence of two or more classes of enzymes that carry out the same chemical process by different chemical strategies is a common feature of biology.

The two-step strategy is also often deployed in phosphoryl-group transfer reactions, particularly those catalyzed by kinases and phosphatases, which, as we shall see in Chapter 3, are among the most important mechanisms for regulating protein function. Unlike the serine proteases, in which a serine (or a cysteine in the case of the related cysteine protease family) is always the attacking nucleophile, phosphoryl-group transfer can occur with a wide variety of attacking groups. Serine, threonine and tyrosine –OH groups, the carboxylate groups of aspartate and glutamate, cysteine –SH and the nitrogen of histidine are all known to participate in various phosphoryl-group transfers. In the bacterial enzyme alkaline phosphatase, which catalyzes the transfer of phosphate to water from many organophosphate substrates, a serine –OH on the enzyme first attacks the phosphorus atom, releasing the organic portion of the substrate and leaving behind a phosphoserine-enzyme intermediate, which is then hydrolyzed by water, a mechanism analogous to that of the serine proteases. Although it is difficult to trap acyl-enzyme intermediates such as those formed in the serine and cysteine protease reactions, phosphoenzyme intermediates can be more stable and have been isolated for many enzymes. For example, phosphoglucomutases catalyze the interconversion of D-glucose 1-phosphate and D-glucose 6-phosphate, a reaction central to energy metabolism in all cells and to the synthesis of cell-wall polysaccharides in bacteria. The reaction proceeds through a phosphoaspartyl anhydride intermediate, which is formed when the phosphate on the aspartyl group of the enzyme attacks the unphosphorylated position on the sugar and transfers the phosphate group to it (Figure 2-41). Subsequent reaction of the phosphoryl group at the other position on the sugar with the aspartate regenerates the phosphoaspartyl-enzyme. The structure of the phosphorylated enzyme from *Lactococcus lactis* clearly shows the phosphate bound to the side chain of aspartic acid 8. The absence of a catalytic base near the aspartyl phosphate group to activate a water molecule or another substrate group accounts for the persistence of the phosphorylated enzyme under physiological conditions.

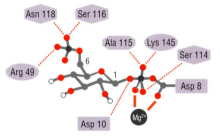

Figure 2-41 The phosphoenzyme–substrate intermediate in the active site of beta-phosphoglucomutase from *Lactococcus lactis* In the active site of beta-phosphoglucomutase, phosphoryl-group transfer occurs via aspartic acid 8, which can be phosphorylated to form a stable phosphoaspartyl-enzyme. In the reaction glucose 6-phosphate (G6P) to glucose 1-phosphate (G1P), for example, the un-phosphorylated C1 OH of the substrate is positioned such that it can become linked to the phosphoaspartyl group as shown. In this intermediate, the phosphate being transferred is pentavalent and is still associated with both aspartic acid 8 (in a mixed anhydride linkage) while becoming covalently linked to the G6P substrate to form the intermediate 1,6-bisphosphoglucose–enzyme complex. Subsequent reaction of the phosphoryl group at the 6 position of the sugar with the aspartic acid 8 is thought to complete the transfer to C1, forming G1P and regenerating the phosphoaspartyl-enzyme.

Definitions

catalytic triad: a set of three amino acids that are hydrogen bonded together and cooperate in catalysis.

nucleophile: a group that is electron-rich, such as an alkoxide ion (–O⁻), and can donate electrons to an electron-deficient center.

oxyanion hole: a binding site for an alkoxide in an enzyme active site. The "hole" is a pocket that fits the –O⁻ group precisely, and has two hydrogen-bond-donating groups that stabilize the oxyanion with –O⁻···H–X hydrogen bonds.

References

Lahiri, S.D. *et al.*: **Caught in the act: the structure of phosphorylated beta-phosphoglucomutase from *Lactococcus lactis*.** *Biochemistry* 2002, **41**:8351–8359.

Silverman, R.B.: *The Organic Chemistry of Enzyme-Catalyzed Reactions* Revised ed. (Academic Press, New York, 2002).

Walsh, C.: *Enzymatic Reaction Mechanisms* Chapter 3 (Freeman, San Francisco, 1979).

Some enzymes can catalyze more than one reaction

In some cases an enzyme may catalyze more than one chemical transformation. Such enzymes may be composed of a single polypeptide chain with one or more active sites, or may be composed of more than one polypeptide chain, each with an active site. In the latter case, each polypeptide chain represents an independently folded subunit which normally does not, however, exist in the absence of the others. Such enzymes are called **bifunctional** (or **multifunctional** if more than two reactions are involved) and they fall into three classes. In the first class, the two reactions take place consecutively at the same active site. In the second, two separate chemical reactions are catalyzed by two distinct active sites, each located in a different domain some distance apart. In the third, two or more reactions are also catalyzed by two or more distinct active sites, but these are connected by internal channels in the protein, through which the product of the first reaction diffuses to reach the next active site, where it undergoes further reaction. In this section we shall look at examples of the first and second classes. In the following section, we deal with multifunctional enzymes with internal channels, and will also consider enzymes that have additional non-enzymatic functions.

Some bifunctional enzymes have only one active site

In bifunctional enzymes that carry out two different reactions using the same active site, the second reaction is simply an inevitable chemical consequence of the first, because the product of the initial reaction is chemically unstable. Although catalysis of the second step, involving the breakdown of this first product, might seem unnecessary to sustain a rapid overall reaction rate, this second reaction is often catalyzed as well. The active site is the same for the two consecutive transformations, as the first product does not dissociate from the enzyme. An example is the reaction catalyzed by isocitrate dehydrogenase (ICDH), an enzyme in the TCA-cycle pathway. ICDH uses the cofactor NADP to oxidize isocitrate to oxalosuccinate, a compound that is unstable because it is a beta-ketoacid. In the same active site, with the assistance of a manganese ion, oxalosuccinate is catalytically decarboxylated, independent of the presence of NADP, to give the final product, alpha-ketoglutarate (Figure 2-42).

Some bifunctional enzymes contain two active sites

A second class of bifunctional enzyme contains two independently folded domains, each of which has a distinct, non-overlapping active site. Although there are some exceptions, it is usually the case that the product of one active site is the substrate for the other active site.

Figure 2-42 The reaction catalyzed by isocitrate dehydrogenase The reaction catalyzed by this enzyme consists of an oxidation reaction and a decarboxylation reaction. In the active site of this TCA-cycle enzyme, isocitrate is oxidized and decarboxylated to form alpha-ketoglutarate. In the first reaction, isocitrate is oxidized to oxalosuccinate and the cofactor NADP is concomitantly reduced to NADPH. Although oxalosuccinate will decarboxylate spontaneously to form the desired product, alpha-ketoglutarate, in the active site of isocitrate dehydrogenase this decarboxylation is catalyzed in a second reaction with the assistance of an enzyme-bound manganese ion. Oxalosuccinate never escapes from the enzyme.

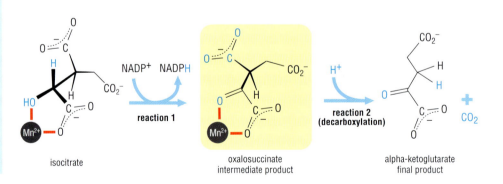

isocitrate

oxalosuccinate
intermediate product

alpha-ketoglutarate
final product

Definitions

bifunctional: having two distinct biochemical functions in one gene product. Bifunctional enzymes catalyze two distinct chemical reactions.

multifunctional: having a number of distinct biochemical functions in one gene product.

References

Greasley, S.E. *et al.*: **Crystal structure of a bifunctional transformylase and cyclohydrolase enzyme in purine biosynthesis.** *Nat. Struct. Biol.* 2001, **8**:402–406.

Hurley, J.H. *et al.*: **Catalytic mechanism of NADP(+)-dependent isocitrate dehydrogenase: implications from the structures of magnesium–isocitrate and NADP+ complexes.** *Biochemistry* 1991, **30**:8671–8678.

Knighton, D.R. *et al.*: **Structure of and kinetic channelling in bifunctional dihydrofolate reductase-thymidylate synthase.** *Nat. Struct. Biol.* 1994, **1**:186–194.

Liang, P.H. and Anderson, K.S.: **Substrate channeling and domain-domain interactions in bifunctional**

thymidylate synthase-dihydrofolate reductase. *Biochemistry* 1998, **37**:12195–12205.

Liu, J. *et al.*: **Thymidylate synthase as a translational regulator of cellular gene expression.** *Biochim. Biophys. Acta* 2002, **1587**:174–182.

Wolan, D.W. *et al.*: **Structural insights into the avian AICAR transformylase mechanism.** *Biochemistry* 2002, **41**:15505–15513.

Unless these two sites were to face each other (which has never been found) or be physically connected in some fashion, the first product must dissociate from the first domain and diffuse through the cellular medium or along the protein surface in order to find the second active site. There would seem to be no catalytic advantage to having the two activities residing in one fused gene product, but the arrangement does at least allow coordinate regulation of the biosynthesis of both enzyme activities. Genome sequencing has provided evidence that any advantage is small: in nearly all instances, at least one organism can be found in which the two different catalytic activities are encoded by two distinct genes.

One example of this type of bifunctional enzyme is dihydrofolate reductase-thymidylate synthase from the parasite *Leishmania major*, which catalyzes two reactions in the biosynthesis of thymidine. In most other organisms, including humans and *E. coli*, these two enzymatic activities are carried by two separate proteins, encoded by separate genes: a thymidylate synthase and a dihydrofolate reductase. In the first active site of the *Leishmania* enzyme, a molecule of 2′-deoxyUMP is methylated by N^5,N^{10}-methylene-tetrahydrofolate to give thymidylate (TMP) and dihydrofolate. In the second active site, located 40 Å away in a separate folded domain, the dihydrofolate is reduced by NADH to generate tetrahydrofolate. (Cofactor regeneration is completed by another, separate, enzyme.) In *Leishmania*, presumably, two genes encoding the two separate enzyme activities were originally arranged in tandem and at some point during evolution the stop codon for the first one was lost, creating a single gene with a bifunctional protein product.

Another example of this is found in the purine biosynthesis pathway. In most organisms AICAR transformylase-IMP cyclohydrolase (ATIC) is a 64 kDa bifunctional enzyme that possesses the final two activities in *de novo* purine biosynthesis, 5-aminoimidazole-4-carboxamide-ribonucleotide (AICAR) transformylase and IMP cyclohydrolase. ATIC forms an intertwined dimer, with each monomer composed of two separate different functional domains. The amino-terminal domain (up to residue 199) is responsible for the IMP cyclohydrolase activity, whereas the AICAR transformylase activity resides in the carboxy-terminal domain (200–593). The active sites of the two domains are approximately 50 Å apart, with no structural evidence of a tunnel connecting the two active sites (Figure 2-43). In the archaeon *Methanopyrus kandleri*, however, these two activities are carried out by two independent proteins, encoded by two distinct genes.

Figure 2-43 The bifunctional enzyme, AICAR transformylase-IMP cyclohydrolase (ATIC) is a single enzyme with two distinct active sites (a) This enzyme occurs as an intertwined homodimer, with each monomer composed of two distinct domains (dark and light blue/dark and light red). The two domains are each involved in catalyzing a different reaction in the biosynthesis of inosine monophosphate (IMP), the initial purine derivative that is the precursor to adenosine and guanosine. One site of each type is indicated by an arrow, with the reaction that takes place there described below. In all there are four active sites on the dimer, one on each domain. **(b)** The active sites for the AICAR transformylase reaction on the two monomers are each formed by residues contributed by both monomers. In the first site a transformylase reaction on the substrate 5-aminoimidazole-4-carboxamide ribonucleotide (AICAR) produces a product, formyl-5-AICAR (FAICAR), that is a substrate for the second active site. 10^f-THF is the cofactor N^{10}-formyltetrahydrofolate. **(c)** In the second active site, FAICAR is cyclized to IMP. In most organisms these two reactions are carried out by the bifunctional enzyme, but in some species the transformylase and the cyclohydrolase reactions are carried out by two distinct proteins (encoded by two separate genes). Each of these resembles in structure the corresponding domain in the bifunctional enzyme.

(a)

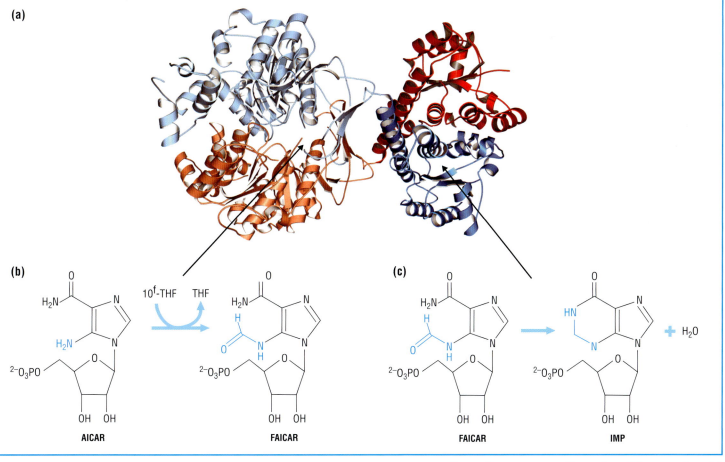

(b) AICAR → FAICAR (10^f-THF, THF) **(c)** FAICAR → IMP + H_2O

Some bifunctional enzymes shuttle unstable intermediates through a tunnel connecting the active sites

A third small class of multifunctional enzymes has been found to possess a remarkable connection between their separate active sites. A physical channel (or channels) allows the product of one reaction to diffuse through the protein to another active site without diffusing out into contact with the cellular medium. Two reasons for such a feature have been identified: either the first reaction product is an uncharged species that might be lost from the cell altogether by diffusion through the membrane if it were allowed to escape from the protein; and/or the reaction product is so unstable in free solution that it would decompose before it had time to find the second active site.

The first enzyme found to have such a tunnel through its structure was the bifunctional tryptophan synthase from *Salmonella typhimurium*. Unlike the enzymes discussed in the previous section, this is composed of two subunits, encoded by separate genes. The tunnel is 25 Å long and connects the active site of the alpha-subunit, in which indole, an uncharged molecule that might diffuse out of the cell, is generated from indole 3-glycerolphosphate, to the second active site, in which the indole is added to a molecule of acrylate, derived from serine, to produce tryptophan (Figure 2-44). Indole remains enzyme-bound throughout the reaction. The tunnel, as expected, is lined with nonpolar side chains to retain the nonpolar indole molecule.

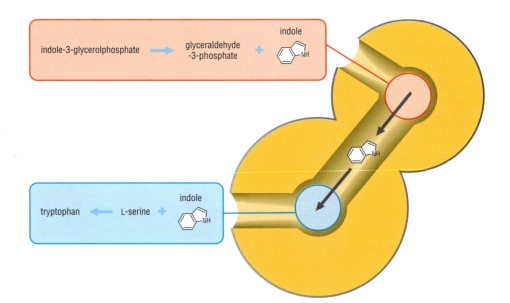

Figure 2-44 The two active sites of the bifunctional enzyme tryptophan synthase are linked by an internal channel In the reaction catalyzed by this enzyme, indole-3-glycerolphosphate (IGP) is first converted to indole and glyceraldehyde-3-phosphate in an active site located in the alpha-subunit. The uncharged indole molecule diffuses through a channel in the enzyme, lined with nonpolar side chains, to the second active site, in the beta-subunit, where a second substrate, serine, has been converted to an acrylate adduct of the cofactor pyridoxal phosphate (PLP). There the indole adds to the acrylate and the product hydrolyzes to give tryptophan and regenerate PLP.

References

Huang, X. *et al.*: **Channeling of substrates and intermediates in enzyme-catalyzed reactions.** *Annu. Rev. Biochem.* 2001, **70**:149–180.

Hyde, C.C. *et al.*:**Three-dimensional structure of the tryptophan synthase alpha 2 beta 2 multienzyme complex from *Salmonella typhimurium*.** *J. Biol. Chem.* 1988, **263**:17857–17871.

James, C.L. and Viola, R.E.: **Production and characterization of bifunctional enzymes. Substrate channeling** in the aspartate pathway. *Biochemistry* 2002, **41**:3726–3731.

Thoden, J.B. *et al.*: **Structure of carbamoyl phosphate synthetase: a journey of 96 Å from substrate to product.** *Biochemistry* 1997, **36**:6305–6316.

Trifunctional enzymes can shuttle intermediates over huge distances

The second example is considerably more complicated because it is a trifunctional enzyme, carbamoyl phosphate synthetase, an enzyme involved in the synthesis of 2'-deoxyUMP and in the urea cycle. The single-chain protein has three separate active sites connected by two tunnels through the interior of the protein, and individual genes corresponding to the different activities have never been found in any organism. This design has presumably evolved because the first reaction produces ammonia, a neutral species, which travels along a tunnel to the second active site where it reacts with carboxyphosphate to give a carbamate intermediate that would be too unstable to survive in aqueous solution. Therefore, it is transported through the interior of the protein to the third active site, where it is phosphorylated by ATP to give the final product. The entire journey from first substrate to carbamoyl phosphate covers a distance of nearly 100 Å: this enzyme is illustrated in Figure 2-45. Bifunctional enzymes are known in which the two reactions catalyzed are not consecutive; in this case a separate enzyme catalyzes the intervening reaction and channels substrate between the two active sites.

Some enzymes also have non-enzymatic functions

The genomes of mammals are not that much larger, in the number of genes, than those of plants, fish, flies and worms. It is becoming clear that one reason for this economy of gene number is that, in multicellular eukaryotes at least, many enzymes have at least a second biochemical function that is unrelated to their catalytic activity. This function is usually regulatory: some enzymes double as transcription factors; others act as signaling proteins; some are cofactors for essential reactions in protein synthesis; and yet others are transported out of the cell to serve as cytokines or growth factors. It is possible that this multiplicity of functions, in addition to allowing the genome to remain relatively compact, also connects various processes. Having a metabolic enzyme double as a repressor, for example, couples the expression of some genes to metabolism in a direct way. For example, the folate-dependent enzyme thymidylate synthase (see section 2-15) also functions as an RNA-binding protein. It interacts with its own mRNA to form a ribonucleoprotein complex, and there is also evidence that it can interact with a number of other cellular mRNAs, including transcripts of the *p53* tumor suppressor gene and the *myc* family of transcription factor genes. The functional consequence of such binding is to repress the translation of these genes. Hence, the metabolic enzyme thymidylate synthase may have a critical role in regulating the cell cycle and the process of programmed cell death through its regulatory effects on the expression of cell-cycle-related proteins. It is also a target for several anti-cancer drugs, and its ability to function as a translational regulator may have important consequences for the development of cellular resistance to such drugs. A list of some proteins with more than one function can be found in Figure 4-47.

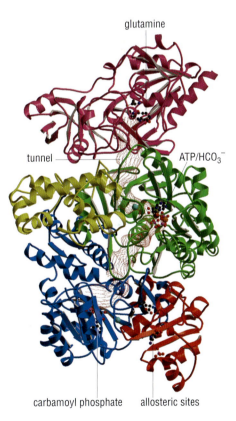

glutamine

tunnel

ATP/HCO$_3^-$

carbamoyl phosphate   allosteric sites

Figure 2-45 Three consecutive reactions are catalyzed by the three active sites of the enzyme carbamoyl phosphate synthetase Carbamoyl phosphate synthetase is a trifunctional enzyme with two long tunnels connecting the three active sites. In the first active site, shown at the top of the figure, glutamine is hydrolyzed to ammonia, which migrates through the first tunnel in the interior of the protein to the second active site, shown in the middle, where it reacts with carboxyphosphate to produce a carbamate intermediate. (The carboxyphosphate is formed by phosphorylation of bicarbonate by ATP.) The carbamate intermediate diffuses through the second tunnel to the third active site, shown at the bottom of the figure, where it is phosphorylated by another ATP molecule to give the final product, carbamoyl phosphate, which escapes from the enzyme. Graphic kindly provided by Frank Raushel and Hazel Holden.

3

Control of Protein Function

In the cell, precise regulation of protein function is essential to avoid chaos. This chapter describes the most important molecular mechanisms by which protein function is regulated in cells. These range from control of a protein's location and lifetime within the cell to the binding of regulatory molecules and covalent modifications such as phosphorylation that rapidly switch protein activity on or off. Also covered here are the nucleotide-driven switches in conformation that underlie the action of motor proteins and that regulate many signal transduction pathways.

Protein function in living cells is precisely regulated

A typical bacterial cell contains a total of about 250,000 protein molecules (comprising different amounts of each of several thousand different gene products), which are packed into a volume so small that it has been estimated that, on average, they are separated from one another by a distance that would contain only a few molecules of water. It is likely that eukaryotic cells are at least as densely packed. In this crowded environment, precise regulation of protein function is essential to avoid chaos. Regulation of protein function *in vivo* tends to occur by many different mechanisms, which fall into several general classes. Protein function can be controlled by localization of the gene product and/or the species it interacts with, by the covalent or noncovalent binding of **effector** molecules, and by the amount and lifetime of the active protein.

Proteins can be targeted to specific compartments and complexes

Not all proteins are absolutely specific, and many also have more than one function. Consequently, it is often undesirable to have such proteins distributed everywhere in the cell, where they may carry out unwanted reactions. A simple way to regulate their activity is to ensure that the protein is only present in its active form in the specific compartment where it is needed, or when bound in a complex with other macromolecules that participate in its function. There are many ways in which specific localization can be achieved. Proteins can be targeted to cellular compartments by so-called signal sequences that are an intrinsic part of the encoded amino-acid sequence, or by attachment of, for example, a lipid tail that inserts into membranes. They can be directed to a complex of interacting proteins by a structural *interaction domain* that recognizes some covalent modification such as phosphorylation on another protein. Localization is a dynamic process and a given protein may be targeted to different compartments at different stages of the cell cycle: many transcription factors, for example, cycle between the nucleus and the cytosol in response to extracellular signals. When the protein is not in the location where it is needed, very often it is maintained in an inactive conformation.

Protein activity can be regulated by binding of an effector and by covalent modification

Protein activity can also be controlled by the binding of effector molecules, which often work by inducing conformational changes that produce inactive or active forms of the protein. Effectors may be as small as a proton or as large as another macromolecule. Effectors may bind noncovalently or may modify the covalent structure of the protein, reversibly or irreversibly. Effectors that regulate activity by binding to the active site usually take the form of inhibitors that compete with the substrate for binding. Often, the product of an enzyme reaction can act as such a **competitive inhibitor**, allowing the enzyme to regulate itself when too much product might be made. Ligands, including reaction products, may also bind to sites remote from the active site and in so doing either activate or inhibit a protein. Proteins regulated in this way tend to be oligomeric and *allosteric*. Allosteric proteins have multiple ligand-binding sites and these show *cooperativity* of binding: in *positive cooperativity* the first ligand molecule to bind is bound weakly, but its binding alters the conformation of the protein in such a way that binding of the second and subsequent ligand molecules is promoted. Cooperativity may also be negative: the first ligand binding weakens and thereby effectively inhibits subsequent binding to the other sites. Metabolic pathways often employ allosteric effectors as part of a feedback control mechanism: the end product of the pathway acts as an allosteric inhibitor of one of the earlier

---

Definitions

competitive inhibitor: a species that competes with substrate for binding to the active site of an enzyme and thus inhibits catalytic activity.

effector: a species that binds to a protein and modifies its activity. Effectors may be as small as a proton or as large as a membrane and may act by covalent binding, noncovalent binding, or covalent modification.

References

Goodsell, D.S.: **Inside a living cell.** *Trends Biochem. Sci.* 1991, **16**:203–206.

Hodgson, D.A. and Thomas, C.M. (eds): **Signals, switches, regulons, and cascades: control of bacterial gene expression.** *61st Symposium of the Society for General Microbiology* (Cambridge, Cambridge University Press, 2002).

Jensen, R.B. and Shapiro, L.: **Proteins on the move: dynamic protein localization in prokaryotes.** *Trends*

Cell Biol. 2000, **10**:483–488.

Kornitzer, D. and Ciechanover, A.: **Modes of regulation of ubiquitin-mediated protein degradation.** *J. Cell Physiol.* 2000, **182**:1–11.

Perutz, M.F.: **Mechanisms of cooperativity and allosteric regulation in proteins.** *Q. Rev. Biophys.* 1989, **22**:139–237.

Sato, T.K. et al.: **Location, location, location: membrane targeting directed by PX domains.** *Science* 2001, **294**:1881–1885.

enzymes in the pathway, so when too much of this product is synthesized it feeds back and shuts off one of the enzymes that help make it, as we shall see later in this chapter.

Binding of effector molecules can be covalent or can lead to covalent changes in a protein. The most common form of post-translational covalent modification is reversible phosphorylation on the hydroxyl group of the side chains of serine, threonine or tyrosine residues, but many other modifications are known, including side-chain methylation, covalent attachment of carbohydrates and lipids, amino-terminal acetylation and limited proteolytic cleavage, in which proteases cut the polypeptide chain in one or more places. Modifications such as phosphorylation or proteolytic cleavage may either activate or inactivate the protein.

Signal amplification is an essential feature in the control of cell function and covalent modification of proteins is the way such amplification is usually achieved. Often, an extra-cellular stimulus is of short duration and involves only a very low concentration, or a small change in concentration, of a hormone or regulatory molecule. Yet the cellular response must not only be very rapid; in many cases it must be massive, including a change in activities of many enzymes and alteration in the transcription of many genes. Covalent modification of a protein provides a simple mechanism by which a regulatory signal can produce a very large output. A single molecule of an enzyme whose catalytic activity is turned on by a covalent modification can process many thousands of substrate molecules. And if that substrate is another enzyme, the amplification is further magnified. When covalent modification of one enzyme causes it to become active so that it can, in turn, covalently modify and activate another enzyme and so on, a regulatory cascade is set up that leads to enormous, rapid changes in the final output. Blood clotting is an example of such an amplification cascade based on proteolysis.

Protein activity may be regulated by protein quantity and lifetime

The activity of a protein can also be regulated by controlling its amount and lifetime in the cell. This control may be exercised at several places in the flow of information from gene to protein. At its simplest, the amount of protein can be set by the level of transcription, which in turn can be controlled by, for example, the strength of the promoter or the action of a transcription factor, which may be a repressor or activator. The level of mRNA may also be adjusted after transcription by varying the rate of RNA degradation. At the level of the protein, quantities are controlled by the lifetime of the molecule, which is determined by its rate of degradation. The rate of turnover varies considerably from protein to protein; there are several specific mechanisms for targeting protein molecules to degradative machinery in the cell, including covalent attachment of the small protein ubiquitin.

A single protein may be subject to many regulatory influences

These various strategies are not mutually exclusive and any one protein may be subject to several of them. Coordination and integration of regulatory signals is achieved largely through signal transduction networks that set the balance of activities and thereby the balance of metabolism and cell growth and division pathways. Integration of signaling pathways is achieved through proteins, such as protein kinases, whose activity is under the control of several different mechanisms that can be independently regulated by incoming signals, as in the example of the protein tyrosine kinases known as cyclin-dependent kinases (CDKs), which control a eukaryotic cell's progression through the cell cycle (Figure 3-1).

Wyrick, J.J. and Young, R.A.: **Deciphering gene expression regulatory networks.** *Curr. Opin. Genet. Dev.* 2002, **12**:130–136.

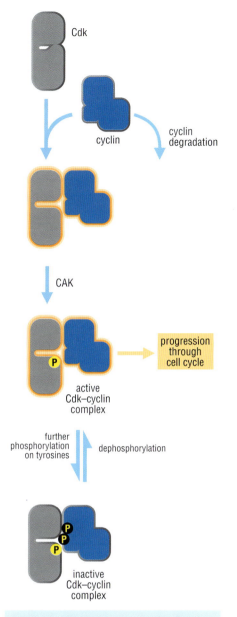

Figure 3-1 The cyclin-dependent protein kinases that control progression through the cell cycle are regulated by a number of different mechanisms Activation of a cyclin-dependent kinase (Cdk) at the appropriate point in the cell cycle requires both binding by its cyclin ligand, which induces a conformational change in the Cdk, phosphorylation of the Cdk in this Cdk–cyclin complex at a particular tyrosine residue, and dephosphorylation at one or two other tyrosines (depending on the particular Cdk). Cyclin binding is determined by the levels of cyclin present in the cell, which are strictly controlled in a temporal fashion by a signaling network that regulates gene transcription and protein degradation. Phosphorylation at the activating site is carried out by a so-called Cdk-activating kinase (CAK) which is present throughout the cell cycle but cannot act until the Cdk has bound cyclin; phosphorylation and dephosphorylation of the inhibitory tyrosines are carried out by another kinase and a phosphatase, respectively, whose activities are also subject to finely tuned regulation.

3-1 Protein Interaction Domains

The flow of information within the cell is regulated and integrated by the combinatorial use of small protein domains that recognize specific ligands

Many proteins that regulate cell behavior are constructed in a modular fashion from a number of different small domains with distinct binding specificities and functions (see Figure 1-46). Signal transduction proteins, regulatory proteins of the cell cycle, proteins that carry out targeted proteolysis, and proteins that regulate secretory pathways, cytoskeletal organization or gene expression all come into this category. Many gene regulatory proteins, for example, bind both to DNA and to another protein to carry out their function; they are composed of a domain that binds a specific DNA sequence and a protein-binding domain, which may target another molecule of the same protein (see Figure 1-41) or another gene regulatory protein. In some enzymes, a catalytic domain is attached to one or a number of protein-binding domains. These **interaction domains** (also called recognition modules) target the attached catalytic domain to a particular multiprotein complex or an appropriate subcellular location (such as the nucleus or the plasma membrane). Some interaction domains, such as the calcium-binding EF hand, often act as regulatory domains for enzymes. In signal transduction pathways, active multiprotein complexes are assembled by interaction domains that target the components to the complex. For example, the WD40 domain of the β-subunit of the heterotrimeric G protein targets this subunit to the α and γ subunits (see Figure 3-15). In some cases the interaction domain of a protein may interact with that protein's own catalytic domain in an autoinhibitory manner. Binding of the interaction domain to another protein then relieves the inhibition and activates the enzyme. Some protein kinases are autoregulated this way (see section 3-13).

Interaction domains are independently folded modules, 35–150 residues in length, which can still bind their target ligands if expressed independently of their "host" protein. Their amino and carboxyl termini are close together in space, with the ligand-binding site being located on the opposite face of the module. This allows an interaction domain to be inserted into a loop region of a catalytic domain, for example, as in the case of mammalian phospholipase Cγ, where two SH2 and one SH3 domains are inserted within the catalytic domain without disturbing its fold or having their own binding sites blocked. This type of domain organization also allows different domains to be strung together in combinatorial fashion and still retain their function.

Interaction domains can be divided into distinct families whose members are related by sequence, structure, and ligand-binding properties (Figure 3-2). Different members of a family recognize somewhat different sequences or structures, providing specificity for the proteins into which they are inserted. Examples of interaction domain families are SH3, WW and EVH1, which recognize proline-rich sequences; SH2 and PTB, which recognize phosphotyrosine-containing sequences; and 14-3-3, FHA, PBD and WD40, which bind to phosphoserine and phosphothreonine motifs. These domains are all common in proteins of signal transduction pathways. The PH and FYVE domains that recognize phospholipids are found in several different pathways, including signal transduction pathways and in proteins that control the traffic between internal membrane-bound compartments of the cell. A given protein may contain one or more copies of several different recognition modules, occurring in a different order in different proteins. Among the GTPase-activating proteins, for example, GAP1 has a PH domain after its GAP domain, while p120 GAP has the opposite order. Some interaction domains also form homo- and hetero-oligomers (typically dimers), which creates yet another level of regulatory versatility and specificity.

Figure 3-2 Interaction domains The name of the particular example shown for each family is given below each structure, along with the function and specificity of the domain.

Definitions

interaction domain: a protein domain that recognizes another protein, usually via a specific recognition motif.

References

Elia, A.E. et al.: **Proteomic screen finds pSer/pThr-binding domain localizing Plk1 to mitotic substrates.** Science 2003, **299**:1228–1231.

Fan, J.S. and Zhang, M.: **Signaling complex organization by PDZ domain proteins.** Neurosignals 2002, **11**: 315–321.

Kuriyan, J. and Cowburn, D.: **Modular peptide recognition in eukaryotic signaling.** Annu. Rev. Biophys. Biomol. Struct. 1997, **26**:259–288.

Macias, M.J. et al.: **WW and SH3 domains, two different scaffolds to recognize proline-rich ligands.** FEBS Lett. 2002, **513**:30–37.

Pawson, A.J. (ed.): Protein Modules in Signal Transduction (Springer, Berlin, New York, 1998).

Stenmark, H. et al.: **The phosphatidylinositol 3-phosphate-binding FYVE finger.** FEBS Lett. 2002, **513**:77–84.

Yaffe, M.B.: **Phosphotyrosine-binding domains in signal transduction.** Nat. Rev. Mol. Cell Biol. 2002, **3**:177–186.

Yaffe, M.B.: **How do 14-3-3 proteins work? Gatekeeper phosphorylation and the molecular anvil hypothesis.** FEBS Lett. 2002, **513**:53–57.

Yaffe, M.B. and Elia, A.E.: **Phosphoserine/threonine-binding domains.** Curr. Opin. Cell Biol. 2001, **13**:131–138.

For a catalog of known intracellular protein interaction domains, see http://www.cellsignal.com/reference

14-3-3

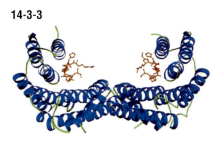

Example: 14-3-3
Function: protein–protein interactions
Specificity: phosphotyrosine

WD40

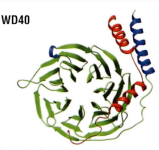

Example: G protein beta subunit
Function: protein–protein interactions;
a stable propeller-like platform to which
proteins bind either stably or reversibly
Specificity: various

EF-hand

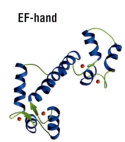

Example: Calmodulin
Function: calcium binding
Specificity: Ca^{2+}

LRR

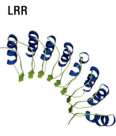

Example: Rpn1
Function: protein–protein
interactions
Specificity: various

Armadillo repeat (ARM)

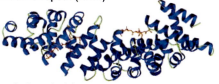

Example: Importin alpha
Function: protein–protein interactions
Specificity: various

SNARE

Example: SNAP-25B
Function: protein–protein interactions
in intracellular membrane fusion
Specificity: other SNARE domains

PTB

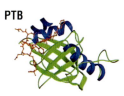

Example: Shc
Function: protein–protein
interactions
Specificity: phosphotyrosine

Death domain (DD)

Example: FADD
Function: protein–protein
interactions in pathway that
triggers apoptosis
Specificity: other DD domains
through heterodimers

ANK (ankyrin repeat)

Example: Swi6
Function: protein–protein
interactions
Specificity: various

C2

Example: PKC
Function: electrostatic switch
Specificity: phospholipids

FHA

Example: Rad53
Function: protein–protein
interactions
Specificity: phosphotyrosine

BH

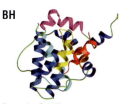

Example: Bcl-Xl
Function: protein–protein
interactions
Specificity: Other BH domains
through heterodimers

SH2

Example: Src
Function: protein–protein
interactions
Specificity: phosphotyrosine

SH3

Example: Sem5
Function: protein–protein
interactions
Specificity: proline-rich sequences

PH

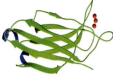

Example: PLC-δ
Function: recruitment of
proteins to the membrane
Specificity: phosphoinositides

SAM

Example: EphA4
Function: protein–protein
interactions via homo- and
heterodimers
Specificity: other SAM domains

Bromo

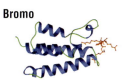

Example: P/CAF
Function: protein–protein
interactions in chromatin remodeling
Specificity: acetylated lysine

PDZ

Example: PSD-95
Function: protein–protein
interactions, often involving
transmembrane proteins or
ion channels
Specificity: -XXXV/I-COOH

GYF

Example: CD2
Function: protein–protein
interactions
Specificity: proline-rich
sequences

Chromo

Example: Mouse modifier protein 1
Function: protein–protein
interactions in chromatin remodeling
Specificity: methylated lysine

FYVE

Example: Vps27p
Function: Regulation of
signaling
Specificity: phosphatidyl-
inositol-3-phosphate

RING finger

Example: c-Cbl
Function: protein–protein
interactions in ubiquitin-
dependent degradation and
transcription regulation
Specificity: various

WW

Example: Pin1
Function: protein–protein
interactions
Specificity: proline-rich
sequences

LIM

Example: CRP2
Function: protein–protein
interactions, usually in
transcription regulation
Specificity: various

F-box

Example: Skp2
Function: protein–protein
interactions in ubiquitin-dependent
protein degradation
Specificity: various

C1

Example: PKC
Function: recruitment of
proteins to the membrane
Specificity: phospholipids

Fibronectin

Example: Fibronectin III
Function: protein–protein
interactions in cell
adhesion to surfaces
Specificity: RGD motif of integrins

Protein function in the cell is context-dependent

Many cellular processes involve the interaction of two or more macromolecules: signal transduction pathways are a good example. But when one estimates the number of gene products apparently involved in such pathways, it frequently appears that there are too few different proteins to account for all the different specific interactions that must be made. Put another way, the genomes of higher organisms seem to contain too few genes to fulfill all the cellular functions required. The logical conclusion is that many proteins participate in more than one cellular process. But if chaos is not to result from all these activities occurring simultaneously, both temporal and spatial control over a protein's activity must be exercised. Temporal control can be achieved partly by regulating gene expression and protein lifetime. However, it is increasingly clear that spatial context, the precise location within the cell at which a gene product exercises its biochemical function, is a major mechanism for regulating function.

It is probable that there is no such thing as a free-floating protein in a eukaryotic cell. Every protein is constrained, whether in a complex with other macromolecules, within a specific organelle, in a cargo vesicle, by attachment to a membrane, or as a passenger on the actin railroads in the cytoskeleton, among others. Moreover, the membrane-bounded vesicles of the cell provide a distinctive environment that can affect protein function, as we discuss in the next section. Prokaryotic cells, which lack organelles and cytoskeletal structures, may be less highly structured, but eukaryotic cells are organized into many compartments (Figure 3-3). In fact, it is increased organization, not increased gene number, that is the real hallmark of eukaryotic cells. For example, the eukaryote fission yeast *Schizosaccharomyces pombe* has fewer genes than the bacterium *Pseudomonas aeruginosa*.

Precise localization of proteins is a central feature of both spatial and temporal organization. In eukaryotic cells, the targeting of proteins to different locations in the cell, such as cytosol versus nucleus, can be regulated according to need, and changes in targeting can modify protein function at the cellular level even when the biochemical function of the protein does not change.

Nowhere is this more evident than in signal transduction pathways. Many of these depend on **protein kinases** which, as discussed later in this chapter, modify protein function by phosphorylating selected amino acids. The number of protein kinase genes in the human genome is large, but is far smaller than the number of potential protein kinase substrates. As we do not appear to possess one kinase for each substrate, kinases must have less than absolute specificity. How then are they prevented from phosphorylating the "wrong" protein at an inappropriate time and place? One answer is to target the kinase to the same location as its "correct" substrate, a location different from that for any other potential substrate; the action of that kinase is then quite specific; and specificity can be altered if required by relocation of kinase and/or substrate.

Another example of control by location is the small monomeric GTPase Tem1 from the budding yeast *Saccharomyces cerevisiae*. Tem1 plays an essential part in terminating the mitotic phase of the cell cycle. The interaction of individual yeast proteins with all the other proteins encoded in the yeast genome has been systematically investigated by two-hybrid analysis (see section 4-4). In such assays Tem1 has been found to interact physically with 24 different yeast gene products. Given its size, no more than about four other proteins could possibly bind to Tem1 at the same time, so what controls which proteins it interacts with at any given time? Differences in the timing of expression of the different target proteins could play some part in determining Tem1's specificity, but binding to specific partners is likely to be mainly controlled by targeting Tem1, and its potential partners, to different locations in the cell at different times.

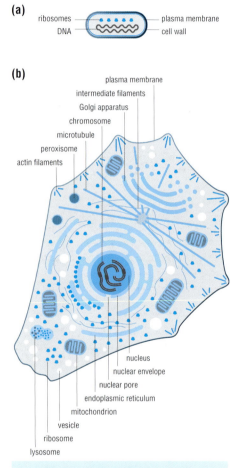

(a)
ribosomes
DNA
plasma membrane
cell wall

(b)
plasma membrane
intermediate filaments
Golgi apparatus
chromosome
microtubule
peroxisome
actin filaments
nucleus
nuclear envelope
nuclear pore
endoplasmic reticulum
mitochondrion
vesicle
ribosome
lysosome

Figure 3-3 The internal structure of cells Schematic diagrams of **(a)** a bacterial cell and **(b)** a typical eukaryotic cell, showing the arrangements of some of the organelles and other internal structures.

Definitions

lipid anchor: lipid attached to a protein that inserts into a membrane thereby anchoring the protein to the bilayer.

protein kinase: enzyme that transfers a phosphate group from ATP to the OH group of serines, threonines and tyrosines of target proteins.

References

Cyert, M.S.: **Regulation of nuclear localization during** signaling. *J. Biol. Chem.* 2001, **276**:20805–20808.

Dhillon, A.S. and Kolch, W.: **Untying the regulation of the Raf-1 kinase.** *Arch. Biochem. Biophys.* 2002, **404**:3–9.

Dorn, G.W. 2nd and Mochly-Rosen, D.: **Intracellular transport mechanisms of signal transducers.** *Annu. Rev. Physiol.* 2002, **64**:407–429.

Garrington, T.P. and Johnson, G.L.: **Organization and regulation of mitogen-activated protein kinase signaling pathways.** *Curr. Opin. Cell Biol.* 1999, **2**:211–218.

Kurosaki, T.: **Regulation of B-cell signal transduction by adaptor proteins.** *Nat. Rev. Immunol.* 2002, **2**:354–363.

Martin, T.F.: **PI(4,5)P(2) regulation of surface membrane traffic.** *Curr. Opin. Cell Biol.* 2001, **13**:493–499.

Penn, R.B. *et al.*: **Regulation of G protein-coupled receptor kinases.** *Trends Cardiovasc. Med.* 2000, **10**:81–89.

Sato, T.K. *et al.*: **Location, location, location: membrane targeting directed by PX domains.** *Science* 2001, **294**:1881–1885.

There are several ways of targeting proteins in cells

Targeting of a protein within the cell is achieved in three main ways: by sequences in the protein itself; by various types of post-translational modification; and by binding to a scaffold protein. Localization to the membrane-bounded compartments of the cell—the nucleus, the endoplasmic reticulum (ER), the Golgi complex, and so on—is usually achieved by specific "localization signals" encoded in the protein sequence. For example, the sequence KDEL targets proteins for retention in the endoplasmic reticulum (by binding to a specific receptor) (Figure 3-4a); lysine/arginine-rich clusters such as KRKR target proteins to the nucleus; and the hydrophobic signal sequence found at the beginning of some proteins directs their secretion from the cell. Depending on the organelle to be targeted, localization sequences may be located at either end of the polypeptide chain or even internally. Intrinsic plasma membrane proteins are also targeted to the membrane via the ER by this type of mechanism.

Localization by sequence motifs is an intrinsic property of the protein. In contrast, post-translational targeting mechanisms are regulatable. The commonest mechanism is covalent chemical modification of the protein. Perhaps the most widely used modification of this type is phosphorylation of tyrosine, serine or threonine residues by protein kinases; such modifications target proteins for binding by other proteins that have a recognition site for the modification (Figure 3-4b). This mechanism is particularly important in the formation of signaling complexes and in biochemical switches, and we discuss it in more detail later in this chapter.

Another important post-translational modification targets proteins to the plasma membrane by covalent attachment of their carboxyl or amino terminus to a lipid molecule, often called a **lipid anchor**, which is then inserted into the phospholipid bilayer (Figure 3-4c). For example, the signaling GTPase Ras, which is found mutated in many human tumors, becomes covalently attached to the isoprenoid farnesyl at its carboxyl terminus via a thioester linkage, and this modification, which targets Ras to the cell membrane and thus brings it close to the signaling complexes attached to receptor proteins, is essential for its function in signal transduction (Figure 3-4c).

Although lipid anchors could insert randomly into membranes solely by virtue of their hydrophobicity, evidence is mounting that the membranes in eukaryotic cells are not just random soups of lipids. Membranes appear to have many patches (sometimes called lipid rafts) where specific lipids congregate. These "islands" target particular lipid anchors or lipid-binding domains on proteins not just to the membrane but to very specific places on the membrane. Control of the location and size of these rafts by enzymatic modification and hydrolysis of phospholipids is yet another level of functional organization. We discuss lipid rafts in more detail later, in the section on lipid modification of proteins (see section 3-19), in which we also describe the various types of lipid anchor.

The third common means of protein targeting is by binding to scaffold proteins. These are proteins that can bind several other proteins simultaneously, thereby promoting their interaction (Figure 3-4d). To carry out this function, scaffold proteins generally contain a number of small domains that recognize specific targets, modifications or sequence motifs. One example is the SH3 domain, which binds to proline-rich regions; another is the SH2 domain, which recognizes sequences containing a phosphorylated tyrosine; yet another is the PH domain, which binds to phospholipids in membranes (see section 3-1).

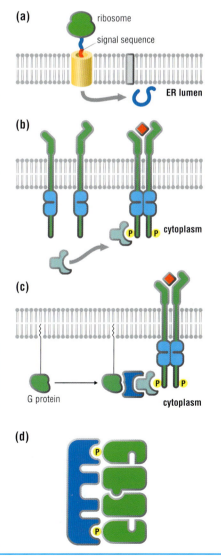

Figure 3-4 Mechanisms for targeting proteins (a) Proteins destined for export from the cell are first co-translationally targeted to the endoplasmic reticulum (ER) by signal sequences (red) at the amino terminus of the protein. The signal sequences are cleaved off when protein synthesis is complete, releasing the protein into the lumen of the ER, from which it is exported to the cell surface in transport vesicles. Transmembrane proteins are targeted in a similar way, but have additional internal hydrophobic signal sequences that retain the protein in the membrane. **(b)** Proteins may be targeted to specific signaling complexes by recognizing phosphorylated sites on their target protein. As shown here, a cell-surface receptor (blue and green) with intrinsic protein kinase activity dimerizes on binding an external ligand (red), and this triggers phosphorylation of the cytoplasmic domains. That generates a target site for a cytoplasmic signaling molecule (light blue) that recognizes the phosphorylated residue and binds to the receptor tail. **(c)** A lipid tail is attached to a small G protein (green), thereby anchoring it in the plasma membrane where it can interact with other membrane-bound signaling proteins. It is shown here interacting with an adaptor protein (dark blue) that is part of the signaling complex that builds up around phosphorylated receptor domains. **(d)** A scaffold protein (blue) binds to several different signaling molecules (green) and thereby targets their activities: the signaling molecules may, for example, be protein kinases that can now sequentially phosphorylate one another as part of a signaling pathway.

Protein function is modulated by the environment in which the protein operates

All proteins are adapted to fold and function optimally in the particular environment of the cellular compartment in which they operate. The cellular aqueous solution is highly viscous and contains many components besides proteins at high concentration, including ions, free polar organic molecules and, most important, the dissociated conjugate acid/conjugate base components of water: the proton and the hydroxide ion. If these two components are present at approximately equal concentration, as is the case in the cytosol of most cells, the solution is neutral with a pH of about 7. If protons are in excess, the solution is acidic, with a lower pH, and ionizable groups tend to be protonated. If hydroxide ions predominate, the solution is alkaline with a pH > 7 and ionizable groups will tend to be deprotonated. Distinct membrane-bounded compartments inside the cell often have a distinct internal microenvironment and the extracellular environment represents a different aqueous environment again from that of the interior. We describe here some examples of the adaptations of proteins for the environments in which they function.

Changes in redox environment can greatly affect protein structure and function

The interiors of cells are for the most part reducing environments: they furnish electrons, often in the form of hydrogen atoms. On the other hand, outside the cell, proteins and small molecules are typically exposed to an oxidizing environment in which electrons can be lost. The chief effect of this difference is that cysteine residues in proteins are usually fully reduced to –SH groups inside the cell, but are readily oxidized to disulfide S–S bridges when a protein is secreted. Cells can exploit this difference to trigger oligomerization by S–S bond formation in secreted proteins, or subunit dissociation and conformational changes when proteins are internalized. For instance, acetylcholinesterase is synthesized as a monomer, but when secreted from muscle and nerve cells self-associates to form dimers and tetramers. These oligomers, which are more stable than the monomer and thus better able to survive outside the cell, are composed of covalently linked subunits. Each monomer has a carboxy-terminal cysteine residue which forms an intersubunit S–S bond with the identical residue on a neighboring polypeptide chain.

Changes in pH can drastically alter protein structure and function

Most cytosolic fluid is maintained at near neutral pH, so that neither acids nor bases predominate. However, there are specialized compartments, such as endosomal vesicles, where the pH is quite acidic. As the surfaces of soluble proteins are chiefly composed of polar side chains, many of which are ionizable, both the net charge on a protein and the distribution of charge over the surface can vary considerably with pH.

If ligand binding depends on electrostatic interactions (see Figure 2-23), changes in the external pH (or ion concentration) can greatly influence binding strength by directly altering the ionization states of groups that interact with the ligand or of groups on the ligand itself. Modulation of the surface charge distribution of a protein by pH changes can also affect the biochemical function indirectly, by changing the extent of ionization of essential functional groups in an active site or binding site through long-range electrostatic interactions. For instance, endosomal proteases, which degrade internalized proteins, are only catalytically active

(a)

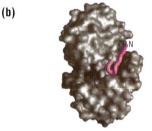

decreasing pH
substrates   increasing pH
inhibitors

(b)

Figure 3-5 Cathepsin D conformational switching Cathepsin D undergoes a pH-dependent conformational change. At the neutral pH of the cytosol **(a)**, the amino-terminal peptide interacts with the active site and blocks substrate access. At the low pH of the endosome **(b)**, in which cathepsin D is active, the amino-terminal segment reorients, opening the active site. At the same time, the low pH sets the protonation states of the catalytic residues for activity. Graphic kindly provided by John Erickson. From Lee, A.Y. et al.: Nat. Struct. Biol. 1998, **5**:866–871, with permission.

References

Authier, F. et al.: **Endosomal proteolysis of internalized insulin at the C-terminal region of the B chain by cathepsin D.** J. Biol. Chem. 2002, **277**:9437–9446.

Authier, F. et al.: **Endosomal proteolysis of internalized proteins.** FEBS Lett. 1996, **389**:55–60.

Lee, A.Y. et al.: **Conformational switching in an aspartic proteinase.** Nat. Struct. Biol. 1998, **5**:866–871.

Weiss, M.S. et al.: **Structure of the isolated catalytic domain of diphtheria toxin.** Biochemistry 1995, **34**:773–781.

at acidic pH, when the charged groups in their active sites are in the proper ionization states. Endosomal proteases are implicated in the degradation of internalized regulatory peptides involved in the control of metabolic pathways and in the processing of intracellular antigens for cytolytic immune responses. Processing occurs in endocytic vesicles whose acidic internal environment is regulated by the presence of an ATP-dependent proton pump. The acidic environment (pH ~ 5) modulates protease activity, protein unfolding and receptor–ligand interactions. The endosomal compartment of liver cells contains an acidic endopeptidase, cathepsin D, that hydrolyzes internalized insulin and generates the primary end-product of degradation. The protease is only active at low pH because its active site contains two essential aspartic acid residues, one of which must be protonated for catalysis to occur.

However, there is an additional aspect to the pH-dependent regulation of cathepsin D activity. The conformation of the protein, particularly the amino terminus, differs at low and high pH. High pH results in a concerted set of conformational changes including the relocation of the amino terminus into the active site, where it blocks substrate binding; also, at this pH both catalytic aspartates are likely to be deprotonated and therefore inactive; their charge state also stabilizes this conformation of the amino terminus. Activation at low pH is due to the protonation of one of the two catalytic aspartates, weakening their electrostatic interactions with the amino-terminal peptide. The released amino-terminal peptide restores the accessibility of the active site to substrate (Figure 3-5).

In some cases, the effect of pH has been exploited to change the structure of a protein completely in an acidic compartment. For example, diphtheria toxin, a protein that is among the deadliest substances known, is synthesized in the neutral pH of the cytosol of the bacterial cell as a single polypeptide chain with three domains: from amino to carboxyl terminus these are designated as domains B, T and A. The B domain binds to a receptor in the target-cell membrane, leading to uptake of the toxin by receptor-mediated endocytosis. The A domain, which is joined to the B domain by a disulfide bond, is an enzyme that kills cells by catalyzing the ADP-ribosylation of elongation factor 2 on the ribosome, leading to a block in protein synthesis. The T domain is the most interesting, because it is responsible for delivering the catalytic domain into the cytoplasm of the target cell.

Endocytosis through binding of the B domain to the cell-surface receptor results in the sequestration of the toxin in the acidic interior of an endocytic vesicle. Exposure to the reducing environment inside the endocytic vesicle breaks the disulfide bond between the A and B domains, releasing the toxic A domain. At the same time, the acidic environment causes a massive conformational change in the T domain, which turns almost inside out, exposing the interior hydrophobic residues. Hydrophobic side chains, as we saw in Chapter 1 (section 1-9) are normally buried in soluble proteins, extensive exposed hydrophobic regions occurring only in the lipid interior of membranes (see section 1-11). Thus the exposure of hydrophobic residues of the diphtheria toxin T domain results in its insertion into the endosomal membrane, where it is thought to create a channel through which the toxic A domain is translocated into the cytoplasm (Figure 3-6).

Figure 3-6 Schematic representation of the mechanism by which diphtheria toxin kills a cell The toxin has a receptor-binding domain (B), a membrane translocation domain (T) and a catalytic subunit (A) that kills cells by ADP-ribosylating a residue on the ribosomal protein EF-2. After receptor binding, the toxin enters the cell by endocytosis. In the low-pH environment of the endosome, a disulfide bond (not shown) in the toxin is reduced and a conformational change occurs that activates the membrane translocation domain, which is thought to release the active catalytic domain from the endosome. Weiss, M.S. *et al.*: *Biochemistry* 1995, **34**:773–781.

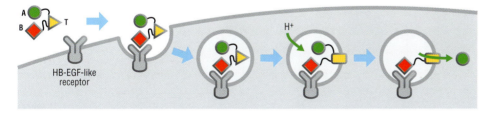

Protein function can be controlled by effector ligands that bind competitively to ligand-binding or active sites

One of the most important ways in which the activity of proteins is controlled is by binding regulatory molecules, termed **effector ligands** or effectors, that alter the activity of the protein with which they interact. Effectors can be as small as a proton (pH-induced conformational changes can be enormous, as in the case of the diphtheria toxin T domain discussed in the preceding section), or as large as another protein. In some instances this regulation can simply be inhibition through competitive binding with the normal ligand. This is a common mechanism for inhibiting enzymes, in which the effector ligand binds to the active site instead of substrate: many metabolic enzymes are feedback-inhibited in this way by their own product or by the product of an enzyme downstream from them in the same metabolic pathway (Figure 3-7). Feedback inhibition ensures that the activity of an enzyme is diminished when there is an over-abundance of its product in the cell. An example is the control by the tripeptide glutathione (GSH) of its own biosynthesis. GSH is ubiquitous in mammalian and other living cells. It has several important functions, including protection against oxidative stress. It is synthesized from its constituent amino acids by the consecutive actions of two enzymes. The activity of the first enzyme is modulated by feedback inhibition of the end product, GSH, ensuring that the level of glutathione does not exceed necessary values.

Cooperative binding by effector ligands amplifies their effects

In the previous example, binding of one molecule of a competitive feedback inhibitor inhibits one molecule of a target enzyme. But many physiological responses need to be rapid and total, which is hard to achieve with a linear system. Amplification would allow a single regulatory molecule to shut down many copies of a target protein or pathway. Amplification can be achieved in either of two ways. One is by covalent modification of the protein, which we discuss later in this chapter. The other is by **cooperativity**. This is a phenomenon of universal importance in biological systems and is as versatile as it is widespread. At the metabolic level, an enzyme in one pathway can cooperate with another pathway by providing a component that can serve as a substrate, enzyme or regulator of that pathway. We are concerned here, on the other hand, with **cooperative binding**, cooperativity between binding sites for the same

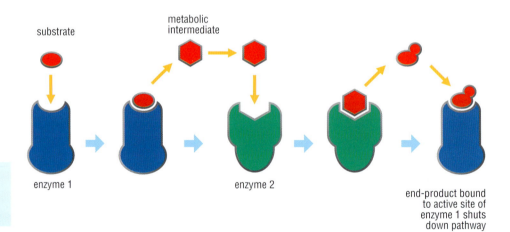

Figure 3-7 Competitive feedback inhibition
In this simple two-enzyme pathway, the end-product of the pathway acts as a competitive inhibitor of the first enzyme.

substrate

metabolic intermediate

enzyme 1

enzyme 2

end-product bound to active site of enzyme 1 shuts down pathway

Definitions

cooperative binding: interaction between two sites on a protein such that the binding of a ligand to the first one affects the properties—usually binding or catalytic—of the second one.

cooperativity: interaction between two sites on a protein such that something that happens to the first one affects the properties of the second one.

effector ligand: a ligand that induces a change in the properties of a protein.

negative cooperativity: binding of one molecule of a ligand to a protein makes it more difficult for a second molecule of that ligand to bind at another site.

positive cooperativity: binding of one molecule of a ligand to a protein makes it easier for a second molecule of that ligand to bind at another site.

ligand on a protein, in which binding of the ligand at one site affects the ease or otherwise of binding of ligand at the other site(s). This type of cooperativity depends on the ligand–protein interaction resulting in a measurable conformational change in other regions, close by or distant, of that protein. Cooperativity is only present in oligomeric proteins, where there are two or more subunits each with a binding site for the ligand.

Cooperativity can be positive or negative: **positive cooperativity** means that binding of one molecule of a ligand to a protein makes it easier for a second molecule of that ligand to bind; **negative cooperativity** means that binding of the second molecule is more difficult. Thus cooperativity can amplify either the activation of a protein or its inhibition. If a protein has, say, four identical subunits, each of which has a binding site for an effector ligand, positive cooperativity occurs if binding of the first effector molecule to one subunit makes binding easier (that is, increases the binding constant) for the second molecule of the effector to the second subunit. In turn, this binding facilitates binding of a third ligand molecule to the third subunit even more, and so on. This is illustrated for a dimer schematically in Figure 3-8. Such positive cooperativity means that activation (or inhibition) of all subunits can be achieved at a lower concentration of the activating or inhibiting ligand than would be the case if each subunit bound the effector independently and with equal affinity. In the extreme case of absolute positive cooperativity, only a few molecules of ligand might be sufficient to activate or inhibit the protein completely. Cooperative binding is also seen in proteins that are not enzymes. The oxygen-transport protein hemoglobin, for example, binds molecular oxygen at its four binding sites with positive cooperativity.

Negative cooperativity in enzymes usually involves the binding of substrates or cofactors, and in this case the binding site in question is the active site. In the extreme case of absolute negative cooperativity, a dimeric enzyme, for example, may never be found with cofactor or substrate bound to both subunits at the same time. For example, the bacterial enzyme ATP-citrate lyase shows strong negative cooperativity with respect to citrate binding. Non-enzymes can display this property as well. For example, the heart drug bepridil binds to the muscle protein troponin C with negative cooperativity. The physiological significance of absolute negative cooperativity is uncertain.

The phenomenon of cooperativity reflects an important consequence of the flexibility of proteins: binding of even a small molecule to a protein surface can induce structural changes at a distance from the binding site. In the case of oligomeric proteins, as we have seen, this can result in communication of a ligand-induced conformational change from one subunit to another. In the next section, we shall see how this ability of proteins to undergo long-range structural adjustments enables regulatory effector ligands to act by binding at sites remote from the normal ligand-binding or active site.

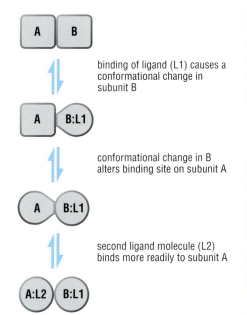

binding of ligand (L1) causes a conformational change in subunit B

conformational change in B alters binding site on subunit A

second ligand molecule (L2) binds more readily to subunit A

Figure 3-8 Cooperative ligand binding Shown here is a simple two-subunit protein displaying cooperative binding of a ligand L. The initial state of the protein (square) has a relatively low affinity for L. When the first molecule of L (L1) binds, it causes a conformational change in the subunit to which it is bound. But interactions between the subunits cause the unoccupied subunit to change its conformation as well, to one that is similar to the ligand-bound state (circle) and therefore has a higher affinity for L. Thus a lower concentration of L is required to bind to the second subunit, and the effect of L is amplified. If the protein is an enzyme, the effect of binding of the first ligand molecule may not only be to increase the affinity of the next subunit for ligand; its enzymatic activity may also be enhanced, increasing the amplifying effect.

References

Koshland, D.E. Jr., and Hamadani, K.: **Proteomics and models for enzyme cooperativity.** *J. Biol. Chem.* 2002, **277**:46841–46844.

Effector molecules can cause conformational changes at distant sites

Because of the close packing of atoms in globular protein structures, even small changes in side-chain and main-chain positions at one site can be propagated through the tertiary structure of the molecule and cause conformational changes at a distant location in the protein. Indeed, the most common type of regulatory effector ligand is one that is different from the normal substrate or functional ligand for the protein and which binds at a site distinct from the enzyme's catalytic site or from the site through which the protein's function is mediated (in the case of a non-enzyme). In hetero-oligomeric enzymes, the regulatory site is often located on a different subunit from the active site.

There are two extreme models for the action of such regulators and evidence exists for each in different cases. In one, the binding of successive effector molecules causes a sequential series of conformational changes from the initial state to the final state (Figure 3-9a). The other model postulates a preexisting conformational equilibrium between a more active and a less active form of the protein. The effector can only bind to one of these forms, and thus its binding shifts the equilibrium in a concerted manner in favor of the bound form (Figure 3-9b). Because structural changes are produced in both models, this form of regulation is called **allostery** from the Greek for "another structure", and a protein that is regulated in this way is called an allosteric protein. A molecule that stabilizes the more active form is an **allosteric activator**; one that stabilizes the less active form is an **allosteric inhibitor**. When the effector ligand is another protein, the protein that is regulated can be, and often is, monomeric, as in the case of the regulation of the monomeric cyclin-dependent kinase Cdk2 by the protein cyclin A, as described later in this chapter. When the effector ligand is a small molecule, however, although allosteric regulation does not formally require cooperative binding, in practice the allosteric protein is nearly always oligomeric, and binding of the effector ligand is usually cooperative (see section 3-4).

ATCase is an allosteric enzyme with regulatory and active sites on different subunits

The allosteric enzyme aspartate transcarbamoylase (ATCase) is a hetero-oligomer made up of six catalytic and six regulatory subunits (Figure 3-10). ATCase catalyzes the formation of *N*-carbamoyl aspartate, an essential metabolite in the synthesis of pyrimidines, from carbamoyl phosphate and L-aspartate. The enzyme is allosterically inhibited by cytidine triphosphate (CTP), the end-product of pyrimidine nucleotide biosynthesis, and allosterically activated by adenosine triphosphate, the end-product of purine nucleotide biosynthesis. Thus, CTP is a feedback inhibitor and shuts down ATCase when pyrimidine levels are high; ATP activates the enzyme when purine levels are high and pyrimidines are needed to pair with them to make nucleic acids. Structural studies have shown that the enzyme exists in at least two states, with very different intersubunit contacts (Figure 3-10). The more compact state of the dodecamer (the tense or T state) is the less active form; it is stabilized by the binding of CTP to sites on the regulatory subunits. The more open arrangement (the relaxed or R state) is the more active enzyme; this structure is stabilized by binding of ATP, also to the regulatory subunits. Both these effectors bind cooperatively, as do the substrates, so there is great potential for amplification of the signals that regulate this enzyme. It is not surprising that the ATCase-catalyzed reaction is the major control step in pyrimidine biosynthesis.

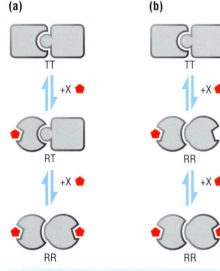

(a)
TT
+X
RT
+X
RR

(b)
TT
+X
RR
+X
RR

Figure 3-9 Two models of allosteric regulation (a) A sequential change in the conformation of subunits of a dimeric protein from a less active state (the "tense" state TT) to the more active state (the "relaxed" state RR) on binding of a regulatory ligand (X). **(b)** The case where the TT and RR states are in equilibrium and the regulatory ligand stabilizes one of them, shifting the equilibrium in favor of this form. To emphasize the conformational change, this diagram shows an asymmetrical protein; in reality, the subunits are usually identical.

Definitions

allosteric activator: a ligand that binds to a protein and induces a conformational change that increases the protein's activity.

allosteric inhibitor: a ligand that binds to a protein and induces a conformational change that decreases the protein's activity.

allostery: the property of being able to exist in two structural states of differing activity. The equilibrium between these states is modulated by ligand binding.

co-activator: a regulatory molecule that binds to a gene activator protein and assists its binding to DNA.

co-repressor: a regulatory molecule that binds to a gene repressor protein and assists its binding to DNA.

References

Kantrowitz, E.R. and Lipscomb, W.N.: *Escherichia coli* **aspartate transcarbamylase: the relation between structure and function.** *Science* 1988, **241**:669–674.

Koshland, D.E. Jr. and Hamadani, K.: **Proteomics and models for enzyme cooperativity.** *J. Biol. Chem.* 2002, **277**:46841–46844.

White, A. *et al.*: **Structure of the metal-ion-activated diphtheria toxin repressor/tox operator complex.** *Nature* 1998, **394**:502–506.

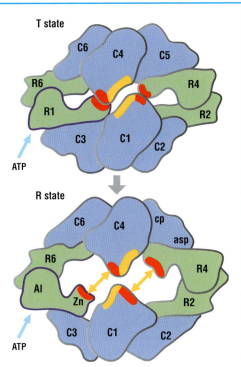

T state

R state

ATP

Figure 3-10 Ligand-induced conformational change activates aspartate transcarbamoylase
Binding of the allosteric activator ATP to its intersubunit binding sites on the regulatory subunits (that between R1, outlined in purple, and R6 is arrowed) of the T state of ATCase (top) causes a massive conformational change of the enzyme to the R state (bottom). In this state the structure of the enzyme is opened up, making the active sites on the catalytic subunits (C) accessible to substrate. Al and Zn in the lower diagram indicate the allosteric regions and the zinc-binding region, respectively; cp and asp indicate the binding sites for the substrates carbamoyl phosphate and aspartate, respectively. The red and yellow regions are the intersubunit interfaces that are disrupted by this allosteric transition.

Disruption of function does not necessarily mean that the active site or ligand-binding site has been disrupted

As we have seen, most proteins are like machines: they have moving parts. And like a machine, the introduction of a monkey wrench into any part of the protein can disrupt any activity that depends on the propagation of a conformational change. The monkey wrench can be a drug that binds directly to the active site or functional ligand-binding site, or a mutation at these sites. But drugs or mutations may also disrupt a protein's function by binding elsewhere and interfering with the conformational transitions necessary for function. For example, binding of anti-viral compounds to the rhinovirus coat protein blocks entry of the virus into host cells. But these drugs do not bind to the site on the virus that binds the cell-surface virus receptor; instead they bind to an unrelated site where they stabilize the structure of the coat protein, thereby preventing the structural rearrangements required for receptor-mediated virus entry into the cell.

The effects of both regulatory ligand binding and mutation are illustrated by ATCase. The T state can be stabilized by the binding of CTP to a site on the regulatory subunit. Exactly the same stabilizing effect can be achieved by mutation of tyrosine 77 to phenylalanine in the regulatory subunit. This mutation stabilizes the T state and thus shuts down catalytic activity, even though it is very far from the active site. Therefore, in the absence of structural information, one should not automatically assume that any mutation that disrupts a protein's function must be in the active site or that any molecule that inhibits function does so because it binds directly in the active site.

Binding of gene regulatory proteins to DNA is often controlled by ligand-induced conformational changes

DNA-binding proteins are not usually intended to interact with the genetic material all the time. Activators and repressors of gene expression are usually under the control of specific regulatory ligands, which may be small molecules, metal ions or proteins, whose binding determines whether or not the activator or repressor can bind to DNA. Such molecules are called **co-activators** and **co-repressors**, respectively. Expression of the gene encoding diphtheria toxin is under the control of a specific repressor, DtxR. Binding of DtxR to its operator sequence is controlled in turn by the concentration of Fe^{2+} in the bacterial cell. Iron acts as a co-repressor by binding to DtxR and inducing a conformational change that allows the helix-turn-helix DNA-recognition motif in the repressor to fit into the major groove of DNA. In the absence of bound iron, the repressor adopts a conformation in which DNA binding is sterically blocked (Figure 3-11).

Figure 3-11 Iron binding regulates the repressor of the diphtheria toxin gene
Comparison of the structures of the aporepressor DtxR (red, left, PDB 1dpr) and the ternary complex (right) of repressor (green), metal ion (Fe^{2+}, orange) and DNA (grey) (PDB 1fst). Iron binding induces a conformational change that moves the recognition helices (X) in the DtxR dimer closer together, providing an optimal fit between these helices and the major groove of DNA. In addition, metal-ion binding changes the conformation of the amino terminus of the first turn of the amino-terminal helix (N) of each monomer. Without this conformational change, leucine 4 in this helix would clash with a phosphate group of the DNA backbone. Thus, DtxR only binds to DNA when metal ion is bound to the repressor.

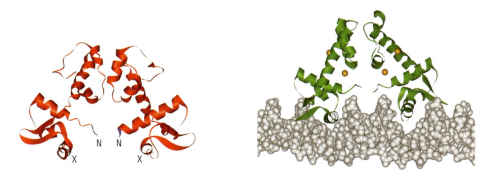

Conformational changes driven by nucleotide binding and hydrolysis are the basis for switching and motor properties of proteins

Not every process in a living cell runs continuously. Many of them must cycle between "on" and "off" states in order to control cell growth and division and responses to extracellular signals. Signal transduction pathways such as those that operate in vision and hormone-based signaling, vesicular transport (which is often called protein or membrane trafficking), polypeptide chain elongation during protein synthesis, and actin- and tubulin-based motor functions are also examples of processes that must be switched on and off under precisely determined circumstances. This cycling is controlled by a special set of proteins that function as molecular switches. Although these proteins vary in structure and in the processes they control, they have a number of common features. The most important common element is the switching mechanism itself: most of these proteins undergo conformational changes induced by the difference between the triphosphate and diphosphate forms of a bound nucleotide. The conformational changes are such that completely different target proteins recognize the two bound states of the switch protein, providing a simple means of altering the output of a signal.

Most protein switches are enzymes that catalyze the hydrolysis of a nucleoside triphosphate to the diphosphate. Most often the nucleotide is guanosine triphosphate, GTP, and the switch protein is a GTPase that hydrolyzes it to GDP. GTPase switches (also commonly called **G proteins** or occasionally **guanine-nucleotide-binding proteins**) are one major class of switch proteins; they control the on/off states of most cellular processes, including sensory perception, intracellular transport, protein synthesis and cell growth and differentiation. The second major class of switch proteins is composed of those ATPases that are usually associated with motor protein complexes or transporters that move material into and out of cells and some organelles. Members of the third major class, the two-component response regulators, a group of switches thus far found only in microbes and plants, are composed of a histidine protein kinase and a second "response regulator" protein. They do not bind GTP or ATP in the same way as the nucleotide switch proteins discussed here, but use a covalently bound phosphate derived from the hydrolysis of ATP by the kinase to trigger a conformational change in the response regulator.

Hydrolyzable nucleotides are used to control many types of molecular switches because the energy derived from hydrolysis of the terminal phosphate of a GTP or ATP is large enough to make the conformational change in the switch effectively irreversible until another protein binds to the switch, displaces the diphosphate, and allows the triphosphate to bind again. Another reason is that using ATP or GTP can couple the switching process to the energy state of the cell and to the synthesis of DNA and RNA, both of which change the levels of nucleoside tri- and diphosphates.

All nucleotide switch proteins have some common structural and functional features

Nucleotide-dependent switch proteins are found in every kingdom of life. The ATPases and the GTPases constitute two protein families that are both characterized by a core domain that carries out the basic function of nucleotide binding and hydrolysis—called the G domain in the GTPases. The core domain structure is conserved within each family, but its secondary structure arrangement is very different in GTPases and ATPases. Remarkably, in spite of the lack of similarity between their protein folds, both the nucleotide-binding site and the switch mechanism are extremely similar for the two families. The standard fold of the G domain in

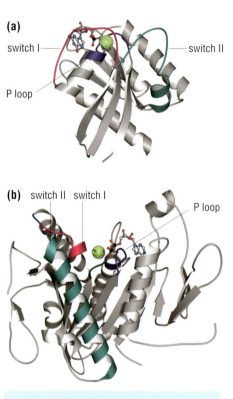

(a)

switch I — — switch II

P loop

(b) switch II switch I

P loop

Figure 3-12 Structure of the core domains of a typical GTPase and an ATPase **(a)** Ribbon diagram of the minimal GTPase G domain, with the conserved sequence elements and the switch regions in different colors. The nucleoside diphosphate is shown in ball-and-stick representation and the bound Mg²⁺ ion as a green sphere. The switch I region is the pink loop that surrounds the ligand in this view. The switch II region is in turquoise and the P-loop in purple. (PDB 4q21) **(b)** Structure of the ATPase domain of the mitotic spindle kinesin Eg5. As in the GTPase, one molecule of Mg-ADP is located in the nucleotide-binding site. The equivalent switch I, switch II and P-loop regions are colored the same as for the GTPase. (PDB 1ii6)

Definitions

G protein: a member of a large class of proteins with GTPase activity that act as molecular switches in many different cellular pathways, controlling processes such as sensory perception, intracellular transport, protein synthesis and cell growth and differentiation. They undergo a large conformational change when a bound GTP is hydrolyzed to GDP.

guanine-nucleotide-binding protein: see **G protein**.

P-loop: a conserved loop in GTPase- and ATPase-based

nucleotide switch proteins that binds to phosphate groups in the bound nucleotide.

phosphate-binding loop: see **P-loop**.

switch I region: a conserved sequence motif in GTPase- and ATPase-based nucleotide switch proteins that, with the **switch II region**, binds the terminal gamma-phosphate in the triphosphate form of the bound nucleotide and undergoes a marked conformational change when the nucleotide is hydrolyzed.

switch II region: a conserved sequence motif in

GTPase- and ATPase-based nucleotide switch proteins that, with the **switch I region**, binds the terminal gamma-phosphate in the triphosphate form of the bound nucleotide and undergoes a marked conformational change when the nucleotide is hydrolyzed.

the GTPases consists of a mixed six-stranded beta sheet with five helices located on both sides (Figure 3-12a). There are three conserved features in nucleotide switches: the **P-loop**, and the **switch I** and **switch II** sequence motifs. The P-loop or **phosphate-binding loop** binds the alpha- and beta-phosphates from the phosphate tail of the nucleotide. Residues from the two switch motifs coordinate the terminal gamma-phosphate in the triphosphate form of the bound nucleotide. A Mg^{2+} ion, which is complexed with the bound nucleotide, is coordinated by the nucleotide phosphate groups and, in the triphosphate form of the switch, by residues from the switch I and II regions.

These regions usually contain four to five conserved sequence elements, which are lined up along the nucleotide-binding site. Additional important contribution to binding is made by the interactions of the nucleotide base with a sequence that has the motif N/TKXD (where X is any amino acid) and confers specificity for guanine. Specificity for guanine is due to the aspartate side chain in this motif, which forms a bifurcated hydrogen bond with the base, and to a main-chain interaction of an invariant alanine (from a short SAK motif), with the guanine oxygen.

Similar conserved sequence motifs are located in equivalent spatial positions in the switch ATPases, even though the core protein fold is completely different (Figure 3-12b). The adenine base interacts with a conserved RXRP or NP motif (equivalent to the N/TKXD in the GTPases) and alpha- and beta-phosphates are bound to an equivalent P-loop with the same consensus sequence GXXXXGKS/T. Furthermore, both switch I and switch II sequence motifs of GTPases (DX_nT and DXXG respectively) have their equivalents in ATPases: NXXSSR and DXXG, respectively.

The GTPases have been most extensively studied and most of the details of the switch mechanism have been determined for them, but the basic switch mechanism is the same for the ATPases. The trigger for the conformational change is most likely universal. The triphosphate-bound state, which is usually the "on" state of the switch, can be considered as "spring-loaded" because of the terminal phosphate group of the bound nucleotide, which makes a number of interactions with the two switch regions (Figure 3-13). It is the loss of this gamma-phosphate group on hydrolysis of GTP to GDP (or ATP to ADP) that provides the trigger for the conformational change.

Changes in conformation between the triphosphate- and diphosphate-bound states are confined primarily to the two switch regions. These regions usually show an increased flexibility relative to the rest of the protein when the structure of a nucleotide-dependent switch is determined. In the triphosphate-bound form, there are two hydrogen bonds from gamma-phosphate oxygens to the main-chain NH groups of the invariant threonine (or serine in the case of the ATPases) and glycine residues in the switch I and II regions, respectively (see Figure 3-13). The glycine is part of the conserved DXXG motif; the threonine (or serine) is from the conserved sequence involved in binding the Mg^{2+} ion that coordinates to the two terminal phosphate groups of the bound nucleotide. Release of the gamma-phosphate after hydrolysis allows the two switch regions to relax into their diphosphate-specific conformations because the bonds that the threonine/serine and glycine residues make with this phosphate are disrupted when it leaves. The extent of the conformational change that results from the rearrangement of the two switch regions is different for different proteins and involves additional structural elements in some of them.

Although the fold of the core domain is conserved, there are many insertions of other domains in individual GTPases. These domains have various functions, from interacting with other proteins to magnifying the conformational change on nucleotide hydrolysis.

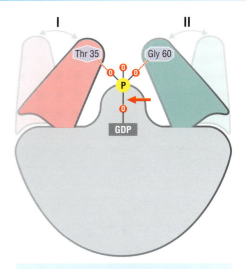

Figure 3-13 Schematic diagram of the universal switch mechanism of GTPases This picture illustrates the GTP-bound state where the switch I (pink) and switch II (turquoise) domains are bound to the gamma-phosphate via the main chain NH groups of the invariant threonine and glycine residues, in a spring-loaded mechanism. Release of the gamma-phosphate after GTP hydrolysis allows the switch regions to relax into a different conformation, as seen in Figure 3-12, which shows the diphosphate-bound forms. Adapted from Vetter, I.R. and Wittinghofer, A.: *Science* 2001, **294**:1299–1304.

References

Sablin, E.P. and Fletterick, R.J.: **Nucleotide switches in molecular motors: structural analysis of kinesins and myosins.** *Curr. Opin. Struct. Biol.* 2001, **11**:716–724.

Spudich, J.A.: **The myosin swinging cross-bridge model.** *Nat. Rev. Mol. Cell Biol.* 2001, **2**:387–392.

Vetter, I.R. and Wittinghofer, A.: **The guanine nucleotide-binding switch in three dimensions.** *Science* 2001, **294**:1299–1304.

An animation showing the conformational changes in the switch regions of Ras can be found at: http://www.mpi-dortmund.mpg.de/departments/dep1/gtpase/ras_gtp_gdp.gif

The switching cycle of nucleotide hydrolysis and exchange in G proteins is modulated by the binding of other proteins

All GTPase switches operate through conformational changes induced by a change in the form of the bound nucleotide, as we saw in the previous section; but this is only the core part of the switching mechanism. Which cellular process a particular G protein controls is determined by the interactions it makes with other proteins in its two conformational states. We will now consider the rest of the switching mechanism in detail in the case of the GTPases, exemplified by the small monomeric GTPase Ras, whose switch function helps to control cell growth and division.

The overall switching mechanism can be viewed as an on–off cycle in which the GTP-bound state is the "on" state and the GDP-bound state is "off". If we start with the switch in the off position, the gamma-phosphate of GTP is not present and the two switch regions are in the relaxed conformations characteristic of the diphosphate-bound state of the protein (see Figure 3-12a). GDP must now dissociate from the protein to allow GTP to bind. The normal rate of dissociation is often very slow, so most G proteins have various **guanine-nucleotide exchange factors** (**GEFs**) that bind to them and facilitate the release of GDP by inducing conformational changes that open up the binding site. A single G protein may be recognized by multiple GEFs, enabling one GTPase to serve as the focal point for integrating signals from different upstream pathways.

After GDP has been released and GTP binds, the protein conformation (that is, the arrangement of the switch I and switch II regions) changes to bind the gamma-phosphate (see Figure 3-13) and the switch is now in the on position. The altered conformation of the switch regions allows the G protein to interact with downstream effectors, activating various enzymes such as phosphatidylinositol-3-kinase and turning on various signaling and other pathways. Several different effectors may recognize the on state of the same G protein, allowing a single switch to control multiple cellular processes. The switch remains on as long as GTP remains bound to the GTPase.

Although GTPases are enzymes and catalyze the hydrolysis of GTP to GDP, they are not very efficient; their intrinsic rate of GTP hydrolysis is very slow. What determines the length of time the switch remains in the on state is the activity of various **GTPase-activating proteins** (**GAPs**) that bind to the GTP-bound conformation and stimulate the catalytic activity: in the case of Ras the binding of a GAP increases the GTPase activity by 100,000-fold (10^5). When GTP has been hydrolyzed, the switch I and switch II regions change to their relaxed conformations and the switch is back in the off state; the cycle is complete. Again, a single G protein can interact with multiple GAPs from different upstream signaling pathways. The on–off cycle is illustrated in Figure 3-14.

How the various GAP proteins facilitate GTP hydrolysis is not known in all cases, but at least some appear to function by stabilizing the transition state (see section 2-9). They insert an arginine side chain into the nucleotide-binding site of the GTPase to which they bind, and the positive charge on this side chain helps to stabilize the negative charge that builds up in the transition state for hydrolysis of the gamma-phosphate group of GTP.

The importance of proper regulation of G-protein signaling is exemplified by the Ras family of small GTPases. There are three human *RAS* genes, *H-*, *N-* and *K-RAS*, all of which code for very similar proteins of around 21 kDa molecular mass. They are post-translationally modified by the covalent attachment of lipophilic groups to the carboxy-terminal end. This modification is necessary for the Ras proteins' biological function as switches because it targets them to the

Definitions

GTPase-activating protein (GAP): a protein that accelerates the intrinsic GTPase activity of switch GTPases.

guanine-nucleotide exchange factor (GEF): a protein that facilitates exchange of GDP for GTP in switch GTPases.

plasma membrane, where their interacting protein partners are found (see Figure 3-4). Interestingly, the pathways controlled by Ras seem to differ in mouse, humans and yeast. Consequently, mutations in the corresponding genes lead to different phenotypes in different organisms. This is also true for some other signal transduction pathways, and it complicates the generalization of data obtained from model organisms such as the mouse.

The lifetime of the signal transduced by Ras is determined by the lifetime of the GTP-bound state. If it is artificially prolonged, the biological response may be unregulated and lead to drastic consequences in the cell. The gene coding for Ras was originally discovered as an oncogene of rodent tumor viruses, the gene responsible for their ability to cause cancer in animals. Mutant forms of the human *RAS* genes that produce a protein with a prolonged on state are found in up to 30% of human tumors. This activated Ras protein is the result of point mutations at amino acids 12, 13 or 61, the biochemical consequence of which is to reduce the rate of hydrolysis of GTP. This leaves Ras in the on state too long, leading to uncontrolled cell growth and proliferation. As many other genes involved in the Ras signal transduction pathway are also found as oncogenes in human or animal tumors, Ras itself and the Ras pathway is considered to be a prime target for anti-tumor therapy.

Figure 3-14 The switching cycle of the GTPases involves interactions with proteins that facilitate binding of GTP and stimulation of GTPase activity The cycle is illustrated here for the small G protein Ras, but the principles are the same for other GTPase switches such as protein synthesis elongation factors and the heterotrimeric G proteins, which are discussed later in this chapter. GTP-binding proteins function as molecular switches by cycling between GDP-bound "off" and GTP-bound "on" states. Exchange of the bound GDP for GTP is facilitated by guanine-nucleotide exchange factors (GEFs) whose binding to the GTPase increases the dissociation rate of the nucleotide by several orders of magnitude. Ras can be switched on by several GEFs, the most important of which are Sos (illustrated here), Ras-GRF and Ras-GRP. Sos is part of the pathway that conveys signals from an activated cell-surface receptor to Ras. In the activated GTP-bound state Ras interacts with and activates several target proteins involved in intracellular signaling pathways. Ras is switched off by hydrolysis of the bound GTP. This reaction is facilitated by the action of specific GTPase-activating proteins (GAPs), the best studied of which are GAP1 (illustrated here), p120GAP and neurofibromin, the product of a tumor suppressor gene.

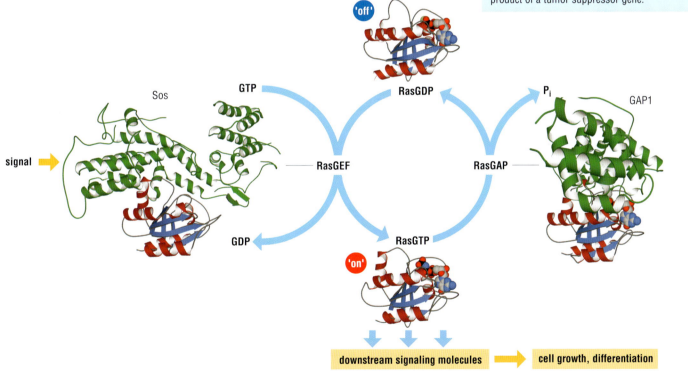

downstream signaling molecules → cell growth, differentiation

References

Geyer, M. and Wittinghofer A.: **GEFs, GAPs, GDIs and effectors: taking a closer (3D) look at the regulation of Ras-related GTP-binding proteins.** *Curr. Opin. Struct. Biol.* 1997, **7**:786–792.

Hamad, N.M. *et al.*: **Distinct requirements for Ras oncogenesis in human versus mouse cells.** *Genes Dev.* 2002, **16**:2045–2057.

Scheffzek, K. *et al.*: **GTPase-activating proteins: helping hands to complement an active site.** *Trends*

Biochem. Sci. 1998, **23**:257–262.

Spoerner, M. *et al.*: **Dynamic properties of the Ras switch I region and its importance for binding to effectors.** *Proc. Natl Acad. Sci. USA* 2001, **98**:4944–4949.

Sprang, S.R.: **G protein mechanisms: insights from structural analysis.** *Annu. Rev. Biochem.* 1997, **66**:639–678.

Takai, Y. *et al.*: **Small GTP-binding proteins.** *Physiol. Rev.* 2001, **81**:153–208.

Vetter, I.R. and Wittinghofer, A.: **Nucleoside triphosphate-binding proteins: different scaffolds to achieve phosphoryl transfer.** *Q. Rev. Biophys.* 1999, **32**:1–56.

Wittinghofer, A.: **Signal transduction via Ras.** *Biol. Chem.* 1998, **379**:933–937.

Heterotrimeric G proteins relay and amplify extracellular signals from a receptor to an intracellular signaling pathway

Most GTPases are either small monomeric proteins like Ras or much larger, heterotrimeric molecules, **heterotrimeric G proteins**, which are composed of an α subunit, whose core polypeptide chain fold is the same as the canonical G-domain fold, and β and γ subunits whose folds bear no relationship to that of the GTPases (Figure 3-15). The α subunit has an extra helical domain compared with the canonical G domain, and this partially projects into the catalytic site. The β and γ subunits are tightly associated with one another, partly by means of a coiled-coil interaction. Heterotrimeric G proteins are associated with the cytoplasmic surface of the cell membrane in complexes with so-called G-protein-coupled receptors (GPCRs), integral membrane proteins with seven transmembrane alpha helices. Affinity of heterotrimeric G proteins for the membrane is aided by prenylation of the β and γ subunits. In general, GPCRs are activated as a result of the binding of specific extracellular ligands to their extracellular or transmembrane domains. Conformational changes in the receptor induced by ligand binding are relayed through the transmembrane domain to the cytoplasmic portion, allowing productive coupling of the receptor with a heterotrimeric G protein, which transduces the signal and relays it onwards by means of its switching function.

The switching cycle of the heterotrimeric G proteins resembles that of the small monomeric GTPases with additional features imparted by the presence of the β and γ subunits. Heterotrimeric G proteins bind to the cytoplasmic domains of GPCRs in their GDP-bound "off" states. When activated by binding of their extracellular ligand, these receptors act as guanine-nucleotide exchange factors (GEFs) for their partner heterotrimeric G proteins, thus triggering GDP dissociation and the GTPase switching cycle. The βγ complex acts as part of the nucleotide-exchange mechanism. In the absence of βγ, α does not bind to GPCR. In addition, βγ does not bind to α when α is in its GTP form. Consequently, βγ acts to guarantee that GPCR only catalyzes replacement of GDP by GTP and not the other way around.

When GDP is released and GTP binds, the heterotrimeric G protein dissociates from the GPCR and the β and γ subunits dissociate as a heterodimer from the α subunit. Both the free GTP-bound α subunit and the βγ heterodimer can now bind to and stimulate their own respective downstream effectors, which are generally ion channels and enzymes. Hydrolysis of GTP in the active site of the free α subunit then results in a conformational change in the switch I and switch II regions, which restores association of α to the βγ heterodimer and causes the reassembled heterotrimeric G protein to rebind to the GPCR, completing the cycle.

Just as GAP proteins switch the state of small monomeric GTPases from GTP-bound to GDP-bound by increasing the GTPase catalytic rate, so-called **regulator of G-protein signaling proteins (RGS proteins)** are responsible for the rapid turnoff of GPCR signaling pathways by functioning as activators of the heterotrimeric G protein GTPase. Structural and mutational analyses have characterized the interaction of the RGS domain of these proteins with Gα in detail. Unlike RasGAPs, the RGS proteins do not directly contribute an arginine residue or any other catalytic residue to the active site of the α subunit to assist GTP hydrolysis. In fact, the α subunit of most heterotrimeric G proteins has a "built-in" arginine residue in the extra helical domain that projects into the catalytic site. RGS proteins probably exert their GAP activity mainly through binding to the switch regions, reducing their flexibility and stabilizing the transition state for hydrolysis in that way. More than 20 different RGS proteins have been isolated, and there are indications that particular RGS proteins regulate particular GPCR signaling pathways. This specificity is probably created by a combination of cell-type-specific

Definitions

heterotrimeric G protein: a GTPase switch protein composed of three different subunits, an α subunit with GTPase activity, and associated β and γ subunits, found associated with the cytoplasmic tails of G-protein-coupled receptors, where it acts to relay signals from the receptor to downstream targets. Exchange of bound GDP for GTP on the α subunit causes dissociation of the heterotrimer into a free α subunit and a βγ heterodimer; hydrolysis of the bound GTP causes reassociation of the subunits.

RGS protein: regulator of G-protein signaling protein; protein that binds to the free GTP-bound α subunit of a **heterotrimeric G protein** and stimulates its GTPase activity.

References

Hamm, H.E. and Gilchrist, A.: **Heterotrimeric G proteins.** *Curr. Opin. Cell Biol.* 1996, **8**:189–196.

Okada, T. *et al.*: **Activation of rhodopsin: new insights**

from structural and biochemical studies. *Trends Biochem. Sci.* 2001, **26**:318–324.

Rodbell, M.: **The complex regulation of receptor-coupled G-proteins.** *Adv. Enzyme Regul.* 1997, **37**:427–435.

Ross, E.M. and Wilkie, T.M.: **GTPase-activating proteins for heterotrimeric G proteins: regulators of G protein signaling (RGS) and RGS-like proteins.** *Annu. Rev. Biochem.* 2000, **69**:795–827.

expression, tissue distribution, intracellular localization, post-translational modifications, and domains other than the RGS domain that link these RGS proteins to other signaling pathways. Many heterotrimeric G proteins operate as complexes that include the RGS protein and the GPCR. This association causes rates of GTP hydrolysis and nucleotide exchange (the deactivation and activation rates, respectively, for the G protein) to become rapid and tightly coupled as long as the ligand for the receptor is present.

GPCRs, which are characterized by seven membrane-spanning alpha helices (Figure 3-15), are the most numerous receptors in all eukaryotic genomes (1–5% of the total number of genes). They transduce extracellular signals as varied as light, odorants, nucleotides, nucleosides, peptide hormones, lipids, and proteins. Rhodopsin, the primary photoreceptor protein in vision, is a prototypical GPCR that contains 11-*cis*-retinal as an intrinsic chromophore. The light-induced *cis–trans* isomerization of this ligand is the primary signaling event in the visual process. It is followed by slower and incompletely defined structural rearrangements leading to an "active state" intermediate of rhodopsin that signals via the heterotrimeric G protein transducin.

As well as relaying the signal onwards, heterotrimeric G proteins can also be one means by which it is amplified. It has been estimated that one rhodopsin molecule activated by a single photon of light activates around 500 molecules of transducin. Each of these activates a molecule of its target phosphodiesterase.

There are at least eight families of GPCRs that show no sequence similarities to each other and that have quite different ligand-binding domains and cytoplasmic domains, but which nevertheless activate a similar set of heterotrimeric G proteins. Homo- and heterodimerization of GPCRs seems to be the rule, and in some cases an absolute requirement, for activation. It has been estimated that there are about 100 "orphan" GPCRs in the human genome for which ligands have not yet been found. Mutations of GPCRs are responsible for a wide range of genetic diseases, including some hereditary forms of blindness. The importance of GPCRs in physiological processes is further illustrated by the fact that they are the targets of the majority of therapeutic drugs and drugs of abuse.

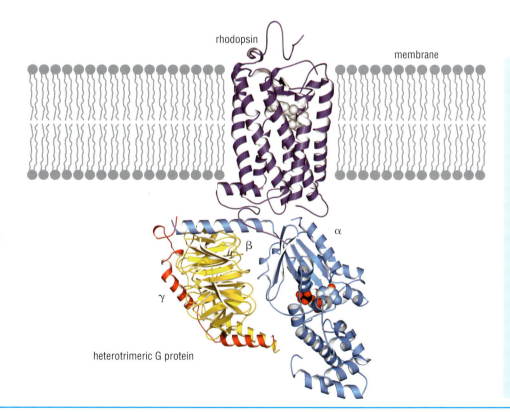

Figure 3-15 Hypothetical model of a heterotrimeric G protein in a complex with its G-protein-coupled receptor The GTPase subunit is shown in blue, the β subunit, which consists of a series of tandem WD40 domains, in yellow, and the γ subunit, which binds to the β subunit by a coiled-coil interaction, in red. The GPCR, shown here as the photoreceptor rhodopsin, is in purple embedded in a schematic of a membrane. A prenylation site on the α subunit that targets the complex to the membrane is not shown. GDP bound to the α subunit is shown in space-filling representation. Rhodopsin is activated by the photoisomerization of 11-*cis*-retinal (grey) bound to the protein. Rhodopsin has quite a short cytoplasmic tail but many other GPCRs have large cytoplasmic domains and additional extracellular domains to which extracellular ligands bind. Because the actual structure of the complex of any heterotrimeric G protein with its receptor is not known, the precise interactions between the components should not be inferred from this model. Adapted from Hamm, H.E. and Gilchrist, A.: *Curr. Opin. Cell Biol.* 1996, **8**:189–196.

EF-Tu is activated by binding to the ribosome, which thereby signals it to release its bound tRNA

One of the central protein components in the machinery of protein synthesis is a guanine-nucleotide-dependent switch. This molecule, called elongation factor Tu or EF-Tu in prokaryotes (the analogous factor in eukaryotes is called EF-1), consists of a core domain that has the canonical switch GTPase fold, and has two other domains that help bind a molecule of transfer RNA. Although there are many different tRNA molecules required for protein synthesis, they have the same overall structure and all can bind to the same EF-Tu. tRNA only binds to the GTP-bound state of EF-Tu; GTP hydrolysis causes dissociation of the complex.

EF-Tu performs a key function in protein synthesis: in its GTP-bound form it escorts aminoacyl-tRNAs coming into the ribosome, where its function is to facilitate codon–anticodon interactions and check their fit. If the fit is correct, this causes conformational changes in the ribosome that stabilize tRNA binding and trigger GTP hydrolysis by EF-Tu. On this, the elongation factor dissociates from the tRNA and leaves the ribosome, leaving the tRNA behind to deliver its amino acid to the growing polypeptide chain (Figure 3-16). If the pairing is incorrect, however, the codon–anticodon interaction is weaker, and this permits the aminoacyl-tRNA–EF-Tu complex to dissociate before hydrolysis of GTP and release of the tRNA can occur. Thus EF-Tu increases the ratio of correct to incorrect amino acids incorporated by providing a short delay between codon–anticodon base pairing and amino-acid incorporation.

tRNA release is driven by the same nucleotide-dependent switch mechanism that operates in the small monomeric and heterotrimeric G proteins. GTP hydrolysis causes a change in the conformations of the two switch regions, which in turn is transmitted to an alpha helix linked to one of them. The helix is part of the interface between the GTPase domain and the other two domains of EF-Tu, and when it moves the entire interface rearranges, which in turn causes a change in the conformations and relative positions of all three domains. The domains, which were closely associated with the tRNA and with each other in the GTP-bound state, now swing apart, one of them moving by almost 40 Å. This opening up of the protein structure dissociates the elongation factor from the bound tRNA.

Thus, in EF-Tu switching it is the correct pairing of the bound tRNA anticodon with the codon in the mRNA on the ribosome that acts as a GAP to promote the GTPase activity. By analogy, one expects to find the equivalent of a GEF to promote exchange of GTP for the bound GDP that remains on EF-Tu after hydrolysis. This function is provided by another protein, elongation factor Ts. When EF-Ts binds to the GDP-bound state of EF-Tu, the nucleoside diphosphate is released and GTP is able to bind. Binding of tRNA to the GTP-bound form completes the cycle (Figure 3-16). EF-Tu resembles the heterotrimeric G proteins in that GTP hydrolysis triggers a large-scale conformational change that leads to a loss of protein–protein interactions, but here the interactions are among domains in the same polypeptide chain rather than between non-identical subunits of an oligomeric protein.

The importance of EF-Tu switching in protein synthesis is underscored by the fact that dozens of antibiotics act by binding to EF-Tu and interfering with its function (although the human homolog EF-1 is similar in structure and function, it differs enough to be unaffected by these molecules). For instance, aurodox is a member of the family of kirromycin antibiotics, which inhibit protein biosynthesis by binding to EF-Tu. When aurodox is bound, the GTP-bound conformation of EF-Tu is observed, even when GDP is bound to the nucleotide-binding site.

References

Abel, K. *et al.*: **An alpha to beta conformational switch in EF-Tu.** *Structure* 1996, **4**:1153–1159.

Hogg, T. *et al.*: **Inhibitory mechanisms of antibiotics targeting elongation factor Tu.** *Curr. Protein Pept. Sci.* 2002, **3**:121–131.

Nyborg, J. and Liljas, A.: **Protein biosynthesis: structural studies of the elongation cycle.** *FEBS Lett.* 1998, **430**:95–99.

Ramakrishnan, V.: **Ribosome structure and the mechanism of translation.** *Cell* 2002, **108**:557–572.

Vogeley, L. *et al.*: **Conformational change of elongation factor Tu (EF-Tu) induced by antibiotic binding. Crystal structure of the complex between EF-Tu.GDP and aurodox.** *J. Biol. Chem.* 2001, **276**:17149–17155.

This suggests that aurodox fixes EF-Tu on the ribosome by locking it in its GTP form. Certain mutations in bacterial EF-Tu genes confer resistance to kirromycin by producing a protein that cannot interact with the antibiotic.

EF-Tu is not the only GTPase switch involved in translation. Another elongation factor, EF-G, is involved at a later step, in the translocation of the mRNA through the ribosome so that it is ready to accept the next aminoacyl-tRNA.

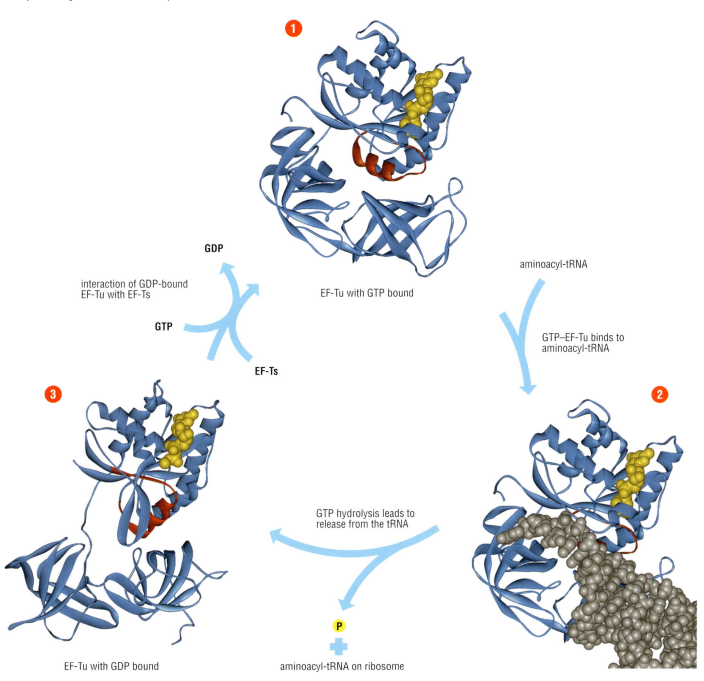

1

GDP

interaction of GDP-bound
EF-Tu with EF-Ts

GTP

EF-Ts

3

EF-Tu with GTP bound

aminoacyl-tRNA

GTP–EF-Tu binds to
aminoacyl-tRNA

2

GTP hydrolysis leads to
release from the tRNA

P
+

EF-Tu with GDP bound

aminoacyl-tRNA on ribosome

Figure 3-16 The switching cycle of the elongation factor EF-Tu delivers aminoacyl-tRNAs to the ribosome EF-Tu in its GTP-bound form (structure 1 at top; GTP shown in yellow bound to the GTPase domain) binds to an aminoacyl-tRNA (structure 2; the tRNA is in grey). This complex binds to the mRNA codon displayed in the A site on the small subunit of the ribosome through pairing of the codon with the anticodon on the tRNA (not shown). Correct pairing triggers GTP hydrolysis by EF-Tu, causing a conformational change that dissociates the factor from the tRNA molecule, which can now deliver its amino acid to the growing polypeptide chain. Generally, only those tRNAs with the correct anticodon can remain paired to the mRNA long enough to be added to the chain. The GDP form of EF-Tu (structure 3; GDP shown in yellow) released from the ribosome then interacts with EF-Ts, its guanine-nucleotide exchange factor (not shown), to facilitate exchange of GDP for GTP. The GTP-bound form of EF-Tu (1) can then restart the cycle. Shown in red is a helical segment in the GTPase domain that undergoes a significant conformational change between the GTP-bound (1) and GDP-bound (3) forms of the enzyme. The two forms also differ markedly in the relative arrangements of their domains.

Figure 3-17 Models for the motor actions of muscle myosin and kinesin Left panel, the sequential conformational changes in muscle myosin result in a rowing "stroke" that moves an actin filament. Muscle myosin is a dimer of two identical motor heads which are anchored in the thick myosin filament (partially visible at the upper edge of the frame) by a coiled-coil tail. At the start of the cycle (top frame of the sequence) ADP and inorganic phosphate (Pi) are bound to the heads as a result of hydrolysis of a bound ATP by the intrinsic ATPase activity of the catalytic cores (blue). The lever-arms that eventually cause the movement are shown in yellow in their "prestroke" conformation. In this conformation the catalytic core binds weakly to actin (bottom of frame). The second frame shows one myosin head docking onto a specific binding site (green) on the actin thin filament (pale grey). (The two myosin heads act independently, and only one attaches to actin at a time.) As shown in the third frame, on actin docking Pi is released from the active site, and there is a conformational change in the head that causes the lever arm to swing to its "poststroke" ADP-bound position (red) while the head remains bound to the actin. This moves the actin filament by approximately 100 Å in the direction shown by the arrow. After completing the stroke, ADP dissociates and ATP binds to the active site and undergoes hydrolysis, reverting the catalytic core domain to its weak-binding actin state, and bringing the lever arm back to its prestroke state. Right panel, kinesin "walks" along a microtubule, its two head domains moving in front of each other in turn. The coiled-coil tail of the kinesin (grey) leads to the attached cargo (not shown). The catalytic heads are shown in blue and purple and the microtubule in green and white. Movement is generated by conformational changes in the linker regions that join the heads to the tail and are colored here according to their conformational state: red, not bound to microtubule; orange, partially bound to microtubule; yellow, tightly bound to microtubule. The direction in which the kinesin will move is shown by the arrows. At the beginning of the movement cycle, the "trailing" head has ADP bound and the "leading" head is empty and neither linker is docked tightly to the microtubule. As shown in the first two frames, when ATP binds to the leading head, its linker (red in the top frame and yellow in the second frame) adopts a conformation that as well as docking it firmly to the microtubule reverses its position and thus throws the trailing head forward by about 160 Å (arrow) towards the next binding site on the microtubule (second frame). This head docks onto the binding site (third frame), which moves the attached cargo forward 80 Å. Binding also accelerates the release of ADP from this head, and during this time the ATP on the other head is hydrolyzed to ADP-Pi (third frame). After ADP dissociates from the new leading head ATP binds in its turn, causing the linker to zipper onto the core (partially docked linker in orange). The new trailing head, which has released its phosphate and detached its neck linker (red) from the core, is being thrown forward. Original illustration kindly provided by Graham Johnson. From Vale, R.D. and Milligan, R.A.: *Science* 2000, **288**:88–95, with permission.

Myosin and kinesin are ATP-dependent nucleotide switches that move along actin filaments and microtubules respectively

Many processes in living organisms and cells require movement. Directed transport of macromolecules, vesicles, organelles and chromosomes within the cytoplasm depends on motors that drive such transport. Bacterial swimming and the movement generated by muscles in higher organisms are examples of processes that also require motor power. Most biological motors move unidirectionally along protein polymers such as actin or microtubules; these polymers can be considered the rails along which protein motor "engines" move cargo; the same sort of machinery is used to slide the actin and myosin filaments past one another in muscle contraction. Defective molecular transport can result in developmental defects as well as cardiovascular and neuronal diseases.

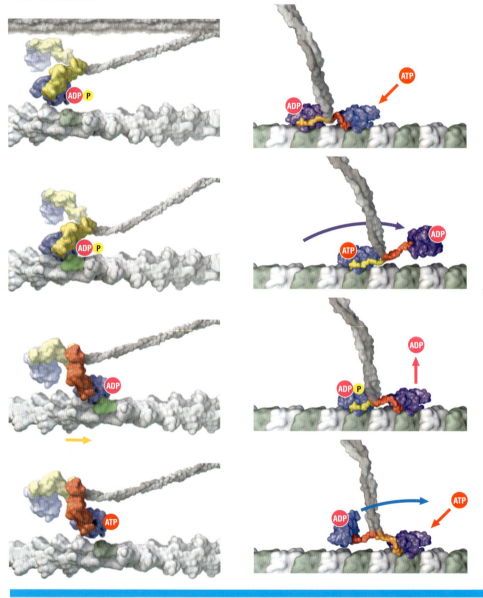

References

Capaldi, R.A. and Aggeler, R.: **Mechanism of the F(1)F(0)-type ATP synthase, a biological rotary motor.** *Trends Biochem. Sci.* 2002, **27**:154–160.

Goldstein, L.S.: **Molecular motors: from one motor many tails to one motor many tales.** *Trends Cell Biol.* 2001, **11**:477–482.

Holmes, K.C. and Geeves, M.A.: **The structural basis of muscle contraction.** *Philos. Trans. R. Soc. Lond. B. Biol. Sci.* 2000, **355**:419–431.

Sablin, E.P. and Fletterick, R.J.: **Nucleotide switches in molecular motors: structural analysis of kinesins and myosins.** *Curr. Opin. Struct. Biol.* 2001, **11**:716–724.

Sindelar, C.V. et al.: **Two conformations in the human kinesin power stroke defined by X-ray crystallography and EPR spectroscopy.** *Nat. Struct. Biol.* 2002, **9**:844–848.

Spudich, J.A.: **The myosin swinging cross-bridge model.** *Nat. Rev. Mol. Cell Biol.* 2001, **2**:387–392.

Vale, R.D. and Milligan, R.A.: **The way things move:**

Nearly all molecular motors share a common feature: a core ATPase domain that binds and hydrolyzes ATP, and in doing so switches between different conformations in a process that is similar to the GTPase molecular switches but which is carried out with a very different protein scaffold. Attached to the core domain are smaller transmission or converter domains that read out the nucleotide-dependent conformation of the core and respond by conformational changes. In turn, these domains relay the conformational change into larger motions by altering the positions of an amplifier region, very often a lever-arm or coiled coil. Large changes in the position of the amplifier in relation to an attached protein are what finally cause very large movements of proteins (Figure 3-17).

Although the overall folds of the ATPase domains of motors and the GTPase domains of G proteins are different, there are striking similarities in the local architecture around the bound nucleotide and in the mechanism of switching. Motor ATPases have switch I and switch II regions that correspond almost exactly to those in the G proteins, and these regions are involved in binding the gamma-phosphate of ATP, just as in the GTPases (Figure 3-18). When ATP is hydrolyzed by the intrinsic ATPase activity of the motor, the switch I and II regions relax into their ADP-bound conformational states, which starts the relay of conformational changes that are ultimately propagated into large movements of attached proteins and domains.

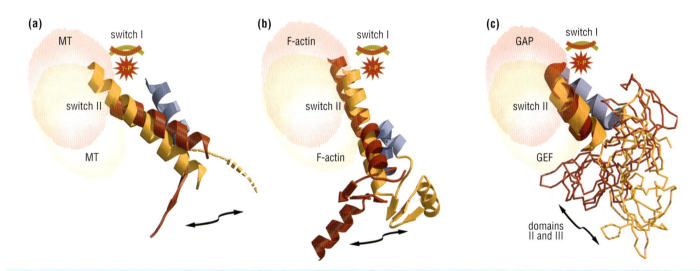

Figure 3-18 Structural and functional similarity between different families of molecular switches In all these graphics, the position of the switch II helix is shown in yellow for the nucleoside diphosphate-bound protein, and in red for the triphosphate-bound conformation. The switch I region is shown schematically with the gamma-phosphate of the nucleotide depicted as a red star. Specific macromolecular partners interact with the switch regions in these two states and are shown as surfaces of the same color. The carboxy-terminal helix of the ATPase or GTPase domain, to which the mechanical elements of molecular motors or additional domains of G proteins are attached, is shown in blue. **(a)** The switch II region of the motor protein kinesin. The nucleotide-driven conformational switching in the kinesin switch II region results in large force-generating rearrangements of the neck domain (see Figure 3-17; not shown here) and movement of kinesin along a microtubule (MT). **(b)** The same mechanism of switching in myosin converts small conformational changes in the nucleotide-binding site into larger movements of the lever-arm helix (see Figure 3-17), part of which is shown here at the bottom of the graphic (ADP state, PDB 2mys; ATP state, PDB 1br1). **(c)** An identical molecular mechanism is used by G proteins for their domain rearrangements and for controlling affinity for their specific macromolecular partners—GEFs, GAPs and effectors. As well as the switch II helix, domains II and III of EF-Tu are shown in GDP state (PDB 1tui) and the GTP state (PDB 1eft). Graphic kindly provided by E.P. Sablin. From Sablin, E.P. and Fletterick, R.J.: *Curr. Opin. Struct. Biol.* 2001, **11**:716–724, with permission.

looking under the hood of molecular motor proteins.
Science 2000, **288**:88–95.

A movie of kinesin moving may be found at:
http://valelab.ucsf.edu/images/mov-procmotconvkin
rev5.mov

Protein function can be controlled by protein lifetime

Proteins not only carry signals that determine their location, they also carry signals that determine their lifetime. Different proteins can have widely differing half-lives within the cell—anywhere from a few minutes to many days—and this time depends not only on the stability of the protein's structure but also on specific cellular machinery for degradation, which in some cases recognizes specific sequence and structural features. Stability in the extracellular environment is also variable, and depends on the intrinsic stability of the protein fold and the presence or absence of both relatively nonspecific and specific proteases. The shortest-lived proteins are usually those that are important in controlling cellular processes, such as enzymes that catalyze rate-determining steps in metabolic pathways or proteins such as cyclins that regulate cell growth and division. Rapid degradation of such proteins makes it possible for their concentrations to be changed quickly in response to environmental stimuli.

Protein degradation in cells is accomplished by machinery that tags both misfolded and folded proteins for specific proteolysis and destroys them. This mechanism depends in part on intrinsic protein stability at physiological temperature. One measure of the intrinsic stability of a protein is its resistance to thermal unfolding. Although there are microorganisms that survive at very high temperatures (so-called thermophilic organisms)—and in which, presumably, all proteins are fairly thermostable—very stable proteins are also found in the cytosol of mesophilic organisms (that is, organisms that live at normal temperatures). There appears to be no correlation between the type of protein fold and the intrinsic stability of a protein; rather, it is the specific sequence that determines stability, through the presence or absence of stabilizing interactions between side chains. Comparison of the structures of thermostable proteins with their mesophilic counterparts shows that many different stabilizing factors can be employed, mostly involving electrostatic interactions of various sorts. As the net free energy of stabilization of most folded proteins is relatively small, about 21–42 kJ/mole, only a few additional interactions can make a large difference in thermal stability.

Temperature-sensitive mutants, which are widely used in genetic research, are mutant organisms that produce a mutated protein that has reduced stability at physiological temperatures. These mutant proteins are more susceptible than the normal forms to degradation by non-specific proteases, and when such mutations occur in humans they can lead to disorders caused by insufficient amounts of the active protein.

Proteins are targeted to proteasomes for degradation

The proteolytic machinery responsible for targeted protein degradation in cells is a giant multiprotein assembly called the **proteasome** (Figure 3-19). In eukaryotes, proteins carrying an appropriate signal for destruction are recognized by an enzyme that then further tags the protein for degradation in the proteasome. The tag consists of many molecules of a small

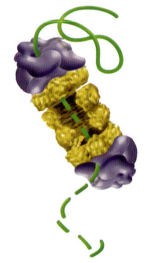

Figure 3-19 The eukaryotic proteasome Proteins targeted for destruction (green) are fed into the multiprotein complex called the proteasome. In prokaryotes, these machines of destruction consist simply of a tunnel-like enzymatic core; in eukaryotes they have an additional cap (here shown in purple) at either or both ends. The core is formed by four stacked rings surrounding a central channel that acts as a degradation chamber. The caps recognize and bind to proteins targeted by the cell for destruction. On entry into the proteasome, proteins are unfolded in a process that uses the energy released by ATP hydrolysis and injected into the central core, where they are enzymatically degraded into small fragments. Graphic kindly provided by US Department of Energy Genomes to Life Program, http://doegenomestolife.org.

Definitions

proteasome: a multiprotein complex that degrades ubiquitinated proteins into short peptides.

stress-response proteins: proteins whose synthesis is induced when cells are subjected to environmental stress, such as heat.

temperature-sensitive mutants: organisms containing a genetic mutation that makes the resulting protein sensitive to slightly elevated temperatures. The temperature at which the mutant protein unfolds is called the restrictive temperature. The term is also used for the protein itself.

ubiquitin: a small protein that when attached to other proteins (**ubiquitination**), targets them for degradation to the **proteasome**. Sometimes ubiquitin tagging targets a protein to other fates such as endocytosis.

ubiquitination: the attachment of ubiquitin to a protein.

protein, **ubiquitin**, which are attached to the protein to be degraded and which are recognized by the cap. **Ubiquitination** is carried out by a multi-enzyme pathway that starts by recognition of an exposed lysine near the amino terminus of the target protein. Once one ubiquitin molecule is covalently attached to this lysine, by enzymes called ubiquitin ligases, additional ubiquitins are attached to the first one, producing a polyubiquitin chain. Polyubiquitinated proteins are the primary substrates for the proteasome, which in an ATP-dependent process degrades these proteins to short peptides (Figure 3-20). In prokaryotes, a homolog of the proteasome acts without such a targeting signal.

Some specific ubiquitin-conjugating systems exist for certain proteins whose regulated destruction is crucial for cell growth, such as cyclins and other cell-cycle control proteins. Defects in ubiquitin-dependent proteolysis have been shown to result in a variety of human diseases, including cancer, neurodegenerative diseases, and metabolic disorders. Proteasome inhibitors are showing promise as treatments for certain cancers.

Proteins are targeted to the ubiquitin/proteasome system by a number of signals. The first to be discovered is remarkably simple, consisting only of the first amino acid in the polypeptide chain. Methionine, serine, threonine, alanine, valine, cysteine, glycine and proline are protective in bacteria; the remaining 12 signal proteolytic attack. Proteins are, however, more commonly targeted to the degradative pathway by a number of other more complex internal signals (for example, phosphorylation of specific residues, denaturation or oxidative damage). For example, the transcription factor NFκB is inhibited and retained in the cytoplasm by another protein IκB, which has a short peptide motif containing two serines. Phosphorylation of these serines, which is the culmination of a complex signal transduction pathway, leads to recognition by a ubiquitin ligase, the ubiquitination of IκB and its degradation, releasing NFκB. Another example of a regulated signal for ubiquitination is seen in the transcription factor hypoxia-inducible factor, HIF-1α, which is active when oxygen levels fall below about 5% O_2. In normal oxygen conditions HIF-1α is rapidly degraded because an oxygen-sensing prolyl hydroxylase modifies it and this modification is recognized by a specific ubiquitin ligase.

Unfolded and damaged proteins are preferential substrates for ubiquitination and degradation, even where the amino-terminal amino acid is protective. Some mechanisms for regulating protein degradation include modification of the protein's amino terminus. All proteins are initially synthesized with methionine as their amino-terminal amino acid. Methionine aminopeptidases and associated enzymes remove the methionine and trim the end of the polypeptide chain until certain residues are reached. However, methionine aminopeptidases will not remove this initiator methionine if it would expose an amino acid that signals destruction.

A class of proteins exists in all cells in all organisms that bind to selected proteins to help solubilize and refold aggregates of misfolded and unfolded forms before they can be degraded. These are known as **stress-response proteins** and their synthesis is upregulated in response to heat shock and other stresses that increase the amount of protein unfolding, misfolding and aggregation in the cell.

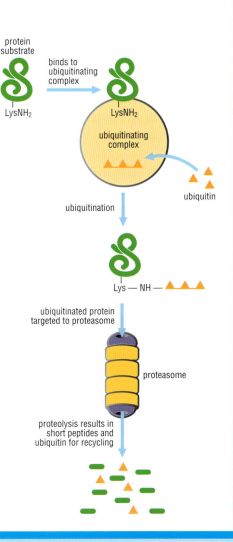

Figure 3-20 Pathway for degradation of ubiquitinated proteins A substrate protein with an exposed lysine side chain near the amino terminus is targeted by binding of a multienzyme ubiquitinating complex which, in this example, recognizes the amino-terminal amino acid of the substrate. The complex attaches polyubiquitin chains to the substrate in an ATP-dependent reaction. The polyubiquitinated substrate is then targeted to the proteasome, whose cap recognizes the ubiquitin tag. After the substrate is chopped up into peptide fragments (which may then be degraded further by other proteases), the ubiquitin is recycled.

References

Benaroudj, N. *et al.*: **The unfolding of substrates and ubiquitin-independent protein degradation by proteasomes.** *Biochimie* 2001, **83**:311–318.

Goldberg, A.L. *et al.*: **The cellular chamber of doom.** *Sci. Am.* 2001, **284**:68–73.

Kisselev, A.F. and Goldberg, A.L.: **Proteasome inhibitors: from research tools to drug candidates.** *Chem. Biol.* 2001, **8**:739–758.

Laney, J.D. and Hochstrasser, M.: **Substrate targeting in the ubiquitin system.** *Cell* 1999, **97**:427–430.

Petsko GA.: **Structural basis of thermostability in hyperthermophilic proteins, or "there's more than one way to skin a cat".** *Methods Enzymol.* 2001, **334**:469–478.

Sherman, M.Y. and Goldberg, A.L.: **Cellular defenses against unfolded proteins: a cell biologist thinks about neurodegenerative diseases.** *Neuron* 2001, **29**:15–32.

Stock D, *et al.*: **Proteasome: from structure to function.** *Curr. Opin. Biotechnol.* 1996, **7**:376–385.

Web site:
http://www.chembio.uoguelph.ca/educmat/chm736/degradat.htm

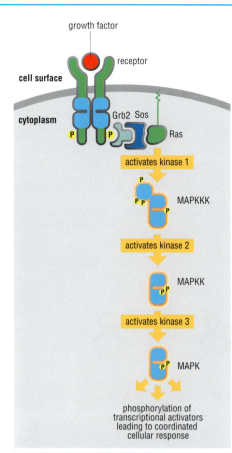

growth factor

receptor

cell surface

cytoplasm

Grb2 Sos

Ras

activates kinase 1

MAPKKK

activates kinase 2

MAPKK

activates kinase 3

MAPK

phosphorylation of
transcriptional activators
leading to coordinated
cellular response

Figure 3-21 A kinase activation cascade in an intracellular signaling pathway that regulates cell growth A growth factor, or mitogen, initiates a kinase cascade by inducing dimerization of its receptor, whose cytoplasmic regions contain tyrosine kinase domains. When these domains are brought together they phosphorylate and activate one another, leading to phosphorylation of tyrosines on other parts of the receptor's cytoplasmic domains, and creating binding sites for an adaptor molecule (such as Grb2), which in turn binds a guanine-nucleotide exchange factor (Sos), bringing this into proximity with the membrane-associated G protein Ras, which is thereby activated (see Figure 3-14). Ras then activates a kinase cascade of three so-called mitogen-activated protein kinases (MAPKs): a mitogen-activated protein kinase kinase kinase (MAPKKK), which phosphorylates and activates a mitogen-activated protein kinase kinase (MAPKK), which in turn phosphorylates and activates a mitogen-activated protein kinase (MAPK). This last kinase phosphorylates a number of different gene regulatory proteins.

Protein function can be controlled by covalent modification

It is estimated that 50–90% of the proteins in the human body are post-translationally modified. Post-translational covalent modification allows the cell to expand its protein structural and functional repertoire beyond the constraints imposed by the 20 naturally encoded amino acids. More than 40 different post-translational covalent modifications have been identified in eukaryotic cells. While some are widespread, others have been observed on only a few proteins. Phosphorylation, glycosylation, lipidation, and limited proteolysis are the most common. Of the remainder, some of the most important are methylation, *N*-acetylation, attachment of the protein SUMO and nitrosylation. In the remaining sections of this chapter, we shall describe the part played by these modifications in controlling protein function.

Most covalent modifications can change the location of the protein, or its activity, or its interactions with other proteins and macromolecules. Limited proteolysis can also be deployed to amplify low-concentration or transient signals, through proteolytic cascades in which the initial stimulus activates a proteolytic enzyme which in turn activates many molecules of the next enzyme in the cascade, and so on. The complement cascade, which is activated in response to microbial cell surfaces and promotes immune defenses, is one example; the blood clotting cascade, discussed later, is another.

The commonest form of covalent modification on proteins is reversible phosphorylation of serine, threonine or tyrosine side chains and we shall discuss this first. Phosphorylation, like limited proteolysis, can produce a very large response to a relatively small signal input. A single molecule of an enzyme that is catalytically activated by phosphorylation can process many thousands of substrate molecules. If the substrate is itself another enzyme, the amplification effects can be magnified further. When phosphorylation of one enzyme causes it to become active so that it can, in turn, covalently modify another enzyme and so on, a regulatory cascade can be set up which produces a large final response very rapidly.

Unlike limited proteolysis, however, phosphorylation is reversible, and this makes it well suited as a regulatory mechanism for intracellular signaling pathways, where it is important that responses can be turned off rapidly as well as turned on (Figure 3-21). In signaling pathways, an important feature of the regulatory cascade may be the opportunity it offers for independent regulation of different downstream targets and thus for a flexible and coordinated response by the cell to the initial signal.

Phosphorylation is the most important covalent switch mechanism for the control of protein function

Post-translational modification by phosphorylation has been found in all living organisms from bacteria to humans. Target proteins are phosphorylated by the action of protein kinases and dephosphorylated by **protein phosphatases**, and this reversible modification provides a switch that controls many diverse cellular processes including metabolic pathways, signaling cascades, intracellular membrane traffic, gene transcription, and movement. The phosphoryl group derives from the terminal phosphate of nucleoside triphosphates, usually ATP. Use of separate enzymes for phosphorylation (kinases) and dephosphorylation (phosphatases) enables independent control of these events by different stimuli. Kinases that phosphorylate proteins on serine, threonine or tyrosine residues constitute the third most common domain encoded in the human genome sequence, with 575 such kinases (about 2% of the genome) identified to date. In prokaryotes, a different type of kinase domain phosphorylates proteins on histidine and aspartate

Definitions

protein phosphatase: enzyme that specifically removes phosphate groups from phosphorylated serines, threonines or tyrosines on proteins.

References

Barford, D. *et al.*: **Structural mechanism for glycogen phosphorylase control by phosphorylation and AMP.** *J. Mol. Biol.* 1991, **218**:233–260.

Cozzone, A.J.: **Regulation of acetate metabolism by protein phosphorylation in enteric bacteria.** *Annu.*

Rev. Microbiol. 1998, **52**:127–164.

Dean, A.M. and Koshland, D.E. Jr.: **Electrostatic and steric contributions to regulation at the active site of isocitrate dehydrogenase.** *Science* 1990, **249**:1044–1046.

Hunter, T.: **Signaling—2000 and beyond.** *Cell* 2000, **100**:113–127.

Hurley, J.H. *et al.*: **Regulation of an enzyme by phosphorylation at the active site.** *Science* 1990, **249**:1012–1016.

Johnson, L.N. and O'Reilly, M.: **Control by phosphoryla-**

tion. *Curr. Opin. Struct. Biol.* 1996, **6**:762–769.

Lee, M. and Goodbourn, S.: **Signalling from the cell surface to the nucleus.** *Essays Biochem.* 2001, **37**:71–85.

Manning, G. *et al.*: **The protein kinase complement of the human genome.** *Science* 2002, **298**:1912–1934.

Wilkinson, M.G. and Millar, J.B.: **Control of the eukaryotic cell cycle by MAP kinase signaling pathways.** *FASEB J.* 2000, **14**:2147–2157.

Zolnierowicz, S. and Bollen, M.: **Protein phosphorylation and protein phosphatases.** *EMBO J.* 2000, **19**:483–488.

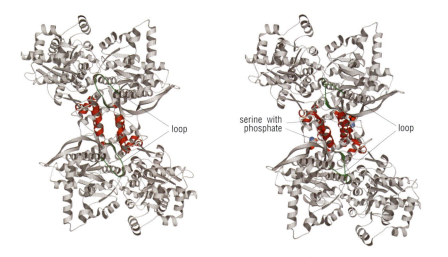

loop

serine with phosphate

loop

Figure 3-22 Conformational change induced by phosphorylation in glycogen phosphorylase The enzyme glycogen phosphorylase (GP) cleaves glucose units from the nonreducing ends of glycogen, the storage polymer of glucose, by phosphorolysis to produce glucose 6-phosphate. Muscle GP is a dimer of identical subunits. In response to hormonal signals the enzyme is covalently converted from the less active phosphorylase b form (left) to the more active phosphorylase a form (right) through phosphorylation of a single serine residue, serine 14, catalyzed by the enzyme phosphorylase kinase. On phosphorylation, a loop (shown in green) that sterically restricts access to the active site in the b form moves out, making the substrate-binding site more accessible. The large changes in conformation that occur at the interface between the two monomers are shown in red. (PDB 1gpa and 1gpb)

residues (although some eukaryotic-like serine/threonine and tyrosine kinases have been found in some bacteria), and the percentage of the genome devoted to regulation by phosphorylation is strikingly similar: about 1.5% of the *E. coli* genome, for example, encodes such proteins.

Covalent addition of a phosphoryl group to the side chain of serine, threonine, tyrosine, histidine or aspartic acid can have profound effects on the function of the protein. Phosphorylated residues have acquired a group that carries a double negative charge and is capable of multiple hydrogen-bonding interactions. Structural studies of phosphorylated proteins have shown that two types of interaction predominate: hydrogen bonding to main-chain amide groups at the positively polarized amino-terminal end of an alpha helix, and salt-bridging to one or more arginine residues. Other residues may also be involved in the recognition and binding of phosphoryl groups in specific cases.

Phosphorylation can affect the target protein in two ways, which are not mutually exclusive. One effect is to change the activity of the target protein, either considerably or subtly. This change in activity may come about solely from the added bulk and charge properties of the phosphoryl group, or may result from a large conformational change in the protein, or both. The second effect of phosphorylation is to provide a new recognition site for another protein to bind (see, for example, the binding of Grb2 to the phosphorylated receptor tail in Figure 3-21). Such protein–protein interactions usually involve specialized interaction domains on the second protein that recognize the phosphorylated peptide segment; the most common domain recognizing phosphotyrosines, for example, is the SH2 domain (see section 3-1).

An instance of phosphorylation activating an enzyme by inducing a large conformational change is seen in muscle glycogen phosphorylase. Covalent attachment of a phosphoryl group to serine 14 results in a rearrangement of the amino-terminal residues in the enzyme such that the serine side chain shifts 50 Å, leading to a change in the subunit–subunit contacts of this dimeric protein, with a concomitant rearrangement of the active-site residues that activates the enzyme (Figure 3-22). The phosphoserine is stabilized in its new position primarily by salt bridges to two arginine side chains.

In contrast, the inactivation of the TCA-cycle enzyme isocitrate dehydrogenase by serine phosphorylation involves no conformational changes. Serine 113, the residue phosphorylated, is located in the active site (Figure 3-23), and attachment of a phosphoryl group inhibits binding of the negatively charged substrate isocitrate by steric exclusion and electrostatic repulsion. In isocitrate dehydrogenase the phosphoserine is stabilized by helix dipole and main-chain hydrogen-bond interactions.

One of the principal functions of reversible phosphorylation carried out by the receptor tyrosine kinases that initiate many eukaryotic signal transduction pathways is the temporary provision of new protein-interaction sites (see Figure 3-21, where phosphorylation of the receptor's own cytoplasmic domains provides a recognition site for an adaptor molecule that relays the signal onward). Another is to phosphorylate and directly activate other downstream signaling and effector molecules. The regulation of these kinases is therefore of central importance in biology, and in the next section we describe the conserved features of their activation mechanism and discuss its regulation in two important kinase families: the Src family and the cyclin-dependent kinases.

(a)

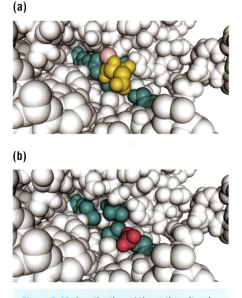

(b)

Figure 3-23 Inactivation of the active site of *E. coli* isocitrate dehydrogenase by phosphorylation The substrate-bound (a) and phosphorylated (b) states of the active site of isocitrate dehydrogenase are shown as space-filling models. The active-site residues include the serine 113 side chain, which is shown in green, the negatively charged portion of the isocitrate substrate in yellow and the phosphoryl group on serine 113 in red (in b). The phosphoryl group occupies almost the same location in the active site as the negatively charged portion of isocitrate. The phosphorylated form is inactive because isocitrate binding is both sterically blocked and electrostatically repelled by the negatively charged phosphoryl group. (PDB 5icd)

Protein kinases are themselves controlled by phosphorylation

Phosphorylation of proteins on serine/threonine or tyrosine residues is probably the single most important regulatory mechanism in eukaryotic signal transduction, and tyrosine phosphorylation lies at the heart of control of the eukaryotic cell cycle. To a first approximation, the protein kinases responsible for phosphorylating proteins on serine, threonine and tyrosine residues all have the same fold for the catalytic domain (Figure 3-24a), although many of them also have other subunits or other domains that serve regulatory functions or target their kinase activity to specific protein substrates (Figure 3-24b). Most, but not all, of these kinases are normally inactive, and before they can phosphorylate other proteins they must themselves be activated by phosphorylation of a threonine or tyrosine residue that is located in a region termed the **activation segment**, also known as the **activation loop**. The conserved mechanism of activation of these kinases is illustrated in Figure 3-25. Kinases controlled by activation-loop phosphorylation represent some of the most important enzymes in signal transduction cascades. Two of the best understood are the Src family of tyrosine protein kinases, which are activated early in eukaryotic cell signaling pathways, and the cyclin-dependent kinases (Cdks) which coordinate the eukaryotic cell cycle. We discuss the part played by regulatory and targeting domains in the activation of Src kinases below, and in the next section we describe the activation and targeting of the cyclin-dependent kinase Cdk2.

Src kinases both activate and inhibit themselves

Src-family kinases are activated early in many signaling pathways and once activated sustain their own activated states by **autophosphorylation**, providing for a large amplification of the signal. The original Src kinase was discovered as an oncogenic variant in a tumor virus that causes sarcomas in chickens (Src is short for sarcoma). Its tumorigenic action is due to a mutation that causes unregulated autophosphorylation, leading to a sustained growth signal, and we now understand the structural basis both for the normal regulation of the kinase and for its oncogenic activation.

Src kinases recognize their target proteins through SH2 and SH3 domains that are joined to one another and to the catalytic domain by flexible linker regions (Figure 3-24b). In the absence of activating signals, these domains bind to the kinase domain, holding it in an inactive conformation (Figure 3-26): the SH2 domain binds to an inhibitory phosphate on a tyrosine

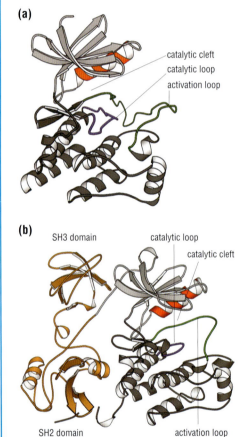

(a)

catalytic cleft
catalytic loop
activation loop

(b)

SH3 domain   catalytic loop
catalytic cleft

SH2 domain   activation loop

Figure 3-24 The conserved protein kinase catalytic domain (a) The catalytic domain of Lck, a signaling tyrosine kinase expressed in the cells of the immune system, is typical of the highly conserved catalytic domain of all protein kinases. This is structurally divided into two lobes, the amino-terminal (upper) lobe (light grey) being formed almost entirely from beta strands but for a single alpha helix (the C helix, red), while the carboxy-terminal (lower) lobe (dark grey) is formed almost entirely from alpha helices. Between the two lobes, which are joined by a flexible hinge, is the catalytic cleft containing the catalytic loop (purple). The activation loop, which is repositioned by phosphorylation and thereby regulates the catalytic action of the kinase, is shown in green. The flexible hinge between the two lobes is important in allowing major conformational changes that accompany activation and inhibition in some kinases. Movement of the single helical element in the amino-terminal lobe is critical to the establishment of the active conformation of the catalytic site. Kindly provided by Ming Lei and Stephen Harrison.
(b) Structure of the Src-family tyrosine protein kinase Hck in its inactive form. The catalytic domain is attached to two small domains (gold), an SH3 domain and an SH2 domain, that target this kinase to specific protein substrates. (PDB 1ad5)

Definitions

activation loop: a stretch of polypeptide chain that changes conformation when a kinase is activated by phosphorylation and/or protein binding. This segment may or may not be the one containing the residue that is phosphorylated to activate the kinase. Usually, in the inactive state, the activation loop blocks access to the active site.

activation segment: see **activation loop**.

autophosphorylation: phosphorylation of a protein

kinase by itself. Autophosphorylation may occur when the active site of the protein molecule to be phosphorylated catalyzes this reaction (*cis* autophosphorylation) or when another molecule of the same kinase provides the active site that carries out the chemistry (*trans* autophosphorylation). Autophosphorylation *in trans* often occurs when kinase molecules dimerize, a process that can be driven by ligand binding as in the receptor tyrosine kinases.

Figure 3-25 Conserved mechanism of kinase activation In all protein kinases, the activation loop (green) plays a central part in regulating catalytic activity. A conserved aspartate in the activation loop is critical to the catalytic action of the kinase, and changes in the position of the loop on phosphorylation lead to repositioning of this and other critical catalytic residues, and in some cases also regulate access of the substrate to the active-site cleft. While the active configuration of the active site (right) is essentially the same for all kinases, the inactive configuration may vary considerably. In the example schematically represented here, the activation loop swings down in the inactive conformation (left), but in some kinases it instead swings up and occludes the catalytic cleft so that substrate cannot gain access. The purple loop is the conserved catalytic loop, found in all kinases, which contains residues that participate in the chemical step of protein phosphorylation.

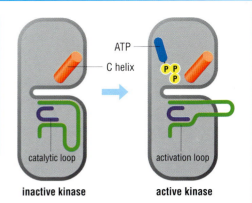

residue close to the carboxyl terminus of the protein, and when it is bound, the linker region joining the SH2 domain to the catalytic domain forms a polyproline helix to which the SH3 domain can bind. This clamps the catalytic domain in an inactive state from which it is released either by dephosphorylation of the tyrosine on the carboxyl tail of the protein, or by binding of the SH2 domain by a phosphotyrosine on the cytoplasmic tail of an activated receptor tyrosine kinase: this causes conformational changes that activate phosphorylation by the Src kinase by rearranging the activation loop in the active site, and at the same time releases the SH3 domain to bind to the target (Figure 3-26). The oncogenic properties of the viral Src kinase are now known to be due to a mutation of the carboxyl tail of the molecule that eliminates the tyrosine residue that is the target of the inhibitory phosphorylation.

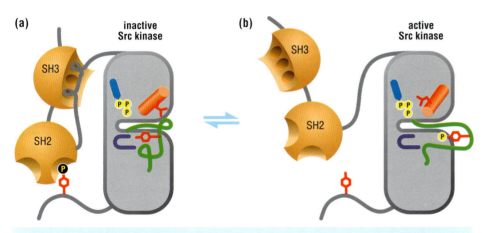

Figure 3-26 Regulation of a Src-family protein kinase In the absence of activating signals the SH2 and SH3 domains hold the kinase in an inactive conformation **(a)** in which the SH2 domain binds to an inhibitory phosphate on a tyrosine residue close to the carboxyl terminus of the protein, while the SH3 domain binds to a polyproline helix in the linker region joining the SH2 domain to the catalytic domain. **(b)** When the SH2 domain releases the carboxyl tail of the protein, either because the inhibitory phosphate is removed, or on binding a target phosphotyrosine, the polyproline helix rearranges, releasing the SH3 domain, and initiating a series of structural changes that propagate to the C helix (red) in the upper catalytic domain and the activation loop (green), which assume new conformations appropriate for substrate binding and autophosphorylation of a tyrosine in the activation loop.

References

Huse, M. and Kuriyan, J.: **The conformational plasticity of protein kinases.** *Cell* 2002, **109**:275–282.

Theodosiou, A. and Ashworth, A.: **MAP kinase phosphatases.** *Genome Biol.* 2002, **3**:reviews3009.

Xu, W. *et al.*: **Crystal structures of c-Src reveal features of its autoinhibitory mechanism.** *Mol. Cell* 1999, **3**:629–638.

Young, M.A. *et al.*: **Dynamic coupling between the SH2**

and SH3 domains of c-Src and Hck underlies their inactivation by C-terminal tyrosine phosphorylation. *Cell* 2001, **105**:115–126.

Cyclin acts as an effector ligand for cyclin-dependent kinases

Cyclin-dependent kinases (Cdks) are the enzymes that drive the cell cycle. Activated periodically during the cycle, they phosphorylate proteins that, for example, move the cell onward from growth phase to DNA replication phase or from DNA replication phase into mitosis. The activation of a Cdk is a two-step process that requires binding of a regulatory protein, the cyclin, and phosphorylation on a threonine in the activation loop of the Cdk by the enzyme Cdk-activating kinase (CAK). In eukaryotes, a Cdk known as Cdk2 regulates passage from the G1 state of the cell cycle, in which cells are growing and preparing to divide, to the S phase in which their chromosomes replicate. Cdk2 is activated by cyclin A. The cyclin subunit of the cyclin–Cdk 2 complex also targets Cdk2 to the downstream targets it phosphorylates.

In its inactive, unphosphorylated state, Cdk2 is autoinhibited by the activation loop (here called the T-loop), which partially blocks the ATP-binding site (Figure 3-27a). Structural studies of phosphorylated Cdk2 in the absence of bound cyclin, a form that has 0.3% of

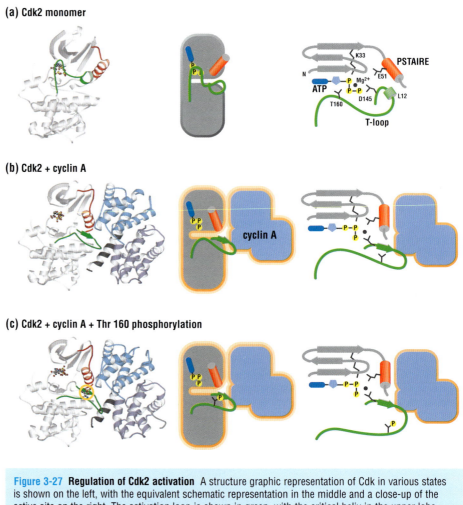

(a) Cdk2 monomer

(b) Cdk2 + cyclin A

(c) Cdk2 + cyclin A + Thr 160 phosphorylation

Figure 3-27 Regulation of Cdk2 activation A structure graphic representation of Cdk in various states is shown on the left, with the equivalent schematic representation in the middle and a close-up of the active site on the right. The activation loop is shown in green, with the critical helix in the upper lobe (here called PSTAIRE) in red. **(a)** In the inactive conformation of the kinase, in the absence of either cyclin or phosphorylation, a small helix (L12) in the activation loop (here called the T loop) displaces the PSTAIRE helix which contains a glutamate residue (E51) critical to catalysis. **(b)** When cyclin binds, the PSTAIRE helix moves inward and the L12 helix melts to form a beta strand, allowing E51 to interact with lysine 33 (K33) and leading to the correct orientation of the ATP phosphates. Cyclin binding also shifts the activation loop out of the active-site cleft. **(c)** Phosphorylation of threonine 160 in the activation loop flattens the activation loop, increasing its interaction with cyclin A, which allows it to interact more effectively with substrates. Taken from David O. Morgan: *The Cell Cycle: Principles of Control* (New Science Press, in the press).

maximal enzymatic activity, show that attachment of a phosphoryl group to threonine 160 causes the activation loop to become disordered, which presumably allows ATP to bind sometimes. No other significant conformational changes are observed. The structure of the complex of unphosphorylated Cdk2 with cyclin A (Figure 3-27b), a form that is also only 0.3% active, shows some significant conformational changes, including reorganization of the activation loop. Cyclin binding and phosphorylation therefore both induce conformational changes in the protein, but these changes are different and neither alone is sufficient to fully activate the enzyme.

The structure of the phospho-Cdk2–cyclin A complex (Figure 3-27c) shows why both modifications are needed for full activity (Figure 3-28). The conformational changes seen on cyclin A binding alone all occur, but in addition the phosphothreonine side chain makes interactions that are not observed without cyclin binding. The phosphothreonine turns into the protein to interact with three arginine residues, leading to a further reorganization of the activation loop beyond that which occurs on either phosphorylation or cyclin binding alone.

This cumulative change in conformation of the activation segment is crucial for proper recognition of the substrate peptides on which this kinase acts. Specific kinases usually phosphorylate specific sequence motifs that incorporate the serine, threonine or tyrosine at which they react. Cdk2 recognizes the sequence SPXR/K. This peptide segment binds in an extended conformation across the catalytic cleft and primarily makes contact with the activation loop. In the unphosphorylated Cdk2–cyclin complex, the conformation of the activation segment does not allow the proline residue to fit properly into the active site because it would lead to a steric clash with valine 163 on the kinase. The phospho-Cdk2–cyclin A structure has a different conformation for the activation loop that repositions valine 163 out of the way. At the same time, the adjacent residue, valine 164, rotates its backbone carbonyl group away from the peptide substrate (Figure 3-28). It is this rotation that gives the kinase absolute specificity for proline in the substrate sequence motif. Proline is the only amino acid that does not have a backbone –N–H group (see Figure 1-3 in Chapter 1). Any other amino acid would have its –N–H pointing into the place where the valine 164 carbonyl group used to be, leading to an unsatisfied hydrogen bond donor in the enzyme–substrate complex, a very unfavorable situation energetically. Proline-containing substrates do not have this problem. Thus, the conformational changes in Cdk2 on phosphorylation and cyclin binding combine not only to activate the enzyme but also to allow specific recognition of the "correct" substrate.

Cdk2 provides a striking example of how protein–protein interactions combine with covalent modification by phosphorylation in regulation. This principle of multiple regulatory mechanisms operating on the same target is widely seen, especially in eukaryotic cells, and provides exquisite control of protein function.

Dephosphorylation by protein phosphatases reverses the effects of phosphorylation and it is assumed that the protein structure simply returns to its original conformation. There are many fewer phosphatases than kinases, implying that they are less specific, but some of them are known to be regulated by phosphorylation or by binding of effector molecules. Since phosphatases are in general able to recognize many phosphorylated protein substrates, it is likely that they are targeted to specific substrates as needed by mechanisms that change protein localization within the cell, as described in other sections of this chapter.

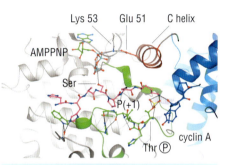

Figure 3-28 The substrate-binding site of Cdk2 The substrate to be phosphorylated (pink) binds to Cdk2 (the activation loop is shown in green) in an extended conformation. The serine to be phosphorylated (Ser) has its –OH group pointed at the gamma phosphate that will be transferred from ATP (here, the nonhydrolyzable analog AMPPNP). Activation of Cdk2 includes a change in the position of the C helix (sequence PSTAIRE, the red helix in Figure 3-27) so that glutamate 51 can interact with lysine 53 to position the ATP properly. This conformational change is induced only by the binding of cyclin A. Sequence specificity for the peptide substrate is provided by the proline residue P(+1) in the substrate. This proline comes close to the backbone of the dipeptide Val 163–Val 164 in the activation loop, which has undergone a conformational change on phosphorylation of Cdk2 to allow the substrate to fit. Graphic kindly provided by Jane Endicott and Martin Noble.

References

Brown, N.R. et al.: **The structural basis for specificity of substrate and recruitment peptides for cyclin-dependent kinases.** Nat. Cell Biol. 1999, **1**:438–443.

Brown, N.R. et al.: **Effects of phosphorylation of threonine 160 on cyclin-dependent kinase 2 structure and activity.** J. Biol. Chem. 1999, **274**:8746–8756.

Cook, A. et al.: **Structural studies on phospho-CDK2/cyclin A bound to nitrate, a transition state analogue: implications for the protein kinase mechanism.** Biochemistry 2002, **41**:7301–7311.

Song, H. et al.: **Phosphoprotein-protein interactions revealed by the crystal structure of kinase-associated phosphatase in complex with phosphoCDK2.** Mol. Cell 2001, **7**:615–626.

Two-component signal carriers employ a small conformational change that is driven by covalent attachment of a phosphate group

G proteins and motor ATPases are generally absent from prokaryotes, which use a different class of molecular switches. Their signaling pathways, referred to as **two-component systems**, are structured around two families of proteins. The first component is an ATP-dependent histidine protein kinase (HK). This is typically a transmembrane protein composed of a periplasmic sensor domain that detects stimuli and cytoplasmic histidine kinase domains that catalyze ATP-dependent autophosphorylation. The second component of the system, a cytoplasmic response regulator protein (RR), is activated by the histidine kinase. Signals transmitted through the membrane from the sensor domain modulate the activities of the cytoplasmic kinase domains, thus regulating the level of phosphorylation of the RR. Phosphorylation of the RR results in its activation and generation of the output response of the signaling pathway. Thus, although ATP hydrolysis is used to drive this switching process, it does so indirectly: a phosphoryl group originally derived from ATP covalently modifies the RR and serves as the trigger.

Two-component system proteins are abundant in most eubacterial genomes, in which they typically constitute ~1% of encoded proteins. For example, the *E. coli* genome encodes 62 two-component proteins, which are involved in regulating processes as diverse as chemotaxis, osmoregulation, metabolism and transport. In eukaryotes, two-component pathways constitute a very small number of all signaling systems. In fungi, they mediate environmental stress responses and hyphal development. In the slime mold *Dictyostelium* they are involved in osmoregulation and development, while in plants they are involved in responses to hormones and light, leading to changes in cell growth and differentiation. To date, two-component proteins have not been identified in animals and do not seem to be encoded by the human, fly or worm genomes. In most prokaryotic systems, the output response is generated directly by the RR; in many systems it functions as a transcription factor whose transcription-activating or repressing activity depends on its state of phosphorylation. Fungi and plants also contain RRs that function as transcription factors. In addition, eukaryotic two-component proteins are found at the beginning of signaling pathways where they interface with more conventional eukaryotic signaling strategies such as mitogen-activated protein (MAP) kinase and cyclic nucleotide cascades.

Similarly to most proteins in signaling pathways, two-component systems are modular in architecture. Different arrangements of conserved domains within proteins and different integration of proteins into pathways provide adaptations of the basic scheme to meet the specific regulatory needs of many different signaling systems. The prototypic prokaryotic two-component pathway (Figure 3-29) illustrates the fundamental phosphotransfer switching mechanism of both simple and more elaborate systems. Stimuli detected by the sensor domain of the histidine kinase regulate the kinase's activities. The kinase catalyzes ATP-dependent autophosphorylation of a specific histidine residue. The RR then catalyzes transfer of the phosphoryl group from this phosphorylated histidine to one of its own aspartate residues, located on the regulatory domain. Phosphorylation of the regulatory domain of the RR activates an effector domain on the same protein (or, on rare occasions, a separate effector protein) that produces the specific output response.

The regulatory domains of RRs have three activities. First, they interact with phosphorylated histidine kinase and catalyze transfer of a phosphoryl group to one of their own aspartate residues. Second, they are phosphatases that catalyze their own dephosphorylation—the counterpart to the GTPase activity of the G proteins. The phosphatase activity varies greatly among different RRs, with half-lives for the phosphorylated state ranging from seconds to hours, a span of four orders

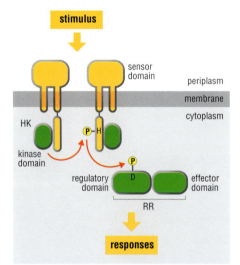

Figure 3-29 Two-component signaling mechanisms The basic two-component phosphotransfer system, found in bacteria, consists of a dimeric transmembrane receptor histidine kinase (HK) and a cytoplasmic response regulator (RR). Information flows between the two proteins in the form of a phosphoryl group (PO$_3$) that is transferred from the HK to the RR. HKs catalyze ATP-dependent autophosphorylation of a specific conserved His residue (H). The activities of HKs are modulated by environmental signals such as nutrients or osmotic stress. The phosphoryl group (P) is then transferred to a specific aspartic acid residue (D) located within the conserved regulatory domain of an RR. Phosphorylation of the RR typically activates an associated (or downstream) effector domain, which ultimately elicits a specific cellular response.

References

Parkinson, J.S. and Kofoid, E.C.: **Communication modules in bacterial signaling proteins.** *Annu. Rev. Genet.* 1992, **26:**71–112.

Stock A.M. *et al.*: **Two-component signal transduction.** *Annu. Rev. Biochem.* 2000, **69:**183–215.

Stock A.M. *et al.*: **Structure of the Mg²⁺-bound form of CheY and mechanism of phosphoryl transfer in bacterial chemotaxis.** *Biochemistry* 1993, **32:**13375–13380.

West, A.H. and Stock, A.M.: **Histidine kinases and response regulator proteins in two-component signaling systems.** *Trends Biochem. Sci.* 2001, **26:**369–376.

Wurgler-Murphy, S.M. and Saito, H.: **Two-component signal transducers and MAPK cascades.** *Trends Biochem. Sci.* 1997, **22:**172–176.

of magnitude. And third, RRs regulate the activities of their associated effector domains (or effector proteins) in a phosphorylation-dependent manner. The different lifetimes of different regulators allow two-component signal systems to regulate a wide variety of cellular processes.

All RRs have the same general fold and share a set of conserved residues (Figure 3-30a). Phosphorylation of the active-site aspartate is associated with an altered conformation of the regulatory domain. A common mechanism appears to be involved in the structural changes that propagate from the active site. The phosphorylated aspartate is positioned by interaction of two of the phosphate oxygens with both a divalent cation (usually a magnesium ion) coordinated to three active-site carboxylate side chains and with the side chain of a conserved lysine residue. Two other highly conserved side chains have distinctly different orientations in the phosphorylated form of the regulatory domain compared to the unphosphorylated form: in the phosphorylated state a serine/threonine side chain is repositioned to form a hydrogen bond with the third phosphate oxygen and a phenylalanine/tyrosine side chain is reoriented towards the interior of the domain, filling the space that is normally occupied by the serine/threonine in the unphosphorylated protein.

Unlike the GTPases, which have large conformational changes in their switching mechanisms, the regulatory domains of two-component signaling pathways undergo less dramatic structural rearrangements on phosphorylation. Structural differences between the unphosphorylated and phosphorylated regulatory domains map to a relatively large surface involving several beta strands, alpha helices and adjacent loops, with backbone displacements ranging from 1 to 7 Å (Figure 3-30b).

In other words, like the G proteins, RR regulatory domains function as generic on–off switch modules, which can exist in two distinct structural states with phosphorylation modulating the equilibrium between them, providing a very simple and versatile mechanism for regulation. Surfaces of the regulatory domain that have altered structures in the two different conformations are exploited for protein–protein interactions that regulate effector domain function.

Not all effector domains are DNA-binding transcriptional regulators. Some function, for example, as regulators of bacterial flagellar rotation. There are many different strategies for regulation of effector domains by RR regulatory domains, including inhibition of effector domains by unphosphorylated regulatory domains, allosteric activation of effector domains by phosphorylated regulatory domains, dimerization of effector domains mediated by domain dimerization, and interaction of RRs with heterologous target effector proteins. Different RRs use different subsets of the regulatory domain surface for phosphorylation-dependent regulatory interactions. Thus, although RRs are fundamentally similar in the design of their phosphorylation-activated switch domains, there is significant versatility in the way these domains are used to regulate effector activity.

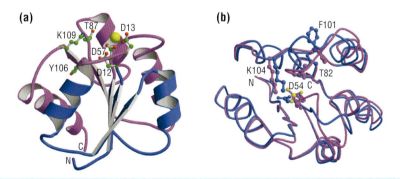

Figure 3-30 Conserved features of RR regulatory domains **(a)** A ribbon diagram of the bacterial RR CheY, whose protein fold is just a bare regulatory domain, is shown with ball-and-stick representations of the side chains of highly conserved residues. Residues that are highly conserved in all RR regulatory domains are clustered in two regions: an active site formed by loops that extend from the carboxy-terminal ends of strands 1, 3 and 5, and a pair of residues that form a diagonal path extending across the molecule from the active site. Three aspartic acid residues (D12, D13 and D57) position a Mg^{2+} ion (yellow) that is required for catalysis of phosphoryl transfer from an HK to aspartate 57. Three additional residues (K109, T87 and Y106) are important in propagation of the conformational change that occurs on phosphorylation. The regions of the backbone that have been observed to differ in unphosphorylated and phosphorylated regulatory domains are shown in magenta. **(b)** The conserved mechanism involved in the phosphorylation-induced conformational change is illustrated by the structures of the unphosphorylated (blue) and phosphorylated (magenta) regulatory domains of the RR protein FixJ. When D54 is phosphorylated, K104 forms an ion pair with the phosphate (yellow). T82 also forms a hydrogen bond with the phosphate and F101 has an inward orientation, positioned in the space occupied by T82 in the unphosphorylated structure. These changes trigger other rearrangements of secondary structure elements, creating a different protein-binding surface in the two states. Graphic kindly provided by Ann Stock. From West, A.H. and Stock, A.M.: *Trends Biochem. Sci.* 2001, **26**:369–376, with permission.

Limited proteolysis can activate enzymes

Both the ubiquitin-dependent proteasome pathway discussed in section 3-11 and digestion by many other nonspecific proteases degrade polypeptide chains to small peptides or individual amino acids. But proteolysis does not have to lead to destruction and inactivation. Many proteins are post-translationally modified by limited proteolytic digestion that produces an active form from their inactive or marginally active precursors.

Limited proteolysis involves the cleavage of a target protein at no more than a few specific sites—commonly just one—usually by a specific protease. The resulting cleaved protein can have one of two fates: either the fragments remain associated, covalently (if they are disulfide-linked) or noncovalently, or they may dissociate to give two or more different polypeptides, each of which may have a completely separate fate and function. Both these outcomes are in fact observed in the maturation of the inactive precursor chymotrypsinogen to active alpha-chymotrypsin, a digestive protease (Figure 3-31). Chymotrypsinogen has the overall fold of an active serine protease with the exception that the active-site region lacks the proper configuration of main chain and catalytic side chains for both catalysis and substrate recognition. The protein is, however, "spring-loaded"; it is only the existence of a covalent bond between arginine 15 and isoleucine 16 and a set of noncovalent interactions between these residues and their neighbors that prevent it from rearranging into the correct conformation for catalysis. Specific proteolytic cleavage between arginine 15 and isoleucine 16 by the protease trypsin releases it from this constraint. After this cleavage, the amino-terminal peptide remains attached through a disulfide linkage, while the newly formed amino terminus (old isoleucine 16) swaps into a new conformation in which it makes a new set of interactions with residues of the active site, causing the active site to assume its correct catalytic configuration. The product is fully active enzymatically, although two more autocatalytic cleavages remove residues 14 and 15 and residues 147 and 148, to yield the mature form of alpha-chymotrypsin that is found in the digestive tract. The function of these latter two modifications is unknown. The activation of most serine proteases of this class from their inactive proenzyme forms occurs in a similar fashion (Figure 3-32).

Polypeptide hormones are produced by limited proteolysis

Limited proteolysis can also produce polypeptides with new functions, as in the production of short polypeptide hormones from long precursor proteins (Figure 3-33). All known polypeptide hormones are synthesized in "prepro" form, with a signal (pre) sequence and additional (pro) sequences that are cleaved out during maturation. Often, a single precursor sequence may contain two or more distinct hormones, each of which is released by additional cleavages. A striking example is the pituitary multihormone precursor pro-opiomelanocortin, which contains sequences for the hormones beta-lipotropin, melanocyte-stimulating hormone (MSH), endorphin, enkephalin, and adrenocorticotropic hormone (ACTH, corticotropin). ACTH is a 39-residue polypeptide that stimulates the synthesis of mineralocorticoids and glucocorticoids in the adrenal cortex, which leads to activation of steroid hormone synthesis. The other hormone products have very different target tissues and different physiological activities. Pro-opiomelanocortin is cleaved at different sites in different cell types so that they produce different spectra of hormones derived from the single precursor. Each of the cleaving enzymes is specific for a particular sequence in the precursor protein, by virtue either of its active-site structure or of the compartment in which it encounters its substrate. Hormone-processing proteases are attractive targets for drugs designed to modulate hormone-sensitive processes such as control of blood pressure and inflammation.

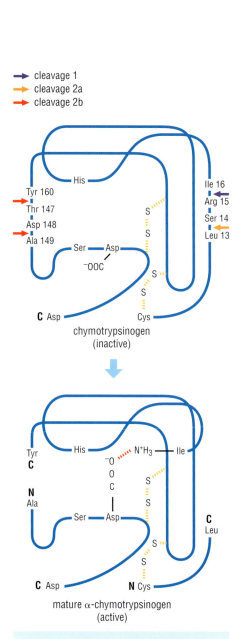

→ cleavage 1
→ cleavage 2a
→ cleavage 2b

chymotrypsinogen
(inactive)

mature α-chymotrypsinogen
(active)

Figure 3-31 Activation of chymotrypsinogen
A schematic view of the activation of chymotrypsinogen. The polypeptide chain (blue) is held together by disulfide (–S–S–) bridges (for clarity only two of the five are shown). Cleavage between residues 15 and 16 (top) results in a rearrangement of part of the polypeptide chain (bottom) and the formation of a new interaction between isoleucine 16 and an aspartate elsewhere in the chain, which helps form the functional active site. After the first cleavage, the first 15 residues remain attached to the rest of the protein, in part by a disulfide linkage. Two more cleavage events remove residues 14–15 and residues 147–148 to produce the mature alpha-chymotrypsin.

Definitions

proteolytic cascade: a series of protein cleavages by proteases, each cleavage activating the next protease in the cascade.

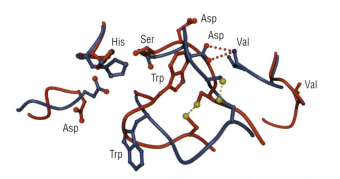

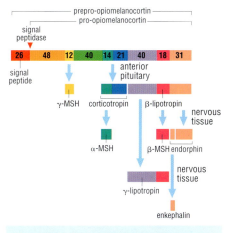

Figure 3-32 Comparison of the active sites of plasminogen and plasmin Plasminogen (red) is the inactive precursor of plasmin (blue), the primary enzyme that dissolves blood clots. Like chymotrypsinogen and many other inactive serine protease precursors, plasminogen is activated by a proteolytic cleavage. The active-site residues in plasminogen must rearrange to form an active enzyme. In chymotrypsinogen, most of the catalytic residues are already in the active configuration, but in plasminogen the entire active site is distorted. The primary cause of this distortion is a tryptophan residue that blocks the substrate-binding pocket and pushes the catalytic residues away from their correct orientations. As in all proteases of this class, the oxyanion hole involved in catalysis (see Figure 2-40) has not yet formed. Cleavage of the arginine–valine bond (not shown) in plasminogen creates a new positively charged amino-terminal amino group that interacts with the central aspartic acid residue of the oxyanion peptide, causing the peptide to rearrange into the correct configuration of the oxyanion hole. At the same time, the tryptophan swings out of the active site, unblocking the substrate-binding pocket and allowing the catalytic residues (His, Ser and Asp at the left of the diagram) to switch to the correct configuration for catalysis.

Figure 3-33 Schematic diagram of prepro-opiomelanocortin and its processing Shown are the domains that comprise the large pre-pro hormone precursor opiomelanocortin, with the number of amino acids in each domain indicated. This primary gene translation product is cleaved into many different smaller hormones by a series of proteolytic steps. First, the signal peptide that directed its secretion is cleaved off by a signal peptidase, generating pro-opiomelanocortin. Then, a number of other cleavages, which are tissue-specific, occur in different cell types to produce different ensembles of hormones. For instance, in the pituitary, cleavage occurs to generate ACTH and beta-lipotropin, while processing in the central nervous system gives endorphin and enkephalin, among other products.

Limited proteolysis can provide for enormous, extremely rapid amplification of a signal if the product of one cleavage is an active protease that can go on to activate other proteins by cleavage. Several such activations may follow one another to generate a **proteolytic cascade** in which the initial activation of a single molecule of an inactive proenzyme produces a huge final output. This is how the blood coagulation pathway works. Regardless of whether the pathway is activated internally, as in thrombosis, or at an open wound, the first step is the production of an active, specific protease from an inactive precursor. A single molecule of this activated protease is capable of activating thousands of molecules of its specific substrate, another inactive protease; each newly activated protease molecule can then activate thousands of molecules of their specific proenzyme substrate, and so on. The final product is millions of activated thrombin molecules, which cleave the soluble blood protein fibrinogen to form insoluble fibrin, which polymerizes to form the clot (Figure 3-34).

Figure 3-34 The blood coagulation cascade Each factor in the pathway is a protein that can exist in an inactive form (green circles) and an active form (blue squares), and many of these are serine proteases (denoted by an asterisk). The cascade of proteolytic activations that ends with the activation of fibrinogen to clotting fibrin is initiated by exposure of blood at damaged tissue surfaces (intrinsic pathway) or from internal trauma to blood vessels (extrinsic pathway). Although the specific initiating factors differ in the two pathways, in both cases the process begins with the conversion of an inactive serine protease to an active one. As little as one molecule of this active protease may be all that is needed to activate thousands of molecules of the next precursor protein in the cascade. Each of these can activate many more of the next, and so on down the cascade. Each protease is highly specific for a cleavage sequence in the next factor in the pathway, preventing unwanted activation. Roman numerals refer to specific proteins called factors.

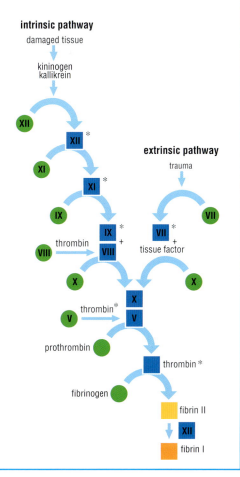

References

Hook, V.Y. *et al.*: **Proteases and the emerging role of protease inhibitors in prohormone processing.** *FASEB J.* 1994, **8**:1269–1278.

Kalafatis, M. *et al.*: **The regulation of clotting factors.** *Crit. Rev. Eukaryot. Gene Expr.* 1997, **7**:241–280.

Peisach, E. *et al.*: **Crystal structure of the proenzyme domain of plasminogen.** *Biochemistry.* 1999, **38**:11180–11188.

Rockwell, N.C. *et al.*: **Precursor processing by kex2/furin proteases.** *Chem. Rev.* 2002, **102**:4525–4548.

Steiner, D.F.: **The proprotein convertases.** *Curr. Opin. Chem. Biol.* 1998, **2**:31–39.

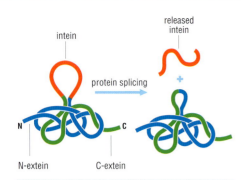

Figure 3-35 Protein splicing Schematic diagram of the process by which an internal segment of a polypeptide chain (intein) excises itself from the protein in which it is embedded, leading to the ligation of the flanking domains to yield a single polypeptide chain. Red indicates sequence to be excised (intein); blue and green are the flanking sequences. Protein splicing does not involve peptide bond breakage and religation but instead occurs entirely by peptide bond rearrangements. This allows it to proceed without a source of metabolic energy such as ATP, which would be required if the precursor protein were cleaved and religated.

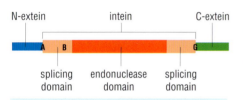

Figure 3-36 Schematic of the organization of intein-containing proteins The intein is represented by the peptide segment labeled A to G. This segment contains two splicing domains separated by a linker region that may be a homing endonuclease. Sites A, B and G represent conserved sequences important for self-splicing. Red indicates sequence to be excised (intein); blue and green are the flanking sequences.

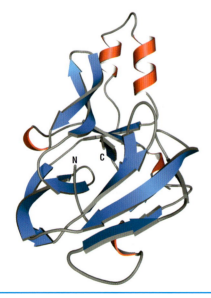

Some proteins contain self-excising inteins

Several genes from prokaryotes and lower eukaryotes contain an in-frame open reading frame encoding an internal domain that is subsequently cleaved out from the protein to form an independent protein known as an **intein**. This process is analogous to the splicing out of introns in messenger RNA, but occurs post-translationally at the level of the protein itself. In protein splicing, internal segments in the translation product (inteins) excise themselves from the protein while ligating the flanking polypeptides (**exteins**) to form the final protein. This process generates two functional proteins from one polypeptide chain, one derived from an internal sequence in the other (Figure 3-35). The novelty of this protein splicing process is the fact that it is self-catalyzed. To date, more than 100 inteins have been discovered, approximately 70% of which reside in proteins involved in DNA replication and repair. Inteins have been found that contain as few as 134 amino acids and as many as 608. Inteins are usually composed of two distinct regions, a protein-splicing domain, which is split into two segments, and an intervening endonuclease domain (Figure 3-36). (In smaller inteins this endonuclease domain may be replaced by a short peptide linker.) The protein-splicing domain catalyzes a series of peptide bond rearrangements that lead to the excision of the entire intein from the protein as well as religation of the flanking exteins. This protein-splicing activity is autoproteolytic and requires no accessory host proteins or cofactors.

As far as is known, inteins are mobile genetic elements with no function other than their own propagation. This function depends on the endonuclease domain. It belongs to a family of enzymes called homing endonucleases. These cleave the DNA sequence corresponding to their own amino-acid sequence from the genome and then mediate its insertion into genes that lack the intein sequence. Why these elements should prefer to exist as inteins in DNA replication and repair proteins in particular is unclear.

Mutant inteins that actually cleave the protein sequence in which they occur, a reaction that does not occur in protein splicing under physiological conditions, are proving useful in protein chemistry and protein engineering. For example, tobacco plants have been genetically engineered to express a mammalian antimicrobial peptide as a fusion protein with a modified intein from vacuolar membrane ATPase. The peptide can then be purified from the plant tissue by taking advantage of the intein-mediated self-cleaving mechanism. Inteins can also be used in the semisynthesis of entire proteins for industrial and medical applications: part of the protein can be synthesized chemically, allowing for the introduction of unnatural amino acids and chemical labels, and part of it can be produced genetically. Intein sequences can then be used to couple these parts together.

The mechanism of autocatalysis is similar for inteins from unicellular organisms and metazoan Hedgehog protein

Both inteins and each of the different types of homing endonucleases are characterized by a few short signature sequence motifs and can be recognized in gene sequences. Certain amino acids are highly conserved within the splicing domain. For instance, the first amino acid on

Figure 3-37 Structure of an intein Ribbon diagram of the polypeptide chain fold of the intein from the gyrase A subunit of *Mycobacterium xenopi*. It contains an unusual beta fold in which the catalytic splice junctions are located at the ends of two adjacent antiparallel beta strands at the amino and carboxyl termini of the intein. (PDB 1am2)

Definitions

intein: a protein intron (intervening sequence). An internal portion of a protein sequence that is post-translationally excised in an autocatalytic reaction while the flanking regions are spliced together, making an additional protein product.

extein: the sequences flanking an **intein** and which are religated after intein excision to form the functional protein.

References

Chevalier, B.S. and Stoddard, B.L.: **Homing endonucleases: structural and functional insight into the catalysts of intron/intein mobility.** *Nucleic Acids Res.* 2001, **29**:3757–3774.

Hall, T.M., *et al.*: **Crystal structure of a Hedgehog autoprocessing domain: homology between Hedgehog and self-splicing proteins.** *Cell* 1997, **91**:85–97.

Jeong, J. and McMahon, A.P.: **Cholesterol modification**

Figure 3-38 Four-step mechanism for protein splicing In step 1, a side-chain oxygen or sulfur (X) of the first intein residue (serine, cysteine or threonine; A in Figure 3-36) attacks the carbonyl group of the peptide bond that it makes with the preceding amino acid. In step 2, this carbonyl, now in the form of an ester or thioester, is attacked by the first residue of the carboxy-terminal extein segment, which is also serine, cysteine or threonine. The last residue of the intein, which is most commonly an asparagine (corresponding to position G in Figure 3-36) then cyclizes internally through its own peptide carbonyl group (step 3), thereby releasing both the intein (with a cyclic asparagine at its carboxyl terminus) and the extein, in which the amino-terminal and carboxy-terminal segments are connected via the side chain of the first residue of the carboxy-terminal segment. The spontaneous rearrangement of this ester or thioester to a normal peptide bond (step 4) completes the splicing process.

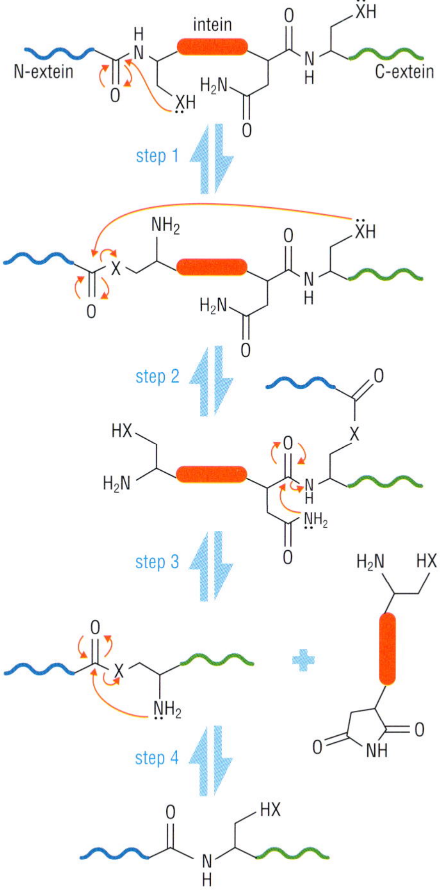

both the amino terminus of the intein domain (indicated by an A in Figure 3-36) and the carboxy-terminal extein fragment is either serine, cysteine or threonine, although an alanine has sometimes been observed. The amino acid on the carboxyl terminus of the intein (indicated by a G in Figure 3-36) is most commonly asparagine, often preceded by histidine. Finally, a ThrXXHis sequence within the amino-terminal splicing domain of the intein (indicated by a B in Figure 3-36) is usually observed. These residues have specific roles in the structure of the intein and the mechanism of protein splicing. The three-dimensional structure of an intein has now been solved for a mycobacterial protein and contains an unusual beta fold with the catalytic splice junctions at the ends of two adjacent beta strands (Figure 3-37). The arrangement of the active-site residues Ser 1 (corresponding to A in Figure 3-36), Thr 72-X-X-His 75 (corresponding to B in Figure 3-36), His 197 and Asn 198 (corresponding to G in Figure 3-36) is consistent with a proposed four-step mechanism for the protein splicing reaction (Figure 3-38).

The reaction carried out by inteins is similar to that performed by the eukaryotic Hedgehog proteins (HH), which play a central part in developmental patterning in vertebrates and invertebrates and are implicated in some important human tumors. Depending on the context, HH signals can promote cell proliferation, prevent programmed cell death, or induce particular cell fates. HH family members can exert their effects not only on cells neighboring the source of the protein signal, but also over considerable distances (up to 30 cell diameters), acting in at least some cases as classic morphogens. Such morphogens are signaling molecules that diffuse from a source to form a concentration gradient over an extended area of the target field and elicit different responses from cells according to their position within the gradient, which reflects the dosage of the ligand they are exposed to. In many cases, the precise biochemical function of the HH protein is unknown.

Like inteins, which autocatalyze their removal from inside other proteins with the concomitant joining of their flanks by a peptide bond, the carboxy-terminal domain of HH proteins autocatalyzes their cleavage from the amino-terminal domain of precursor proteins with a concomitant covalent attachment of a cholesterol molecule to the cleavage point on the amino-terminal domain. This targets the HH protein to the membrane, a process that is essential to its signaling function. The autocatalytic reactions are similar, with the cleaved peptide bond (the amino-terminal one in inteins) first changed to an ester/thioester bond that is cleaved by a nucleophilic attack from the carboxy-terminal flanking sequence in inteins or the cholesterol molecule in HH. Although these two types of self-splicing proteins are found in different types of organisms (single-celled organisms for inteins and metazoans for HH proteins), they have significant sequence similarity and their core three-dimensional structures are very similar (compare Figures 3-37 and 3-39).

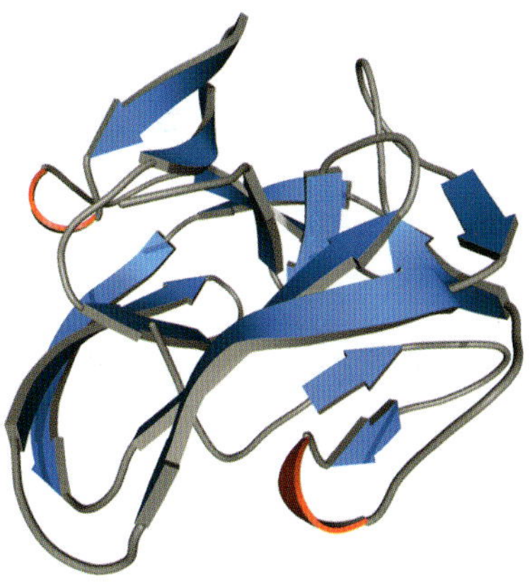

Figure 3-39 Structure of part of the Hedgehog carboxy-terminal autoprocessing domain The carboxy-terminal domain of *Drosophila* Hedgehog protein (HH-C) possesses an autoprocessing activity that results in an intramolecular cleavage of full-length Hedgehog protein and covalent attachment of a cholesterol molecule to the newly generated amino-terminal fragment. The 17 kDa fragment of HH-C (HH-C17) shown here is active in the initiation of autoprocessing. The HH-C17 structure comprises two homologous subdomains that appear to have arisen from tandem duplication of a primordial gene. Both the protein structure and the chemical features of the reaction mechanism are conserved between HH-C17 and the self-splicing regions of inteins (see Figure 3-37). (PDB 1at0)

of Hedgehog family proteins. *J. Clin. Invest.* 2002, **110**:591–596.

Klabunde, T. *et al.*: **Crystal structure of GyrA intein from *Mycobacterium xenopi* reveals structural basis of protein splicing.** *Nat. Struct. Biol.* 1998, **5**:31–36.

Paulus, H.: **Protein splicing and related forms of protein autoprocessing.** *Annu. Rev. Biochem.* 2000, **69**:447–496.

Perler, F.B.: **Protein splicing of inteins and hedgehog**

autoproteolysis: structure, function, and evolution. *Cell* 1998, **92**:1–4.

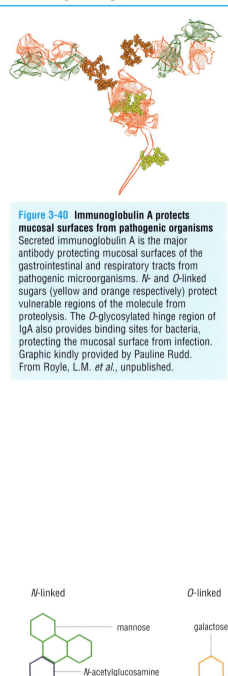

Figure 3-40 Immunoglobulin A protects mucosal surfaces from pathogenic organisms Secreted immunoglobulin A is the major antibody protecting mucosal surfaces of the gastrointestinal and respiratory tracts from pathogenic microorganisms. *N*- and *O*-linked sugars (yellow and orange respectively) protect vulnerable regions of the molecule from proteolysis. The *O*-glycosylated hinge region of IgA also provides binding sites for bacteria, protecting the mucosal surface from infection. Graphic kindly provided by Pauline Rudd. From Royle, L.M. *et al.*, unpublished.

N-linked

O-linked

mannose

galactose

N-acetylglucosamine

N-acetylgalactosamine

HN

O

CH₃ or H

O

Asn

Ser or Thr

Glycosylation can change the properties of a protein and provide recognition sites

Perhaps the most complex and diverse post-translational modifications to proteins are those that attach carbohydrate chains (oligosaccharides) to asparagine, serine and threonine residues on protein surfaces in the process known as **glycosylation**. Almost all secreted and membrane-associated proteins of eukaryotic cells are glycosylated, with oligosaccharides attached to the polypeptide chain at one or more positions by either *N*- or *O*-glycosidic bonds. There are two major types of functions for sugars attached to proteins. First, specific oligosaccharides provide recognition sites that tag glycoproteins for recognition by other proteins both inside and outside the cell and thus direct their participation in various biological processes. One important example of this is the targeting of leukocytes to sites of inflammation. Second, since oligosaccharides can be as large as the protein domains to which they are attached, they shield large areas of the protein surface, providing protection from proteases (Figure 3-40) and nonspecific protein–protein interactions. Inside the cell, oligosaccharides increase the solubility of nascent glycoproteins and prevent their aggregation.

The nature of the modification on any given protein varies from species to species and often from one tissue to the next. Glycoproteins exist as populations of glycoforms in which the same amino-acid sequence is modified by a glycosidic core with a range of different derivatives at each glycosylation site. Protein glycosylation is almost exclusively a eukaryotic property and the complexity of the modification increases as one proceeds up the evolutionary tree. Simple eukaryotes like yeast attach only a simple set of sugars; mammals modify their proteins with highly branched oligosaccharides composed of a wider range of carbohydrates. The commonest modifications are shown in Figure 3-41. The most common *N*-linked attachment consists of two *N*-acetylglucosamines with three mannoses attached; the most common *O*-linked attachment involves an *N*-acetylgalactosamine, beta-galactose core attached at serine or threonine.

For *N*-linked sugars, the core in most eukaryotes is the same, and derives from a larger preformed precursor, Glc3Man9GlcNAc2. As the protein is being synthesized in the endoplasmic reticulum (ER) the precursor is transferred to asparagine residues (N) in the glycosylation signal sequence NXS (Figure 3-42). To produce the mature oligosaccharides, the *N*-glycosylation core is first trimmed in the ER by glucosidases and mannosidases, and additional processing of the oligomannose structures can then occur in the Golgi complex (Figure 3-42), giving rise to diverse complex sugars.

In contrast to *N*-linked sugars, all of which have the same core structure, *O*-linked sugars can have at least eight different core structures, attached through *N*-acetylgalactosamine to the side chains of serine or threonine residues; these are extended by the addition of monosaccharide units.

Some viruses and bacteria use carbohydrates on cell-surface glycoproteins as Trojan horses to gain entry into the cell: they have proteins that bind specifically to these oligosaccharides, a process that leads to internalization. Antibiotics that change the structure of these carbohydrates prevent this binding, and a number of antibiotics act by blocking enzymes in glycoprotein biosynthesis, including 1-deoxymannojirimycin, which inhibits mannosidases in both the Golgi complex and the ER.

Figure 3-41 Schematic representation of the core *N*-linked oligosaccharide and a representative *O*-linked core oligosaccharide

Definitions

glycosylation: the post-translational covalent addition of sugar molecules to asparagine, serine or threonine residues on a protein molecule. Glycosylation can add a single sugar or a chain of sugars at any given site and is usually enzymatically catalyzed.

References

Bruckner, K. *et al.*: **Glycosyltransferase activity of Fringe modulates Notch-Delta interactions.** *Nature* 2000, **406**:411–415.

Imperiali, B. and O'Connor, S.E.: **Effect of *N*-linked glycosylation on glycopeptide and glycoprotein structure.** *Curr. Opin. Chem. Biol.* 1999, **3**:643–649.

Parekh, R.B. and Rohlff, C.: **Post-translational modification of proteins and the discovery of new medicine.** *Curr. Opin. Biotechnol.* 1997, **8**:718–723.

Petrescu, T. *et al.*: **The solution NMR structure of glycosylated *N*-glycans involved in the early stages of glycoprotein biosynthesis and folding.** *EMBO J.* 1997, **16**:4302–4310.

Rudd, P.M. and Dwek, R.A.: **Glycosylation: heterogeneity and the 3D structure of proteins.** *Crit. Rev. Biochem. Mol. Biol.* 1997, **32**:1–100.

Wells, L. *et al.*: **Glycosylation of nucleoplasmic proteins: signal transduction and *O*-GlcNac.** *Science* 2001, **291**:2376–2378.

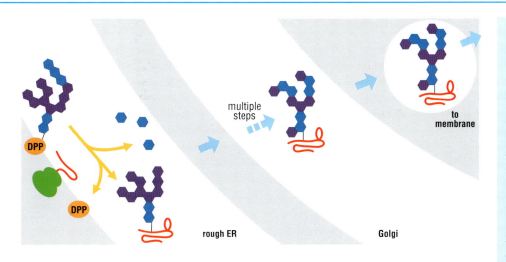

Figure 3-42 Oligosaccharide processing
Schematic pathway of oligosaccharide processing on newly synthesized glycoproteins. The oligosaccharide is first transferred from its membrane-bound dolichol pyrophosphate (DPP) carrier to a polypeptide chain while the latter is still being synthesized on the ribosome. As the polypeptide chain grows within the endoplasmic reticulum (ER), monosaccharides may be cleaved from the nonreducing ends of the oligosaccharide. After completion of the synthesis of the polypeptide chain, the immature glycoprotein is carried in a transport vesicle to the Golgi apparatus where further modification occurs. New monosaccharides may be added and others removed in a multi-step process involving many different enzymes. The completed glycoprotein is finally transported to its ultimate destination.

The importance of the proper glycosylation of glycoproteins can be seen in mice lacking *N*-acetylglucosaminyltransferase I (the initial enzyme in the processing of hybrid or complex *N*-linked sugars), which die at mid-gestation, and in patients with congenital deficiencies in glycosylation, who have many pathologies ranging from neurological disorders to motor and ocular problems. Oligosaccharides attached to proteins have many roles. They may modify the activity of an enzyme, mediate the proper folding and assembly of a protein complex or facilitate the correct localization of glycoproteins; and they can be immunogenic. Glycosylation serves as an identity badge for proteins, especially those on the cell surface, indicating what type of cell or organism they come from, distinguishing mammals from their pathogens. So characteristic is this badge that many viruses bind to specific oligosaccharides as a mechanism for targeting the type of cell they infect. Because asparagine, serine and threonine-linked carbohydrates are nearly always on the outside of a folded protein they contribute to protein–protein recognition in many processes, including cell adhesion during development, immune surveillance, inflammatory reactions, and the metastasis of cancer cells. Yet there are many examples of glycoproteins that can be enzymatically stripped of their attached sugars with no effect on their biochemical function *in vitro*; in other cases, mutational loss of the oligosaccharide attachment sites has no apparent cellular consequence, and many eukaryotic proteins can be expressed in bacteria, which perform no glycosylation, and yet retain full activity *in vitro*. While this may suggest that not all glycosylations are important, it should be borne in mind that a protein enters into many interactions during its lifetime and usually only one functional assay is used to determine whether or not the sugars are important. One physical property that is dramatically affected by glycosylation is solubility: many glycoproteins become very insoluble when their sugar residues are cleaved off.

The structures of some of the most important oligosaccharide complexes have been determined by X-ray diffraction or, more commonly, by NMR. For instance, the structure of Glc3Man9GlcNac2, the precursor oligosaccharide in protein glycosylation, is shown in Figure 3-43. No obvious rules for the structures of these attached carbohydrates have yet been deduced, and in many glycoproteins whose structures have been determined, the oligosaccharides are disordered.

Another, simpler modification deserves special mention because its function has recently been elucidated. This modification is the direct enzymatic attachment of a single *N*-acetyl-glucosamine residue to a serine or threonine residue after the protein has already folded. This glycosylation is reversible by enzymatic cleavage. The sites of monoglycosylation are often identical to the sites of phosphorylation, but the effect on activity of the modified protein is usually completely different. Thus, monoglycosylation appears to transiently block phosphorylation, and is likely to be important in regulating that process. It is important to recognize that mutations that abolish phosphorylation sites may often also remove a site of monoglycosylation, and if the effects of these modifications are different then the observed phenotype will result from blocking both of them. In many breast cancer cells, monoglycosylation levels are significantly reduced owing to an increase in the synthesis of the cleaving enzyme. Less monoglycosylation would lead to more protein phosphorylation, which is also observed in most tumors and is correlated with uncontrolled growth.

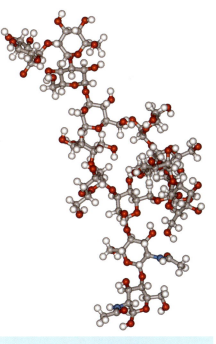

Figure 3-43 The structure of Glc3Man9GlcNac2 Average solution structure determined by NMR of the oligosaccharide involved in the early stage of glycoprotein biosynthesis. This is the 14-mer first transferred from dolichol pyrophosphate to the nascent polypeptide chain. Graphic kindly provided by Mark R. Wormald.

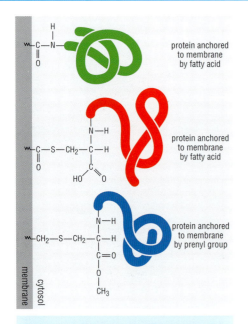

Figure 3-44 Membrane targeting by lipidation Covalent attachment of a lipid molecule to either the amino-terminal or carboxy-terminal end of a protein provides a hydrophobic anchor that localizes the protein to a membrane. The linkage is an amide (top), a thioester (middle) or a thioether (bottom) bond.

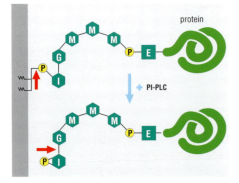

Figure 3-45 Glycosylphosphatidylinositol anchoring Schematic diagram of the reversible modification of a protein by a GPI anchor. The anchor consists of an oligosaccharide chain. The protein is connected through an amide linkage to a phosphoethanolamine molecule (E-P) that is attached to a core tetrasaccharide composed of three mannose sugars (M) and a single glucosamine sugar (G). The tetrasaccharide is in turn attached to phosphatidylinositol (I). It is the fatty-acid residues of this phosphatidylinositol group that anchor the protein to the membrane. These fatty acids may vary somewhat with the protein being modified, providing membrane specificity.

Covalent attachment of lipids targets proteins to membranes and other proteins

In signal transduction pathways and the control of protein traffic through intracellular membranes there are often many variants of one type of protein, each with a similar biochemical function. For example, more than 50 mammalian Rab proteins are known, and each of these small GTPases plays a part in directing the vesicular traffic that transports proteins between membrane-bounded compartments of the cell. To prevent chaotic cross-talk from their similar activities, each of these proteins is targeted to specific intracellular membranes at specific times in the cell cycle and during processes such as metabolism and secretion, as the need for switching functions in different compartments of the cell changes. Membrane targeting is mediated by the covalent attachment of a specific lipid group to an end of the polypeptide chain.

Lipid attachment is one of the most common post-translational modifications in eukaryotic cells. The process is sequence-specific, always involves residues either at or near either the carboxyl terminus or the amino terminus of the protein, and may require several enzymatic steps. There are four broad types of lipid modifications with distinct functional properties, classified according to the identity of the attached lipid.

These are: **myristoylation**, in which a 14-carbon fatty-acid chain is attached via a stable amide linkage to an amino-terminal glycine residue (Figure 3-44, top); **palmitoylation**, in which a 16-carbon fatty-acid chain is attached via a labile thioester linkage to cysteine residues (other fatty-acid chains can sometimes substitute for the palmitoyl group, so this modification is often called **S-acylation**) (Figure 3-44, middle); **prenylation**, in which a prenyl group (either a farnesyl or geranylgeranyl group) is attached via a thioether linkage to a cysteine residue initially four positions from the carboxyl terminus that becomes carboxy-terminal after proteolytic trimming and methylation of the new carboxyl terminus (Figure 3-44, bottom); and modification by a **glycosylphosphatidylinositol (GPI) anchor**, which is attached through a carbohydrate moiety (Figure 3-45).

N-myristoylated proteins include select alpha subunits of heterotrimeric G proteins, a number of non-receptor tyrosine kinases, and a few monomeric G proteins, among others. The myristoyl moiety is attached to the protein co-translationally. S-acylated proteins include most alpha subunits of heterotrimeric G proteins, members of the Ras superfamily of monomeric G proteins, and a number of G-protein-coupled receptors, among others. S-acylation is post-translational and reversible, a property that allows the cell to control the modification state, and hence localization and biological activity, of the lipidated protein. Lymphoma proprotein convertase, an enzyme that activates proproteins in the secretory pathway, occurs in both the palmitoylated and unmodified form. The palmitoylated form is degraded much faster than the unmodified one, allowing the lifetime of this enzyme to be regulated by **lipidation**. There is some overlap between N-myristoylated and S-acylated proteins such that many contain both lipid modifications. Such dual modification can have important consequences, most notably targeting of the dually modified species to distinct membrane subdomains termed lipid rafts or to caveolae; this provides a way of sublocalizing proteins to microdomains that also contain specific protein–protein interaction partners.

S-prenylation occurs in two classes, those with a single prenyl group on a cysteine residue at or near the carboxyl terminus, and those that are modified on two cysteine residues at or near the carboxyl terminus. In both cases, attachment is stable and post-translational. The difference between the two types of attachment is based on the recognition motifs recognized by the

Definitions

glycosylphosphatidylinositol anchor: a complex structure involving both lipids and carbohydrate molecules that is reversibly attached to some proteins to target them to the cell membrane.

lipidation: covalent attachment of a fatty-acid group to a protein.

myristoylation: irreversible attachment of a myristoyl group to a protein via an amide linkage.

palmitoylation: reversible attachment of a palmitoyl group to a protein via a thioester linkage.

prenylation: irreversible attachment of either a farnesyl or geranylgeranyl group to a protein via a thioether linkage.

S-acylation: reversible attachment of a fatty-acid group to a protein via a thioester linkage; **palmitoylation** is an example of S-acylation.

prenylating enzyme. One important protein whose activities are controlled by prenylation is the small GTPase Ras (see section 3-7). A number of *S*-prenylated proteins are also subject to *S*-acylation at a nearby cysteine residue. This type of dual modification does not apparently target the protein to the same type of membrane subdomain as dual acylation does.

A number of mammalian proteins are modified by attachment of GPI to the amino terminus. This anchor is an alternative to the lipid tails described above. GPI-anchored proteins participate in such processes as nutrient uptake, cell adhesion and membrane signaling events. All GPI-anchored proteins are destined for the cell surface via transport through the secretory pathway, where they acquire the preassembled GPI moiety. GPI modification is reversible, as the anchored protein can be released from the membrane by the action of phospholipases. In some cases, this enzyme-catalyzed release can activate an enzyme. A number of human parasites have GPI-anchored enzymes on their cell surfaces that are inactive in this state but become activated when the parasite encounters host phospholipases. By this mechanism the parasite can sense and respond to the host environment.

The enzymes involved in lipid attachment and hydrolysis are potential targets for anti-cancer, anti-fungal and other types of drugs. Several genetic diseases have been identified that involve protein lipidation. Mutations in genes encoding proteins essential for the geranyl-geranyl modification of Rab proteins are responsible for the diseases choroideremia (an incurable X-linked progressive retinal degeneration leading to blindness) and Hermansky-Pudlak syndrome (a rare disorder characterized by oculocutaneous albinism, a bleeding tendency and eventual death from pulmonary fibrosis). These diseases result from a failure to localize Rab properly.

The GTPases that direct intracellular membrane traffic are reversibly associated with internal membranes of the cell

We have already mentioned the Rab GTPases that direct transport vesicles between the membrane-bounded compartments of the cell. Two other small GTPases—Ser1 and ADP ribosylation factor (ARF)—also play an essential part in the control of vesicular traffic by recruiting the specialized coat proteins that are required for vesicle budding from the donor membrane. These small GTPases associate reversibly with their target membranes through the exposure of a covalently attached myristoyl group which occurs on exchange of GDP for GTP. We can illustrate the mechanism of vesicle formation with the ARF protein, which participates in the recruitment of vesicle coat proteins to the membrane of the Golgi complex.

ARF is myristoylated at its amino terminus and when GDP is bound this hydrophobic tail is sequestered within the protein, which therefore exists in soluble form in the cytoplasm. ARF is itself recruited to the Golgi membrane by a GTP-exchange protein, and on binding GTP, ARF undergoes a conformational change that releases the amino-terminal tail so that the protein becomes anchored to its target membrane (Figure 3-46). It is then thought to recruit preassembled coatomers—complexes of coat proteins—to the membrane, causing the coated membrane to bud off from the Golgi complex, capturing specific membrane and soluble proteins for delivery to another membrane-bounded compartment. One of the complex of coat proteins is thought to serve as a GTPase-activating protein, causing ARF to hydrolyze GTP back to GDP, withdraw its tail from the vesicle membrane, and return to the cytoplasm, with the dispersal of the coat components of the vesicle.

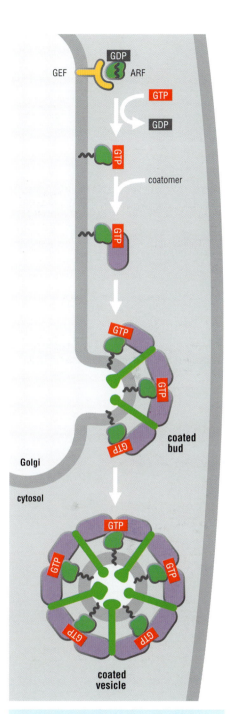

Figure 3-46 Working model for vesicular transport between Golgi compartments A key role in the process is played by the small GTPase, ARF. In its GDP-bound form, ARF is a soluble protein with a hydrophobic myristoylated tail that is tucked into the protein. When ARF is recruited to the Golgi membrane by a GTP-exchange factor, GDP is exchanged for GTP and the amino-terminal region of ARF rearranges, releasing the tail which targets ARF to the membrane. Membrane-bound ARF then recruits coat proteins necessary for vesicle budding and transport, trapping membrane and soluble proteins (not shown) in the vesicle as it pinches off.

References

Butikofer, P. *et al.*: **GPI-anchored proteins: now you see 'em, now you don't.** *FASEB J.* 2001, **15**:545–548.

Casey, P.J.: **Lipid modifications** in *Encyclopedic Reference of Molecular Pharmacology.* Offermanns, S. and Rosenthal, W. eds (Springer-Verlag, Heidelberg, 2003) in the press.

Casey, P.J.: **Protein lipidation in cell signaling.** *Science* 1995, **268**:221–225.

Chatterjee, S. and Mayor, S.: **The GPI-anchor and protein sorting.** *Cell. Mol. Life Sci.* 2001, **58**:1969–1987.

Park, H.W. and Beese, L.S.: **Protein farnesyltransferase.** *Curr. Opin. Struct. Biol.* 1997, **7**:873–880.

Pereira-Leal, J.B. *et al.*: **Prenylation of Rab GTPases: molecular mechanisms and involvement in genetic disease.** *FEBS Lett.* 2001, **498**:197–200.

Spiro, R.G.: **Protein glycosylation: nature, distribution, enzymatic formation, and disease implications of glycopeptide bonds.** *Glycobiology* 2002, **12**:43R–56R.

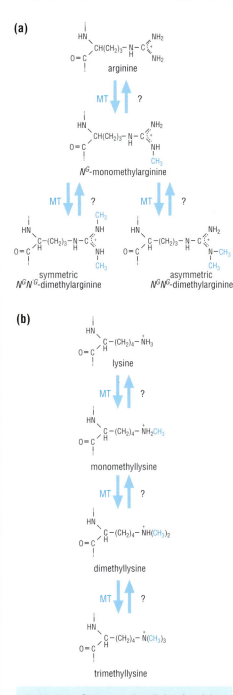

(a)

arginine

MT ↓↑ ?

N^G-monomethylarginine

MT ↓↑ ?   MT ↓↑ ?

symmetric
$N^G N^{'G}$-dimethylarginine

asymmetric
$N^G N^G$-dimethylarginine

(b)

lysine

MT ↓↑ ?

monomethyllysine

MT ↓↑ ?

dimethyllysine

MT ↓↑ ?

trimethyllysine

Figure 3-47 Structures of methylated arginine and lysine residues (a) Arginine may be mono- or dimethylated, either symmetrically or asymmetrically; **(b)** lysine may be mono-, di- or trimethylated. MT, methyltransferase.

Fundamental biological processes can also be regulated by other post-translational modifications of proteins

Phosphorylation, glycosylation, lipidation, and limited proteolysis are the commonest post-translational covalent modifications of proteins. However, important regulatory functions are also performed by methylation, *N*-acetylation, attachment of SUMO and nitrosylation.

Methylation occurs at arginine or lysine residues and is a particularly common modification of proteins in the nuclei of eukaryotic cells. Methylation of eukaryotic proteins is performed by a variety of methyltransferases, which use *S*-adenosylmethionine as the methyl donor. Three main forms of methylarginine have been found in eukaryotes: N^G-monomethylarginine, N^G,N^G-asymmetric dimethylarginine, and $N^G,N^{'G}$-symmetric dimethylarginine (Figure 3-47a); lysine may be mono-, di-, or trimethylated (Figure 3-47b). In contrast to phosphorylation, methylation appears to be irreversible: methylated lysine and arginine groups are chemically stable, and no demethylases have been found in eukaryotic cells. Thus regulation of this modification must occur through regulation of methyltransferase activity, and removal of the methylated proteins themselves. Arginine methylation most commonly occurs at an RGG sequence, and less often at other sites such as RXR and GRG. Although methylation does not change the overall charge on an arginine residue, it greatly alters the steric interactions this group can make and eliminates possible hydrogen-bond donors. It is not surprising, therefore, that methylation has been shown to alter protein–protein interactions. For example, asymmetric methylation of Sam68, a component of signaling pathways that recognizes proline-rich domains, decreases its binding to SH3-domain-containing but not to WW-domain-containing proteins. Arginine methylation is particularly prevalent in heterogeneous nuclear ribonucleoproteins (hnRNPs), which have roles in pre-mRNA processing and nucleocytoplasmic RNA transport. A second centrally important group of nuclear proteins to undergo methylation are the histones that package chromosomal DNA in the DNA–protein complex known as **chromatin**. Histone methylation on lysine by histone methyltransferases changes the functional state of chromatin in the region of the modification, with important effects on gene expression and DNA replication and repair. Some of these are thought to be due to effects on the compaction of the chromatin, but it is clear that others depend on the recruitment to the DNA of "silencing" proteins that recognize specific modifications to specific lysines through chromodomains characteristic of these proteins (see Figure 3-2) and suppress gene expression. Histones can also undergo phosphorylation, ubiquitination and acetylation (see below), which also affect the functions directed by chromosomal DNA.

***N*-acetylation**, which is catalyzed by one or more sequence-specific *N*-acetyltransferases, usually modifies the amino terminus of the protein backbone with an acetyl group derived from acetyl-CoA. It has been estimated that more than one-third of all yeast proteins may be so modified. This modification has a number of roles, including blocking the action of aminopeptidases and otherwise altering the lifetime of a protein in the cell. *N*-acetylation of

lysine   acetylation by HATs →
← deacetylation by HDs

Figure 3-48 *N*-acetylation Acetylation by histone acetyltransferase (HAT) of amino-terminal lysine residues is an important regulatory modification of histone proteins. Deacetylation is catalyzed by histone deacetylases (HDs).

Definitions

chromatin: the complex of DNA and protein that comprises eukaryotic nuclear chromosomes. The DNA is wound around the outside of highly conserved histone proteins, and decorated with other DNA-binding proteins.

methylation: modification, usually of a nitrogen or oxygen atom of an amino-acid side chain, by addition of a methyl group. Some bases on DNA and RNA can also be methylated.

***N*-acetylation:** covalent addition of an acetyl group from acetyl-CoA to a nitrogen atom at either the amino terminus of a polypeptide chain or in a lysine side chain. The reaction is catalyzed by *N*-acetyltransferase.

nitrosylation: modification of the –SH group of a cysteine residue by addition of nitric oxide (NO) produced by nitric oxide synthase.

sumoylation: modification of the side chain of a lysine residue by addition of a small ubiquitin-like protein (SUMO). The covalent attachment is an amide bond between the carboxy-terminal carboxylate of SUMO and

the NH2 on the lysine side chain of the targeted protein.

References

Hochstrasser, M.: **SP-RING for SUMO: New functions bloom for a ubiquitin-like protein.** *Cell* 2001, **107**:5–8.

Jenuwein, T. and Allis, C.D.: **Translating the histone code.** *Science* 2001, **293**:1074–1080.

Kouzarides, T.: **Histone methylation in transcriptional control.** *Curr. Opin. Genet. Dev.* 2002, **12**:198–209.

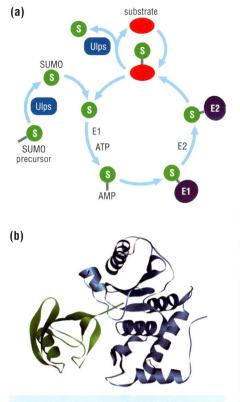

the amino terminus is usually irreversible, but the epsilon amino group on the side chain of lysine residues can also be acetylated by other, specific acetyltransferases, and this reaction can be reversed by deacetylases (Figure 3-48). In contrast to methylation, which maintains the charge on the amino group, acetylation does not. Reversible *N*-acetylation at lysine in histone proteins has a major role in the control of gene expression and other chromosomal functions: while histone methylation may induce either an active or an inactive state of chromatin, depending on the position and the nature of the methyl group, histone acetylation is always associated with an active state of chromatin: this is promoted by chromatin-remodeling enzymes recruited to the DNA by proteins containing bromodomains (see Figure 3-2) that specifically recognize acetylated lysines. Thus, in contrast to methylation, which regulates chromatin by creating non-binding surfaces for regulatory proteins on core histones, acetylation may influence genome function in part through affecting higher-order protein structure. Deregulation of chromatin modification pathways is widely observed in cancer.

Covalent attachment of one protein to another is a very common post-translational modification. In addition to ubiquitination (see section 3-11), another such modification is the attachment of the ubiquitin-like protein SUMO (small ubiquitin-related modifier), called **sumoylation** (Figure 3-49). The consensus sequence for sumoylation is ψKXE (where ψ is a hydrophobic amino acid and X is any amino acid); this is in marked contrast to ubiquitination, where no consensus sequence has ever been found. In yeast, the gene coding for the sole SUMO-like protein, Smt3, is essential for progression through the cell cycle. Septin, a GTP-binding protein that is essential for cell separation, is sumoylated during the G2/M phase of the cell cycle. Like ubiquitin, SUMO is attached to the amino group of lysine residues by specific SUMO-activating and -conjugating enzymes. Attachment of SUMO to proteins has been shown to change their subcellular localization, transcriptional activity and stability: for example, the SUMO conjugate of the RanGAP1 protein binds preferentially to the nuclear pore complex and so sumoylation appears to localize this protein. Other functions of SUMO attachment are uncertain at present.

Nitrosylation is one of only two post-translational modifications conserved throughout evolution—phosphorylation being the other. Yet it has been much less studied, in part because it has been difficult to understand how specificity of action is achieved for the reversible modification of proteins by NO groups. In general, NO modifies the –SH moiety of cysteine residues (Figure 3-50) and reacts with transition metals in enzyme active sites. Over 100 proteins are known to be regulated in this way. The majority of these are regulated by reversible *S*-nitrosylation of a single critical cysteine residue flanked by an acidic and a basic amino acid or by a cysteine in a hydrophobic environment. Cysteine residues are important for metal coordination, catalysis and protein structure by forming disulfide bonds. Cysteine residues can also be involved in modulation of protein activity and signaling events via other reactions of their –SH groups. These reactions can take several forms, such as redox events (chemical reduction or oxidation), chelation of transition metals, or *S*-nitrosylation. In several cases, these reactions can compete with one another for the same thiol group on a single cysteine residue, forming a molecular switch composed of redox, NO or metal ion modifications to control protein function. For example, the JAK/STAT signaling pathway is regulated by NO at multiple loci. NO is a diffusible gas, which modifies reactive groups in the vicinity of its production, which is catalyzed by the enzyme nitric oxide synthase. Thus, the localization of this enzyme is a key factor in which proteins will be susceptible to modification by NO. The mechanism by which this modification is reversed is unknown in many cases; some redox enzymes have been implicated; free glutathione may also be involved.

Figure 3-49 Sumoylation (a) In the SUMO cycle, a SUMO precursor is processed to SUMO by Ulp proteins. SUMO is then derivatized with AMP by E1, the SUMO-activating enzyme, before transfer of SUMO to a cysteine of E1 to form an E1–SUMO thioester intermediate. SUMO is passed to the SUMO-conjugating enzyme, E2, to form an E2–SUMO thioester intermediate. This latter complex is the proximal donor of SUMO to a substrate lysine in the ψKXE target sequence in the final substrate protein. SUMO can also be cleaved from sumoylated proteins by Ulp proteins. **(b)** The structure of the complex of the SUMO-binding domain of a SUMO-cleaving enzyme, Ulp1, with the yeast SUMO protein Smt3. SUMO is the small domain on the left. The size of the active-site cleft of Ulp1 allows even large SUMO–protein conjugates to bind and be cleaved.

McBride, A.E. and Silver, P.A.: **State of the Arg: protein methylation at arginine comes of age.** *Cell* 2001, **106**:5–8.

Stamler, J.S. *et al.*: **Nitrosylation: the prototypic redox-based signaling mechanism.** *Cell* 2001, **106**:675–683.

Turner, M.: **Cellular memory and the histone code.** *Cell* 2002, **111**:285–291.

Workman, J.L. and Kingston, R.E.: **Alteration of nucleosome structure as a mechanism of transcriptional regulation.** *Annu. Rev. Biochem.* 1998, **67**:545–579.

Web resource on nitrosylation:

http://www.cell.com/cgi/content/full/106/6/675/DC1

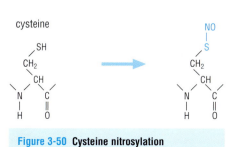

Figure 3-50 Cysteine nitrosylation

From Sequence to Function:
Case Studies in Structural and Functional Genomics

One of the main challenges facing biology is to assign biochemical and cellular functions to the thousands of hitherto uncharacterized gene products discovered by genome sequencing. This chapter discusses the strengths and limitations of the many experimental and computational methods, including those that use the vast amount of sequence information now available, to help determine protein structure and function. The chapter ends with two individual case studies that illustrate these methods in action, and show both their capabilities and the approaches that still must be developed to allow us to proceed from sequence to consequence.

Genomics is making an increasing contribution to the study of protein structure and function

The relatively new discipline of **genomics** has great implications for the study of protein structure and function. The genome-sequencing programs are providing more amino-acid sequences of proteins of unknown function to analyze than ever before, and many computational and experimental tools are now available for comparing these sequences with those of proteins of known structure and function to search for clues to their roles in the cell or organism. Also underway are systematic efforts aimed at providing the three-dimensional structures, subcellular locations, interacting partners, and deletion phenotypes for all the gene products in several model organisms. These databases can also be searched for insights into the functions of these proteins and their corresponding proteins in other organisms.

Sequence and structural comparison can usually give only limited information, however, and comprehensively characterizing the function of an uncharacterized protein in a cell or organism will always require additional experimental investigations on the purified protein *in vitro* as well as cell biological and mutational studies *in vivo*. Different experimental methods are required to define a protein's function precisely at biochemical, cellular, and organismal levels in order to characterize it completely, as shown in Figure 4-1.

In this chapter we first look at methods of comparing amino-acid sequences to determine their similarity and to search for related sequences in the sequence databases. Sequence comparison alone gives only limited information at present, and in most cases, other experimental and structural information is also important for indicating possible biochemical function and mechanism of action. We next provide a summary of some of the genome-driven experimental tools for probing function. We then describe computational methods that are being developed to deduce the protein fold of an uncharacterized protein from its sequence. The existence of large families of structurally related proteins with similar functions, at least at the biochemical level, is enabling sequence and structural motifs characteristic of various functions to be identified. Protein structures can also be screened for possible ligand-binding sites and catalytic active sites by both computational and experimental methods.

As we see next, predicting a protein's function from its structure alone is complicated by the fact that evolution has produced proteins with almost identical structures but different functions, proteins with quite different structures but the same function, and even multifunctional proteins which have more than one biochemical function and numerous cellular and physiological functions. We shall also see that some proteins can adopt more than one stable protein fold, a change which can sometimes lead to disease.

The chapter ends with two case histories illustrating how a range of different approaches were combined to determine aspects of the functions of two uncharacterized proteins from the genome sequences of *E. coli* and yeast, respectively.

Figure 4-1 Time and distance scales in functional genomics The various levels of function of proteins encompass an enormous range of time (scale on the left) and distance (scale on the right). Depending on the time and distance regime involved, different experimental approaches are required to probe function. Since many genes code for proteins that act in processes that cross multiple levels on this diagram (for example, a protein kinase may catalyze tyrosine phosphorylation at typical enzyme rates, but may also be required for cell division in embryonic development), no single experimental technique is adequate to dissect all their roles. In the age of genomics, interdisciplinary approaches are essential to determine the functions of gene products.

Definitions

genomics: the study of the DNA sequence and gene content of whole genomes.

Time		Process	Example System	Example Detection Methods	Distance
10^{-15} sec		electron transfer	photosynthetic reaction center	optical spectroscopy	1 Å
10^{-9} sec		proton transfer	triosephosphate isomerase	fast kinetics	
10^{-6} sec		fastest enzyme reactions	catalase, fumarase, carbonic anhydrase	kinetics	2–10 Å
10^{-3} sec		typical enzyme reactions	trypsin, protein kinase A, ketosteroid isomerase	kinetics, time-resolved X-ray, nuclear magnetic resonance	
sec		slow enzyme reactions/cycles	cytochrome P450, phosphofructokinase	kinetics, low T X-ray, nuclear magnetic resonance, mass spectroscopy	Å – nm
min/ hour		protein synthesis/ cell division	budding yeast cell	light microscopy, genetics, optical probes	nm – μm
day/ year		embryonic development	mouse embryo	genetics, microscopy, microarray analysis	μm – m

References

Brazhnik, P. et al.: **Gene networks: how to put the function in genomics.** Trends Biotechnol. 2002, **20**:467–472.

Chan, T.-F. et al.: **A chemical genomics approach toward understanding the global functions of the target of rapamycin protein (TOR).** Proc. Natl Acad. Sci. USA 2000, **97**:13227–13232.

Guttmacher A.E. and Collins, F.S.: **Genomic medicine— a primer.** N. Engl. J. Med. 2002, **347**:1512–1520.

Houry, W.A. et al.: **Identification of in vivo substrates of the chaperonin GroEL.** Nature 1999, **402**:147–154.

Koonin E.V. et al.: **The structure of the protein universe and genome evolution.** Nature 2002, **420**:218–223.

O'Donovan, C. et al.: **The human proteomics initiative (HPI).** Trends Biotechnol. 2001, **19**:178–181.

Oliver S.G.: **Functional genomics: lessons from yeast.** Philos. Trans. R. Soc. Lond. B. Biol. Sci. 2002, **357**:17–23.

Quevillon-Cheruel, S. et al.: **A structural genomics initiative on yeast proteins.** J. Synchrotron. Radiat. 2003, **10**:4–8.

Tefferi, A. et al.: **Primer on medical genomics parts I–IV.** Mayo. Clin. Proc. 2002, **77**:927–940.

Tong, A.H. et al.: **Systematic genetic analysis with ordered arrays of yeast deletion mutants.** Science 2001, **294**:2364–2368.

von Mering, C. et al.: **Comparative assessment of large-scale data sets of protein–protein interactions.** Nature 2002, **417**:399–403.

4-1 Sequence Alignment and Comparison

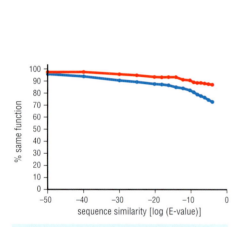

```
S.c. Kss1 INNQNSGFSTLSDDHVQYFTYQILRALKSIHSAQVI
H.s. Erk2 LKTQH-----LSNDHICYFLYQILRGLKYIHSANVL

          HRDIKPSNLLLNSNCDLKVCDFGLARCLASSSDSRET
          HRDLKPSNLLLNTTCDLKICDFGLARVA----DPDHD
```

Figure 4-2 Pairwise alignment Part of an alignment of the amino-acid sequences of the kinase domains from two ERK-like kinases of the MAP kinase superfamily, Erk2 from humans and Kss1 from yeast. The region shown covers the kinase catalytic loop and part of the activation loop (see Figure 3-24). Identical residues highlighted in purple show the extensive similarity between these two homologous kinases (their evolutionary relationship can be seen in Figure 4-5). To maximize similarity, a small number of gaps have had to be inserted in the human sequence.

Figure 4-3 Plot of percentage of protein pairs having the same biochemical function as sequence changes When a series of sequences of homologous proteins are compared, it is observed that as sequence similarity (measured by the E-value from a sequence comparison) decreases, the probability that homologs will have the same function also decreases. The red curve corresponds to single-domain proteins, the blue curve to multidomain proteins. Up to an E-value of approximately 10^{-10}, the likelihood of an identical function is reasonably high, but then it starts to decrease substantially, especially for multidomain proteins.

Sequence comparison provides a measure of the relationship between genes

The comparison of one nucleotide or amino-acid sequence with another to find the degree of similarity between them is a key technique in present-day biology. A marked similarity between two gene or protein sequences may reflect the fact that they are derived by evolution from the same ancestral sequence. Sequences related in this way are called **homologous** and the evolutionary similarity between them is known as **homology**. Unknown genes from newly sequenced genomes can often be identified by searching for similar sequences in databases of known gene and protein sequences using computer programs such as BLAST and FASTA. Sequences of the same protein from different species can also be compared in order to deduce evolutionary relationships. Two genes that have evolved fairly recently from a common ancestral gene will still be relatively similar in sequence to each other; those that have a more distant common ancestor will have accumulated many more mutations, and their evolutionary relationship will be less immediately obvious, or even impossible to deduce from sequence alone.

Alignment is the first step in determining whether two sequences are similar to each other

A key step in comparing two sequences is to match them up to each other in an **alignment** that shows up any similarity that is present. Alignments work on the general principle that two homologous sequences derived from the same ancestral sequence will have at least some identical residues at the corresponding positions in the sequence; if corresponding positions in the sequence are aligned, the degree of matching should be statistically significant compared with that of two randomly chosen unrelated sequences.

As a quantitative measure of similarity, a pairwise alignment is given a score, which reflects the degree of matching. In the simplest case, where only identical matched residues are counted, the fraction of identical amino acids or nucleotides gives a similarity measure known as **percent identity**. When protein sequences are being compared, more sophisticated methods of assessing similarity can be used. Some amino acids are more similar to each other in their physical-chemical properties, and consequently will be more likely to be substituted for each other during evolution (see section 1-1). Most of the commonly used alignment programs give each aligned pair of amino acids a score based on the likelihood of that particular match occurring. These scores are usually obtained from reference tables of the observed frequencies of particular substitutions in sets of known related proteins (see Figure 1-6). The individual scores for each position are summed to give an overall similarity score for the alignment.

In practice, insertions and deletions as well as substitutions will have occurred in two homologous sequences during their evolution. This usually results in two gene or protein sequences of different lengths in which regions of closely similar sequence are separated by dissimilar regions of unequal length. In such cases, portions of the sequence are slid over each other when making the alignment, in order to maximize the number of identical and similar amino acids. Such sliding creates gaps in one or other of the sequences (Figure 4-2). Experience tells us that closely related sequences do not, in general, have many insertions or deletions relative to each other. Because any two sequences could be broken up randomly into as many gaps as needed to maximize matching, in which case the matching would have no biological significance, gaps are subject to a penalty when scoring sequence relatedness.

Definitions

alignment: procedure of comparing two or more sequences by looking for a series of characteristics (residue identity, similarity, and so on) that match up in both and maximize conservation, in order to assess overall similarity.

conserved: identical in all sequences or structures compared.

E-value: the probability that an **alignment** score as good as the one found between two sequences would

be found in a comparison between two random sequences; that is, the probability that such a match would occur by chance.

evolutionary distance: the number of observed changes in nucleotides or amino acids between two related sequences.

Hidden Markov Model: a probabilistic model of a sequence **alignment**.

homologous: describes genes or proteins related by divergent evolution from a common ancestor.

homology: the similarity seen between two gene or protein sequences that are both derived by evolution from a common ancestral sequence.

multiple sequence alignment: alignment of more than two sequences to maximize their overall mutual identity or similarity.

pairwise alignment: alignment of two sequences.

percent identity: the percentage of columns in an **alignment** of two sequences that contain identical amino acids. Columns that include gaps are not counted.

```
                                    Motif 1              Motif 2
H.s. Wee1 409-457    QVGRGLRYIHSMS-LVHMDIKPSNIFISRTSIPNAASEEGDEDDWASNK----
H.s. Ttk  614-659    NMLEAVHTIHQHG-IVHSDLKPANFLIVDG-----MLKLIDFGIANQMQPD--
S.c. Ste7 313-358    GVLNGLDHLYRQYKIIHRDIKPSNVLINSK----GQIKLCDFGVSKKLI----
S.c. Mkk1 332-376    AVLRGLSYLHEKK-VIHRDIKPQNILLNEN----GQVKLCDFGVSGEAV----
S.p. Byr1 168-213    SMVKGLIYLYNVLHIIHRDLKPSNVVVNSR----GEIKLCDFGVSGELV----
S.c. St20 722-767    ETLSGLEFLHSKG-VLHRDIKSDNILLSME----GDIKLTDFGFCAQINE---
S.c. Cc15 129-172    QTLLGLKYLHGEG-VIHRDIKAANILLSAD----NTVKLADFGVSTIV-----
S.p. Byr2 505-553    QTLKGLEYLHSRG-IVHRDIKGANILVDNK----GKIKISDFGISKKLELNST
S.c. Spk1 302-348    QILTAIKYIHSMG-ISHRDLKPDNILIEQDD--PVLVKITDFGLAKVQG----
S.p. Kin1 249-293    QIGSALSYLHQNS-VIHRDLKIENILISKT----GDIKIIDFGLSNLYR----
S.p. Cdr1 111-156    QILDAVAHCHRFR-FRHRDLKLENILIKVN----EQQIKIADFGMATVEP---
M.m. K6a1 507-556    TISKTVEYLHSQG-VVHRDLKPSNILYVDESGNPECLRICDFGFAKQLRA---
R.n. Kpbh 136-180    SLLEAVNFLHVNN-IVHRDLKPENILLDDN----MQIRLSDFGFSCHLE----
H.s. Erk2 132-176    QILRGLKYIHSAM-VLHRDLKPSNLLLNTT---CLSCKICDFGLARVA-----
S.c. Kss1 137-182    QILRALKSIHSAQ-VIHRDIKPSNLLLNSN------CKVCDFGLARCLASSS-
```

Various algorithms have been used to align sequences so as to maximize matching while minimizing gaps. The most powerful is the **Hidden Markov Model**, a statistical model that considers all possible combinations of matches, mismatches and gaps to generate the "best" alignment of two or more sequences. Use of such models provides a third score to go along with percent identity and the similarity score. This score is usually expressed as the probability that the two sequences will have this degree of overall similarity by chance; the lower the score, the more likely the two sequences are to be related. Two virtually identical sequences tend to have probability scores (known in this context as **E-values**) of 10^{-50} or even lower. When the E-value for a sequence comparison is greater than about 10^{-10}, the two sequences could still be related and could have similar structures, but the probability that the two proteins will differ in function increases markedly, especially for multidomain proteins (Figure 4-3).

Multiple alignments and phylogenetic trees

The alignment process can be expanded to give a **multiple sequence alignment**, which compares many sequences (Figure 4-4). Such multiple sequence alignments are arrived at by successively considering all possible pairwise alignments. In effect, one mutates one sequence into all the others to try and determine the most likely evolutionary pathway, given the likelihoods of the various possible substitutions. As more sequences are added to the multiple alignment, such a model becomes "trained" by the evolutionary history of the family of proteins being compared. From this alignment one can see that certain residues are identical in all the sequences. Any residue, or short stretch of sequence, that is identical in all sequences in a given set (such as that of a protein family) is said to be **conserved**. Multiple alignments tend to give a better assessment of similarity than pairwise alignments and can identify distantly related members of a gene family that would not be picked up by pairwise alignments alone.

Multiple sequence alignments of homologous proteins or gene sequences from different species are used to derive a so-called **evolutionary distance** between each pair of species, based, in this instance, on the degree of difference (rather than similarity) between each sequence pair. Given that sequences that diverged earlier in time will be more dissimilar to each other than more recently diverged sequences, these distances can be used to construct **phylogenetic trees** that attempt to reflect evolutionary relationships between species, or, as in the tree illustrated here (Figure 4-5), individual members of a protein superfamily. The tree that emerges, however, will be influenced by the particular tree-building algorithm used and the evolutionary assumptions being made. As the rates of change of protein sequences can vary dramatically, depending on, among other things, the function of the proteins in question and large-scale genomic rearrangements, these specific assumptions are crucial to evaluating the results of phylogenetic analysis.

Figure 4-4 Multiple alignment A small part of a large multiple alignment of more than 6,000 protein kinase domains in the Pfam database (http://pfam.wustl.edu), displaying part of the region shown in Figure 4-2. Residues identical in all or almost all sequences in the complete alignment are highlighted in red, the next most highly conserved in orange and those next most conserved in yellow. The alignment reveals residues and sequence motifs that are common to all protein kinase catalytic domains and can be used to identify additional members of the family. One is motif 1, which identifies the catalytic loop and contains a conserved aspartic acid (D) important to catalytic function. H.s.: human; S.c.: *Saccharomyces cerevisiae*; S.p.: *Schizosaccharomyces pombe*; M.m.: *Mus musculus*, mouse; R.n.: *Rattus norvegicus*, rat.

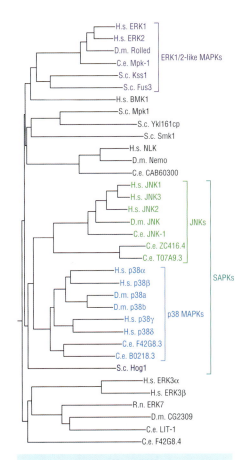

Figure 4-5 Phylogenetic tree comparing the three major MAP kinase subgroups The three major subgroups of MAP kinases (ERKs, JNKs and p38) are well conserved throughout evolution. This dendrogram shows the evolutionary relationships between the ERKs, JNKs and p38 in the budding yeast *S. cerevisiae* (S.c.), the nematode worm *Caenorhabditis elegans* (C.e.), the fruit fly *Drosophila melanogaster* (D.m.) and humans (H.s.). The mammalian MAPK ERK7 was isolated from the rat (R.n.). No human homolog has yet been identified. SAPK stands for stress-activated protein kinases, a general name for the JNK and p38 families. (Kindly provided by James E. Ferrell Jr.)

phylogenetic tree: a branching diagram, usually based on the evolutionary distances between sequences, that illustrates the evolutionary history of a protein family or superfamily, or the relationships between different species of organism.

References

Gerstein, M. and Honig, B.: **Sequences and topology.** *Curr. Opin. Struct. Biol.* 2001, **11**:327–329.

Mount, D.W.: *Bioinformatics: sequence and genome analysis* (Cold Spring Harbor Laboratory Press, New York, 2001).

Wilson, C. *et al.*: **Assessing annotation transfer for genomics: quantifying the relations between protein sequence, structure and function through tradition and probabilistic scores.** *J. Mol. Biol.* 2000, **297**:233–249.

The Pfam database : http://pfam.wustl.edu

Some Examples of Small Functional Protein Domains

Domain	Function
SH2	binds phosphotyrosine
SH3	binds proline-rich sequences
Pleckstrin homology (PH)	binds to G proteins and membranes
WD40	protein–protein interaction
DH	guanine nucleotide exchange
EF-hand	binds calcium
Homeobox	binds DNA
TRBD	binds tRNA
Helix-turn-helix	binds DNA
PUA	RNA modification

Figure 4-6 Representative examples of small functional domains found in proteins These domains are characterized by degenerate sequence motifs that extend over the whole domain. For the structures of some of these domains see Figure 3-2. See Figure 1-46 for an indication of how these domains are combined in proteins.

Structural data can help sequence comparison find related proteins

Some sequence-comparison methods also try to include secondary and tertiary structural information. Because different secondary structural elements can be formed from very similar segments of sequence (see section 4-14), using structural information in the description of the reference protein could, in theory at least, help exclude proteins with somewhat similar sequences but very different structures. It is also known that even similar proteins can have shifts in the relative positions of sequence segments, dictated by differences in secondary-structure packing and the positioning of functionally important groups. This makes the similarity at the sequence level very difficult to determine. For example, there are prokaryotic SH3 domains which, like their eukaryotic relatives, bind to proline-rich sequences. Straightforward sequence alignment does not indicate any relationship between the prokaryotic and eukaryotic domains; however, when the alignment is performed by comparing residues in the corresponding secondary structure elements of the prokaryotic and eukaryotic domains, some regions of sequence conservation appear. A number of small functional domains that can be characterized in this way are listed in Figure 4-6.

Prediction of secondary structure and tertiary structure from sequence alone, by methods such as that of Chou-Fasman and profile-based threading (see sections 1-8 and 4-7), is more accurate when multiple sequences are compared. Both secondary and tertiary structures are determined by the amino-acid sequence; however, there is an interplay between the intrinsic secondary structure propensities of the amino acids and the energetics of the local interactions within a tertiary structure. Tertiary interactions can override a preferred conformation for a residue or segment of residues, and this effect can differ within different local structural contexts. This effect can be taken into account if multiple structures resulting from multiple sequences are available for a *superfamily* of proteins. Therefore, knowledge of the variability of a sequence that can form closely similar structures can improve the performance of prediction methods based on statistical analysis of sequences. Interestingly, all methods for predicting protein structure from sequence seem to have a maximum accuracy of about 70%. The reason for this barrier is unclear.

Sequence and structural motifs and patterns can identify proteins with similar biochemical functions

Sometimes, only a part of a protein sequence can be aligned with that of another protein. Such **local alignments** can identify a functional module within a protein. These function-specific blocks of sequence are called **functional motifs**. There are two broad classes. Short, contiguous motifs usually specify binding sites and can be found within the context of many structures (Figure 4-7). Discontinuous short binding motifs also occur but are often harder to identify by sequence comparisons. Discontinuous or non-contiguous motifs are composed of short stretches of conserved sequence, or even individual conserved residues, separated by stretches

Figure 4-7 Representative examples of short contiguous binding motifs These motifs are determined by comparison of numerous different versions of the given motif from different proteins. Each motif represents a so-called consensus sequence reflecting the residue most likely to occur at each position. Where two or more residues are equally likely at the same position they are shown in square brackets. X can be any amino acid. The subscript numbers represent repeated residues.

Some Examples of Short Sequence Motifs and Their Functions

Contiguous motif	Consensus sequence	Function
Walker (P loop)	[A/G]XXXXGK[S/T]	binds ATP or GTP
Zn finger	$CX_{2-4}CX_{12}HX_{3-5}H$	binds Zn in a DNA-binding domain
Osteonectin	$CX[D/N]XXCXXG[K/R/H]XCX_{6-7}PXCXCX_{3-5}CP$	binds calcium and collagen
DEAD box helicase	XXDEAD[R/K/E/N]X	ATP-dependent RNA unwinding
MARCKS	GQENGNV[K/R]	substrate for protein kinase C
Calsequestrin	[E/Q][D/E]GL[D/N]FPXYDGXDRV	binds calcium

Definitions

BLAST: a family of programs for searching protein and DNA databases for sequence similarities by optimizing a specific similarity measure between the sequences being compared.

functional motif: sequence or structural motif that is always associated with a particular biochemical function.

local alignment: alignment of only a part of a sequence with a part of another.

profile: a table or matrix of information that characterizes a protein family or superfamily. It is typically composed of sequence variation or identity with respect to a reference sequence, expressed as a function of each position in the amino-acid sequence of a protein. It can be generalized to include structural information. Three-dimensional profiles express the three-dimensional structure of a protein as a table which represents the local environment and conformation of each residue.

References

Aitken, A.: **Protein consensus sequence motifs.** *Mol. Biotechnol.* 1999, **12**:241–253.

Altschul, S.F. *et al.*: **Gapped BLAST and PSI-BLAST: a new generation of protein database search programs.** *Nucleic Acids Res.* 1997, **25**:3389–3402.

Elofsson, A. *et al.*: **A study of combined structure/sequence profiles.** *Fold. Des.* 1996, **1**:451–461.

Falquet, L. *et al.*: **The PROSITE database, its status in**

of non-conserved sequence. Such discontinuous patterns can also represent catalytic sites; examples are the motifs characterizing the serine proteases and glycosyltransferases. For example, catalases, which are heme-containing enzymes that degrade hydrogen peroxide, can be identified by the discontinuous motifs RXFXYXD[A/S/T][Q/E/H] where the bold Y is the heme iron ligand tyrosine and [I/F]X[R/H]X$_4$[E/Q]RXXHX$_2$[G/A/S], where the bold H is an essential catalytic histidine. Finally, there are some motifs that extend over the entire sequence of a domain and are highly degenerate. These characterize small protein domains such as SH2 and SH3 (see Figure 4-6). A web server that can be used to find all types of motifs is the PROSITE database, which as of early 2003 contained 1,585 different recognizable motifs.

Protein-family profiles can be generated from multiple alignments of protein families for which representative structures are known

Because functionally important residues must necessarily be conserved over evolution, when multiple sequences from different organisms can be aligned, the probability of recognizing related proteins or a similar biochemical function even at very low overall sequence identity increases dramatically. Specialized computer programs such as PSI-BLAST have been developed for this purpose. This looks for a set of particular sequence features—a **profile**—that characterizes a protein family. Such profiles are obtained from a multiple alignment as described in Figure 4-8. A profile is derived from a position-specific score matrix (PSSM) and this method is used in PSI-BLAST (position-specific iterated **BLAST**). Motif or profile search methods are frequently much more sensitive than pairwise comparison methods (such as ordinary BLAST) at detecting distant relationships. PSI-BLAST may not be as sensitive as the best available dedicated motif-search programs, but its speed and ease of use has brought the power of these methods into more common use.

During evolution, certain positions in a sequence change more rapidly than others. Functionally and structurally important residues tend to be conserved, although the former can change if the specificity or biochemical activity of a protein changes over time; this is how new families branch off from old ones, building up a large superfamily. The concept of a position-based matrix of information to represent a sequence can be generalized to include structural information, which changes more slowly, as well as sequence similarity. This information is used to refine PSSMs such as the one shown in Figure 4-8 to provide a more accurate profile. Profile-based comparison methods differ in two major respects from other methods of sequence comparison. First, any number of known sequences can be used to construct the profile, allowing more information to be used to test the target sequence than is possible with pairwise alignment methods. This is done in PSI-BLAST, where the number of sequences grows with each iteration as more distantly related sequences are found, increasing the informational content of the profile on each iteration. The profile can include penalties for insertion or deletion at each position, which enables one to include information derived from the secondary structure and other indicators of tertiary structure such as the pattern of hydrophobicity or even the local environment around each residue in the comparison. Evolutionary information can also be incorporated.

Profile construction allows the identification of sequences that are compatible with a specific tertiary structure even when sequence identity is too low to be detected with statistical significance. This is the theoretical basis for the profile-based threading method of assigning folds to sequences of unknown proteins (see section 4-7). However, if such a match is not found, it is not an indication that the sequence is incompatible with the protein fold or that two sequences do not have the same structure. False negatives are common in profile-based methods.

Constructing a Family Profile

Position	1	2	3	4	5
	C	C	G	T	L
	C	G	H	S	V
	G	C	G	S	L
	C	G	G	T	L
	C	C	G	S	S

Position	1	2	3	4	5
Prob(C)	0.8	0.6	-	-	-
Prob(G)	0.2	0.4	0.8	-	-
Prob(H)	-	-	0.2	-	-
Prob(S)	-	-	-	0.6	0.2
Prob(T)	-	-	-	0.4	-
Prob(L)	-	-	-	-	0.6
Prob(V)	-	-	-	-	0.2

Figure 4-8 Construction of a profile In this simple example, five homologous sequences five residues long are compared and a matrix is constructed that expresses, for each position, the probability of a given amino acid being found at that position in this family, which is simply a fraction representing the frequency of occurrence. This position-specific score matrix (PSSM) represents the "profile" of this sequence family. An unknown sequence can be scanned against this profile to determine the probability that it belongs to the family by multiplying the individual probabilities of each residue in its sequence, selected from the profile, to obtain a total probability. This can be compared to the value generated by scanning random sequences against the same profile, to assess the significance of the value. For example, the sequence CCHTS would have a probability score of $0.8 \times 0.6 \times 0.2 \times 0.4 \times 0.2 = 0.0077$, comparable to the score of the aligned sequence CGHSV ($0.8 \times 0.4 \times 0.2 \times 0.6 \times 0.2$). The sequence CLHTG would have a score of zero ($0.8 \times 0.0 \times 0.2 \times 0.4 \times 0.0$).

2002. *Nucleic Acids Res.* 2002, **30**:235–238.

Gaucher, E.A. *et al.*: **Predicting functional divergence in protein evolution by site-specific rate shifts.** *Trends Biochem. Sci.* 2002, **27**:315–321.

Gribskov, M. *et al.*: **Profile analysis: detection of distantly related proteins.** *Proc. Natl Acad. Sci. USA* 1987, **84**:4355–4358.

Kawabata, T. *et al.*: **GTOP: a database of protein structures predicted from genome sequences.** *Nucleic Acids Res.* 2002, **30**:294–298.

Kunin, V. *et al.*: **Consistency analysis of similarity between multiple alignments: prediction of protein function and fold structure from analysis of local sequence motifs.** *J. Mol. Biol.* 2001, **307**:939–949.

Jones, D.T. and Swindells, M.B.: **Getting the most from PSI-BLAST.** *Trends Biochem. Sci.* 2002, **27**:161–164.

Pellegrini, M. *et al.*: **Assigning protein functions by comparative genome analysis: protein phylogenetic profiles.** *Proc. Natl Acad. Sci. USA* 1999, **96**:4285–4288.

Snel, B. *et al.*: **The identification of functional modules from the genomic association of genes.** *Proc. Natl Acad. Sci. USA* 2002, **99**:5890–5895.

von Mering, C. *et al.*: **STRING: a database of predicted functional associations between proteins.** *Nucleic Acids Res.* 2003, **31**:258–261.

The PROSITE database: http://ca.expasy.org/prosite

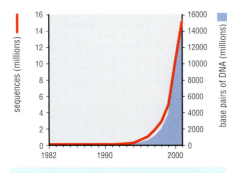

Figure 4-9 The growth of DNA and protein sequence information collected by GenBank over 20 years There has been an exponential increase in both base pairs of DNA sequence and coding sequences, especially since 1994 when various genomics projects were initiated. (Information from http://www.ncbi.nlm.nih.gov/Genbank/genbankstats.html)

Sequence information is increasing exponentially

During the past decade, more than 800 organisms have been the object of genome-sequencing projects. We now know the complete DNA sequences of the genomes of over 100 species of bacteria and archaea, including some important pathogens, and three yeasts, and have partial or complete genome sequences of a number of protozoan parasites. Among multicellular organisms, the genomes of the nematode worm (*Caenorhabditis*), the fruit fly (*Drosophila*) and the plants *Arabidopsis thaliana* and rice have also been completely sequenced. The human genome sequence is now completely finished and a draft mouse genome sequence has also been completed. The growth of sequence information is exponential, and shows no sign of slowing down (Figure 4-9). However, in all these organisms the biochemical and cellular functions of a large percentage of the proteins predicted from these sequences are at present unknown.

It is hardly surprising, therefore, that much effort is being expended on the attempt to define the structures and functions of proteins directly from sequence. Such efforts are based on comparison of sequences from many different organisms using computational tools such as BLAST to retrieve related sequences from the databases (see section 4-1). Attempts to derive function from sequence depend on the basic assumption that proteins that are related by sequence will also be related by structure and function. In this chapter, we will show that the assumption of structural relatedness is usually valid, but that function is less reliably determined by such methods. Structure and function can be derived in this way only for sequences that are quite closely related to those encoding proteins of known structure and function, and sometimes not even then.

As one proceeds from prokaryotes to eukaryotes, and from single-celled to multicellular organisms, the number of genes increases markedly (Figure 4-10), by the addition of genes such as those involved in nuclear transport, cell–cell communication, and innate and acquired immunity. The number of biochemical functions also increases. With increasing evolutionary distance, sequences of proteins with the same structure and biochemical function can diverge so greatly as to render any relationship extremely difficult to detect. Consequently, defining functions for gene products from higher organisms by sequence comparisons alone will be difficult until even more sequences and structures are collected and correlated with function.

In some cases function can be inferred from sequence

If a protein has more than about 40% sequence identity to another protein whose biochemical function is known, and if the functionally important residues (for example, those in the active site of an enzyme) are conserved between the two sequences, it has been found that a reasonable working assumption can be made that the two proteins have a common biochemical function (Figure 4-11). The 40% rule works because proteins that are related by descent and have the same function in different organisms are likely still to have significant sequence similarity, especially in regions critical to function. Sequence comparison will not, however, detect proteins of identical structure and biochemical function from organisms so remote from one another on the evolutionary tree that virtually no sequence identity remains. Moreover, identity of biochemical function does not necessarily mean that the cellular and other higher-level

Genome Sizes of Representative Organisms

Organism	Genome size (base pairs)	Number of genes
Mycoplasma genitalium	45.8×10^5	483
Methanococcus jannaschii	1.6×10^6	1,783
Escherichia coli	4.6×10^6	4,377
Pseudomonas aeruginosa	6.3×10^6	5,570
Saccharomyces cerevisiae	1.2×10^7	6,282
Caenorhabditis elegans	1.0×10^8	19,820
Drosophila melanogaster	1.8×10^8	13,601
Arabidopsis thaliana	1.2×10^8	25,498
Homo sapiens	3.3×10^9	~30,000 (?)

Figure 4-10 Table of the sizes of the genomes of some representative organisms The first four organisms are prokaryotes. A continuous update on sequencing projects, both finished and in progress, may be found at http://ergo.integratedgenomics.com/GOLD/

References

Brenner, S.: **Theoretical biology in the third millennium.** *Philos. Trans. R. Soc. Lond. B. Biol. Sci.* 1999, **354**: 1963–1965.

Brizuela, L. *et al.*: **The FLEXGene repository: exploiting the fruits of the genome projects by creating a needed resource to face the challenges of the post-genomic era.** *Arch. Med. Res.* 2002, **33**:318–324.

Domingues, F.S. *et al.*: **Structure-based evaluation of sequence comparison and fold recognition align-** ment accuracy. *J. Mol. Biol.* 2000, **297**:1003–1013.

Hegyi, H. and Gerstein, M.: **Annotation transfer for genomics: measuring functional divergence in multi-domain proteins.** *Genome Res.* 2001, **11**:1632–1640.

Genomic and protein resources on the Internet:

http://bioinfo.mbb.yale.edu/lectures/spring2002/show/index_2

http://ergo.integratedgenomics.com/GOLD/

http://www.ncbi.nlm.nih.gov/Genbank/genbankstats.html

Figure 4-11 Relationship of sequence similarity to similarity of function The percentage of protein pairs with the same precise biochemical function is plotted against the sequence identity (enzymes, blue curve; non-enzymes, green curve). The orange area represents proteins whose fold and function can be reliably predicted from sequence comparison. The yellow area represents proteins whose fold can reliably be predicted from sequence but whose precise function cannot. The blue area represents proteins for which neither the fold nor the function can reliably be predicted from sequence. Note that below about 40% identity, the probability of making an incorrect functional assignment increases dramatically. Adapted from an analysis by Mark Gerstein (http://bioinfo.mbb.yale.edu/lectures/spring2002/show/index_2).

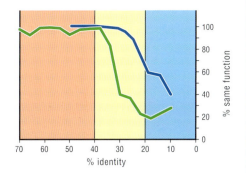

functions of the proteins will be similar. Such functions are expressed in a particular cellular context and many proteins, such as hormones, growth factors and cytokines, have multiple functions in the same organism (see section 4-13).

Local alignments of functional motifs in the sequence (see section 4-2) can often identify at least one biochemical function of a protein. If the sequence motif is large enough and contiguous, it can identify an entire domain or structural module with a recognizable fold and function. For example, helix-turn-helix motifs (see Figure 1-50) and zinc finger motifs (see Figure 1-49) are

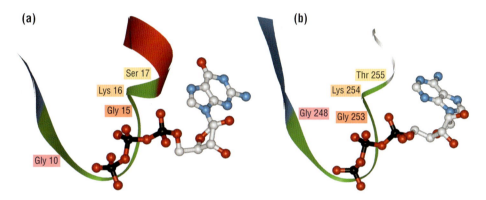

often recognizable in the sequence and are diagnostic for, respectively, small secondary structure elements and small domains that potentially bind DNA. The SH2 and SH3 domains present in many signal transduction proteins can also often be recognized by characteristic stretches of sequence. When present, such sequences usually indicate domains that are involved in the recognition of phosphotyrosines or proline-rich sequences, respectively, in dynamic protein–protein interactions. The so-called Walker motif, which identifies ATP- and GTP-binding sites, is also easily identified at the level of sequence, although its presence does not reveal what the nucleotide binding is used for and it is found in many different protein folds. The Walker motif is actually three different, non-contiguous stretches of sequence, labeled Walker A, B, and C. Of these, the Walker A motif, or P loop, which defines the binding site for the triphosphate moiety, is the easiest to recognize (Figure 4-12 and see Figure 4-7). The B and C motifs interact with the base of the nucleotide.

Sequence comparison is such an active area of research because it is now the easiest technique to apply to a new protein sequence. Figure 4-13 shows an analysis of the functions of all the known or putative protein-coding sequences in the yeast genome: some of these are experimentally established, but a large proportion are inferred only by overall sequence similarity to known proteins (labeled homologs in the figure) or by the presence of known functional motifs, and 32% of them are unknown. Similar distributions are observed for many other simple organisms. For more complex organisms, the proportion of proteins of unknown function increases dramatically. Current efforts are focused on ways of identifying structurally and functionally similar proteins when the level of sequence identity is significantly below the 40% threshold. As we shall see, identification of structural similarity is easier and more robust than the identification of functional similarity.

Figure 4-12 The P loop of the Walker motif A contiguous sequence block, the so-called Walker A block or P loop, is a stretch of sequence with a consensus pattern of precisely spaced phosphate-binding residues; this is found in a number of ATP- or GTP-binding proteins, for example ATP synthase, myosin heavy chain, helicases, thymidine kinase, G-protein alpha subunits, GTP-binding elongation factors, and the Ras family. The consensus sequence is: [A or G]XXXXGK[S or T]; this forms a flexible loop between alpha-helical and beta-pleated-sheet domains of the protein in question. The proteins may have quite different overall folds. The triphosphate group of ATP or GTP is bound by residues from the P loop. Shown are the interactions **(a)** of GTP with the P loop of the signaling protein H-Ras (PDB 1qra) and **(b)** of ATP with the P loop of a protein kinase (PDB 1aq2).

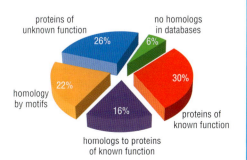

Figure 4-13 Analysis of the functions of the protein-coding sequences in the yeast genome Some are known experimentally, some are surmised from sequence comparison with proteins of known function in other organisms, and some are deduced from motifs that are characteristic of a particular function. Some of these surmised functions may not be correct, and a large percentage of the coding sequences cannot at present be assigned any function by any method.

Gene function can sometimes be established experimentally without information from protein structure or sequence homology

The explosive growth of sequence information has driven the development of new experimental methods for obtaining information relevant to the function of a gene. Many of these methods are high throughput: they can be applied to large numbers of genes or proteins simultaneously. Consequently, databases of information about the expression level, cellular localization, interacting partners and other aspects of protein behavior are becoming available for entire genomes. Such data are then combined with the results of more classical biochemical and genetic experiments to suggest the function of a gene of interest. The order of experiments is flexible and many will be carried out in parallel. Here we review some of the most common techniques. Most of them require either cloned DNA or protein samples (which sometimes must be purified) for the gene(s) of interest. The rest of the chapter discusses methods, both experimental and computational, that attempt to derive functional information primarily from either protein sequence or protein structure data.

One valuable clue to function is the expression pattern of the gene(s) in question. Experience suggests that genes of similar function often display similar patterns of expression: for example, proteins that are involved in chromosome segregation tend to be expressed at the same phase of the cell cycle, while proteins involved in response to oxidative stress usually are expressed—or their expression levels are greatly increased—when cells are subjected to agents that produce oxidative damage (hydrogen peroxide, superoxide, nitric oxide, and so on). Expression can be measured at the level of mRNA or protein; the mRNA-based techniques, such as **DNA microarrays** (Figure 4-14) and SAGE (serial analysis of gene expression), tend to be easier to carry out, especially on a genome-wide scale. Microarray technology, in particular, can provide expression patterns for up to 20,000 genes at a time. It is based on the fidelity of hybridization of two complementary strands of DNA. In its simplest form, the technique employs synthetic "gene chips" that consist of thousands of oligonucleotide spots on a glass slide, one for each gene of the genome. Complementary DNA, labeled with a fluorescent dye, is then made from the mRNA from two different states of the cells being analyzed, one labeled with a red probe (the test state) and the other labeled with a green probe (the reference state). Both are mixed and applied to the chip, where they hybridize to the DNA in the spots. If the level of mRNA for a particular gene is increased in the test cells relative to the reference cells that spot will show up as red; if the level is unchanged the spot will be yellow, and if the mRNA has decreased the spot will be green. In theory, differences of 3–4-fold or greater in mRNA level can be detected reliably with this technique, but in practice the threshold for significance is often 5–10-fold. Any important change in expression must always be verified by **northern blot** analysis.

Protein expression in the cell can be monitored by antibody binding, but this method is only useful for one protein at a time. High throughput can be achieved by two-dimensional gel electrophoresis, which can separate complex protein mixtures into their components, whose identity can be determined by cutting out the bands and measuring the molecular weight of each protein by mass spectrometry (Figure 4-15). In addition to the amount of protein present, this method can also detect covalent modifications of the protein. The technique is powerful but is also relatively slow and expensive, cannot resolve all the proteins in a cell extract, and can fail to detect proteins that are only present in a few copies per cell. Experience suggests that mRNA levels determined by microarray are good predictors of relative protein levels as determined by two-dimensional gels for the most abundant proteins in a cell; the correlation breaks down for scarcer proteins. Efforts are underway to develop protein microarrays (so-called "protein chips") that can rapidly measure the levels of larger numbers of proteins.

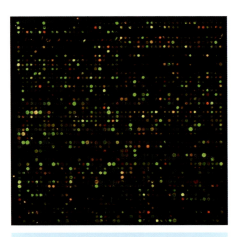

Figure 4-14 DNA microarray Part of a microarray chip showing changes in gene expression when yeast cells are treated with a drug. Genes whose expression increases on drug treatment appear as red spots; those that decrease are green; those that do not change are yellow. Some genes do not appear because they are not expressed under these conditions. Each spot represents a single gene.

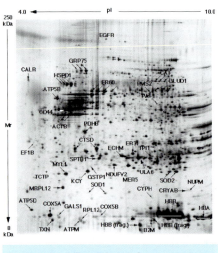

Figure 4-15 2-D protein gel Some spots have been identified and labeled.

Definitions

DNA microarray: an ordered array of nucleic acid molecules, either cDNA fragments or synthetic oligonucleotides, where each position in the array represents a single gene.

gene knockout: inactivation of the function of a specific gene in a cell or organism, usually by recombination with a marker sequence but sometimes by antisense DNA, RNA interference, or by antibody binding to the gene product. The phenotype resulting from the knockout can often provide clues to the function of the gene.

northern blot: technique for detecting and identifying individual RNAs by hybridization to specific nucleic acid probes, after separation of a complex mixture of mRNAs by electrophoresis and blotting onto a nylon membrane.

RNA interference (RNAi): Abolition of the expression of a gene by a small (~22 base pair) double-stranded RNA.

yeast two-hybrid: a method for finding proteins that interact with another protein, based on activation of a reporter gene in yeast.

References

Colas, P. and Brent, R.: **The impact of two-hybrid and related methods on biotechnology.** *Trends Biotechnol.* 1998, **16**:355–363.

Gasch, A.P. *et al.*: **Genomic expression programs in the response of yeast cells to environmental changes.** *Mol. Biol. Cell* 2000, **11**:4241–4257.

Kallal, L. and Benovic, J.L.: **Using green fluorescent proteins to study G-protein-coupled receptor localization and trafficking.** *Trends Pharmacol. Sci.* 2000, **21**:175–180.

The phenotype produced by inactivating a gene, a **gene knockout**, is highly informative about the cellular pathway(s) in which the gene product operates (Figure 4-16). Knockouts can be obtained by classical mutagenesis, targeted mutations, **RNA interference (RNAi)**, the use of antisense message RNA, or by antibody binding. Microarray analysis on the knockout, comparing the pattern of gene expression in the presence and absence of the gene, will often provide a wealth of information about how the cell responds to its expression, as will studies of changes in protein expression and modification. Of course, the phenotype is an overall response to the loss of the gene product, not a direct readout of biochemical or cellular function. In addition, expressing the gene at high levels in tissues or organisms where it is normally not expressed significantly (ectopic expression) frequently also produces an interesting, and informative, phenotype.

The location of a protein in the cell often provides a valuable clue to its functions. If a gene product is nuclear, cytoplasmic, mitochondrial, or localized to the plasma membrane, for example, and especially if that localization changes in different states of the cell, then inferences about the pathways in which the protein participates can be drawn. A number of techniques exist for determining location, all dependent on attachment of a tag sequence to the gene in question. A commonly used method is to fuse the sequence encoding green fluorescent protein (GFP) to one end of the gene sequence for the protein in question, and then use the intrinsic fluorescence of GFP to monitor where the protein is in the cell (Figure 4-17). Of course, care must be taken that the fusion does not interfere with folding or localization of the gene product.

Many proteins do not function on their own; they are part of a complex of two or more gene products. If the function of one of the interacting proteins is known, then the fact that it binds to a given protein will help reveal the latter's function. Interacting proteins can be found by the **yeast two-hybrid** system. This exploits the fact that transcriptional activators are modular in nature. Two physically distinct domains are necessary to activate transcription: (1) a DNA-binding domain (DBD) that binds to the promoter; and (2) an activation domain that binds to the basal transcription apparatus and activates transcription. In the yeast two-hybrid system, the gene for the target protein is cloned into a "bait" vector next to a sequence encoding the DBD of a given transcription factor. cDNAs encoding potential interactor proteins (the "prey") are cloned separately into another set of plasmids in-frame with the sequence encoding the activation domain of the transcription factor. A bait plasmid and a prey plasmid are introduced together into yeast cells, where the genes they carry are translated into proteins (all combinations of bait and prey are tested in parallel experiments). To form a working transcription factor within the yeast cell, the DBD and the activation domain must be brought together, and this can only happen if the protein carrying the activation domain interacts with the protein fused to the DBD (Figure 4-18). The complete transcription factor can then activate a reporter gene, producing enzyme activity, for example, or cell growth in the absence of a nutrient. Although the two-hybrid screen is a powerful and rapid method for detecting binding partners, it is plagued by false positives and irreproducibility, so any putative interaction must be verified by direct methods such as isolation of the protein complex and identification of its components by antibody binding.

Many other techniques exist and can be employed as needed. Among them are techniques for identifying possible substrates and regulatory molecules. Some of the most popular of these are surface plasmon resonance to detect ligand binding, and purification and direct assay of possible biochemical function *in vitro*. More are being developed as the need for methods to probe function increases. Many of these, like the techniques described here, will produce large databases, so computational analysis of and correlation between such databases will be of great importance for functional genomics.

Figure 4-16 The phenotype of a gene knockout can give clues to the role of the gene The mouse on the right is normal; the mouse on the left lacks the gene that encodes pro-opiomelanocortin (POMC), which, among other things, affects the regulation of energy stores and has been linked to obesity. Photograph kindly provided by Ute Hochgeschwender. (From Yaswen, L. *et al.*: **Obesity in the mouse model of pro-opiomelanocortin deficiency responds to peripheral melanocortin.** *Nat. Med.* 1999, **5**:1066–1070.)

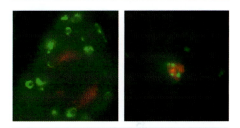

Figure 4-17 Protein localization in the cell The protein has been fused to GFP (green); the nucleus is stained red. In different stages of the cell cycle the protein is either cytoplasmic (left) or localized to the nucleus (right). Photographs kindly provided by Daniel Moore and Terry Orr-Weaver. (From Kerrebrock, A.W. *et al.*: *Cell* 1995, **83**:247–256.)

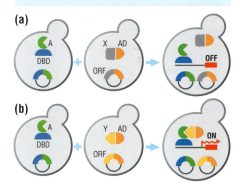

Figure 4-18 Two-hybrid system for finding interacting proteins The "bait" vector expresses a transcription factor DNA-binding domain (DBD, blue) fused to the test protein (protein A, green). The "prey" expression vectors each contain an individual open reading frame (ORF) of interest placed adjacent to the sequence encoding the activation domain (AD) of the same transcription factor. **(a)** When a bait and a prey vector are introduced into a yeast cell, the DBD and its attached protein A binds to the reporter gene (red). If the protein encoded by the ORF (protein X, grey) does not interact with the bait, the reporter gene is not activated. **(b)** If the prey protein (Y, yellow) does interact with protein A, the two parts of the transcription factor are reunited and the reporter gene is expressed.

Kerrebrock, A.W. *et al.*: **Mei-S332, a *Drosophila* protein required for sister-chromatid cohesion, can localize to meiotic centromere regions.** *Cell* 1995, **83**:247–256.

Patterson, S.D.: **Proteomics: the industrialization of protein chemistry.** *Curr. Opin. Biotechnol.* 2000, **11**:413–418.

Phizicky, E.M. and Fields, S.: **Protein–protein interactions: methods for detection and analysis.** *Microbiol. Rev.* 1995, **59**:94–123.

Reymond, M.A. *et al.*: **Standardized characterization** of gene expression in human colorectal epithelium by two-dimensional electrophoresis. *Electrophoresis* 1997, **18**:2842–2848.

Schena, M. *et al.*: **Quantitative monitoring of gene expression patterns with a complementary DNA microarray.** *Science* 1995, **270**:467–470.

Sherlock, G. *et al.*: **The Stanford Microarray Database.** *Nucleic Acids Res.* 2001, **29**:152–155.

A movie of protein transport in a live cell is at: http://elab.genetics.uiowa.edu/ELABresearch.htm

Evolution has produced a relatively limited number of protein folds and catalytic mechanisms

Although the total number of different enzymatic activities in any living cell is large, they involve a smaller number of classes of chemical transformation (see, for example, section 2-10). For each of these transformations, there is an even smaller number of different catalytic mechanisms by which they can be achieved. This all suggests that most enzymes should be related in both sequence and structure to many others of similar mechanism, even where their substrates are different. Such structural relatedness has indeed been observed: there are only a limited number of protein structural *superfamilies* and the proteins in the same superfamily often share some features of their mechanisms. In practice, however, detecting these structural and functional relationships from sequence alone is fraught with complications.

As described in section 4-1, two proteins with high sequence identity throughout can be assumed to have arisen by **divergent evolution** from a common ancestor and can be predicted to have very similar, if not identical, structures. In general, if the overall identity between the two sequences is greater than about 40% without the need to introduce an inordinate number of gaps in the alignment, and if this identity is spread out over most of the sequence, then the expectation is that they will code for proteins of similar overall fold (Figure 4-19). However, problems in deducing evolutionary relationships and in predicting function from sequence and structure arise when the situation is less clear-cut. And even proteins with greater than 90% sequence identity, which must have very similar structures and active sites, can in rare cases operate on quite different substrates.

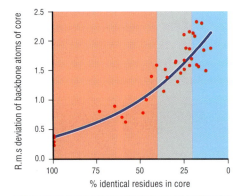

Figure 4-19 Relationship between sequence and structural divergence of proteins The percent identity of the protein cores of 32 pairs of proteins from eight different structural families was plotted against their structural divergence as measured by the root-mean-square difference in spatial positions of backbone atoms. A striking relationship is found, which holds for all the families studied. As the sequences diverge, the structures diverge, but not at the same rate. Small differences in sequence have little effect on structure, but structural divergence increases exponentially as sequence divergence becomes greater. Sequences with greater than 40% identity are generally considered to be homologous and the probability that they will have the same overall structure is also very high. For proteins with sequence identities below about 20%, evolution has usually altered much of the structure, and homology cannot be determined with any certainty. In between is a "grey area", where the overall identity between two sequences is less than 40% but greater than about 20%, and when it may be impossible from sequence comparisons alone to determine that two proteins are related. Data from Lesk, A.M., *Introduction to Protein Architecture* (Oxford University Press, Oxford, 2001).

Proteins that differ in sequence and structure may have converged to similar active sites, catalytic mechanisms and biochemical function

The structure of the active site determines the biochemical function of an enzyme, and in many homologous proteins active-site residues and structure are conserved even when the rest of the sequence has diverged almost beyond recognition. One might therefore suppose that all proteins with similar active sites and catalytic mechanisms would be homologs. This is, however, not the case. If two such proteins have quite distinct protein folds as well as low sequence similarity, it is likely that they are examples of **convergent evolution**: that is, they did not diverge from a common ancestor but instead arose independently and converged on the same active-site configuration as a result of natural selection for a particular biochemical function. Clear examples of convergent evolution are found among the serine proteases and the aminotransferases, which include proteins of quite different structure and fold, but with similar catalytic sites and biochemical function; these are considered in detail later in the chapter (see sections 4-8 and 4-12, respectively).

Proteins with low sequence similarity but very similar overall structure and active sites are likely to be homologous

It can be difficult to discern homology from sequence gazing alone, because sequence changes much more rapidly with evolution than does three-dimensional structure (Figure 4-20). In fact, proteins with no detectable sequence similarity at all, but with the same structures and biochemical functions, have been found. Among numerous examples are the glycosyltransferases, which transfer a monosaccharide from an activated sugar donor to a saccharide, protein, lipid, DNA or small-molecule acceptor. Some glycosyltransferases that operate on different

Definitions

convergent evolution: evolution of structures not related by ancestry to a common function that is reflected in a common structure.

divergent evolution: evolution from a common ancestor.

References

Bullock, T.L. et al.: **The 1.6 angstroms resolution crystal structure of nuclear transport factor 2 (ntf2).** *J. Mol. Biol.* 1996, **260**:422–431.

Hasson, M.S. et al.: **The crystal structure of benzoylformate decarboxylase at 1.6Å resolution: diversity of catalytic residues in thiamine diphosphate dependent enzymes.** *Biochemistry* 1998, **37**:9918–9930.

Irving, J.A. et al.: **Protein structural alignments and functional genomics.** *Proteins* 2001, **42**:378–382.

Kim, S.W. et al.: **High-resolution crystal structures of delta5-3-ketosteroid isomerase with and without a reaction intermediate analogue.** *Biochemistry* 1997, **36**:14030–14036.

Lesk, A.M.: *Introduction to Protein Architecture* (Oxford University Press, Oxford, 2001).

Lundqvist, T. et al.: **Crystal structure of scytalone dehydratase—a disease determinant of the rice pathogen, *Magnaporthe grisea*.** *Structure* 1994, **2**:937–944.

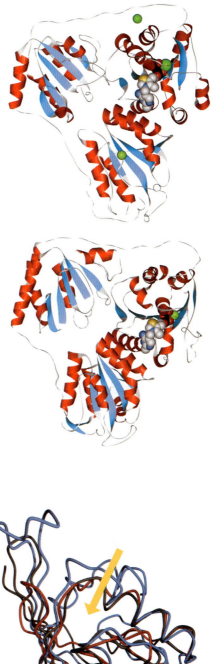

Figure 4-20 Ribbon diagram of the structure of a monomer of benzoylformate decarboxylase (BFD) and pyruvate decarboxylase (PDC) BFD (top) and PDC (bottom) share a common fold and overall biochemical function, but they recognize different substrates and have low (21%) sequence identity. The bound thiamine pyrophosphate cofactor is shown in space-filling representation in both structures. The green spheres are metal ions. (PDB 1bfd and 1pvd)

substrates and show no significant sequence identity nevertheless contain a structurally very similar catalytic domain and are thought to have a common ancestor.

In some—probably most—cases, low sequence homology combined with high structural similarity reflects selective conservation of functionally important residues in genuinely homologous, but highly diverged, sequences. Mandelate racemase, muconate lactonizing enzyme and enolase display very little overall sequence identity but have similar structures and active sites (see section 4-11). The reactions they catalyze share a core step and this step is catalyzed in the same way by all three enzymes, implying that they have probably diverged from a common ancestor.

Convergent and divergent evolution are sometimes difficult to distinguish

In other cases, however, there is spatial equivalence at the functional site, but little or no sequence conservation of the functionally important residues. In such cases, distinguishing between convergent and divergent evolution may be difficult. For example, the enzymes benzoylformate decarboxylase (BFD) and pyruvate decarboxylase (PDC) have only about 21% overall sequence identity but have essentially identical folds (Figure 4-20). The catalytic amino-acid side chains are conserved in spatial position in the three-dimensional structure but not in the sequence. It is possible that the two proteins evolved independently and converged to the same chemical solution to the problem of decarboxylating an alpha-ketoacid. But their great similarity in overall structure would seem to indicate that they diverged from a common ancestor. The level of sequence identity between them is, however, too low to distinguish between these two possibilities with confidence.

Divergent evolution can produce proteins with sequence and structural similarity but different functions

Conversely, there are proteins with very different biochemical functions but which nevertheless have very similar three-dimensional structures and enough sequence identity to imply homology. Such cases suggest that structure also diverges more slowly than function during evolution. For example, steroid-delta-isomerase, nuclear transport factor-2 and scytalone dehydratase share many structural details (Figure 4-21) and are considered homologous, yet the two enzymes—the isomerase and the dehydratase—have no catalytically essential residue in common. This suggests that it is general features of the active-site cavity of this enzyme scaffold that have the potential ability to catalyze different chemical reactions that proceed via a common enolate intermediate, given different active-site residues. The third protein in this homologous set—nuclear transport factor-2—is not an enzyme at all, as far as is known, but its active-site-like cavity contains residues that are present in the catalytic sites of both enzymes. Thus, determination of function from sequence and structure is complicated by the fact that proteins of similar structure may not have the same function even when evolutionarily related.

Murzin, A.G.: **How far divergent evolution goes in proteins.** *Curr. Opin. Struct. Biol.* 1998, **8**:380–387.

Patthy, L.: *Protein Evolution* (Blackwell Science, Oxford, 1999).

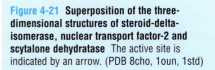

Figure 4-21 Superposition of the three-dimensional structures of steroid-delta-isomerase, nuclear transport factor-2 and scytalone dehydratase The active site is indicated by an arrow. (PDB 8cho, 1oun, 1std)

4-6 Structure from Sequence: Homology Modeling

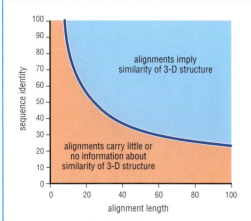

Figure 4-22 The threshold for structural homology Sequence space, plotted as a function of length of the segment being aligned and the percent identity between the two sequences, can be divided into two regions. The upper region (above the curve) shows where sequence similarity is likely to yield enough structural similarity for homology modeling to work. The lower region is highly problematic. At present 25% of known protein sequences fall in the safe area, implying 25% of all sequences can be modeled reliably.

(a)

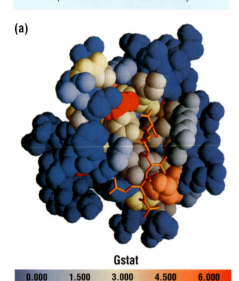

Gstat

| 0.000 | 1.500 | 3.000 | 4.500 | 6.000 |

(b)

Structure can be derived from sequence by reference to known protein folds and protein structures

Because structure changes more slowly than sequence, if there is a high degree of sequence identity between two proteins, their overall folds will always be similar. But at sequence identity of less than around 40% (see Figure 4-19), structures can be markedly different from each other. In practice, however, structural similarity often extends to lower levels of sequence identity, depending on how the identical residues are distributed. And there are many cases of two proteins having virtually identical overall folds and closely related functions despite having no statistically significant degree of sequence identity/similarity. The real problem in deducing structure from sequence is how to treat these difficult cases.

There are at present about 20,000 entries in the Protein Data Bank representing, depending on how one classifies them, 1,000–2,000 distinct structural "domains", that is unique folds. It has been estimated that the total number of unique folds will be at most several thousand. One of the major goals of work in structural genomics is to determine structures representative of all unique folds so that the structure of any unknown sequence can be modeled. Currently, the known protein structures and canonical protein folds are used to derive structure from sequence by two quite different approaches. The first is described here; the second is the subject of the next section.

Homology modeling is used to deduce the structure of a sequence with reference to the structure of a close homolog

The technique of **homology modeling** aims to produce a reasonable approximation to the structure of an unknown protein by comparison with the structure of a known sequence homolog (a protein related to it by divergent evolution from a common ancestor). Structures that have diverged too far from each other cannot be modeled reliably; the arrangements in space of their secondary structure elements tend to shift too much. In practice, a sequence with greater than about 40% amino-acid identity with its homolog, and with no large insertions or deletions having to be made in order to align them (Figure 4-22), can usually produce a predicted structure equivalent to that of a medium-resolution experimentally solved structure.

Higher-resolution models can be obtained, in principle at least, when there are a number of aligned sequences. To exploit such information better, a technique was developed that uses evolutionary data for a protein family to measure statistical interactions between amino-acid positions. The technique is based on two hypotheses that derive from empirical observation of

Figure 4-23 Evolutionary conservation and interactions between residues in the protein-interaction domain PDZ and in rhodopsin (a) Highly conserved regions of the PDZ domain were determined using a representative known structure plus information from a structure-based multiple alignment of 274 PDZ-domain sequences, which show a low degree of sequence similarity. This analysis shows that the peptide-binding groove is the most conserved portion of this protein family. Evolutionary conservation is measured by Gstat, a statistical "energy" function: the larger the value of Gstat for a position, the more highly conserved the position is. These data are plotted onto the three-dimensional structure to show the protein interaction surface of the fold, which has a co-crystallized peptide ligand (orange wire model). The high Gstat values for the residues in the groove are consistent with the intuitive expectation that functionally important sites on a protein tend to have a higher than average degree of conservation. **(b)** The structure of the integral membrane protein rhodopsin with the cluster of conserved interacting residues shown in red surrounded by brown van der Waals spheres. This connected network of coevolving residues connects the ligand-binding pocket (green) with known protein-binding regions through a few residues mediating packing interactions between the transmembrane helices. Graphics kindly provided by Rama Ranganathan.

Definitions

homology modeling: a computational method for modeling the structure of a protein based on its sequence similarity to one or more other proteins of known structure.

References

Al-Lazikani, B. *et al.*: **Protein structure prediction.** *Curr. Opin. Chem. Biol.* 2001, **5**:51–56.

Baker, D. and Sali, A.: **Protein structure prediction and structural genomics.** *Science* 2001, **294**:93–96.

Cardozo, T. *et al.*: **Estimating local backbone structural deviation in homology models.** *Comput. Chem.* 2000, **24**:13–31.

Cline, M. *et al.*: **Predicting reliable regions in protein sequence alignments.** *Bioinformatics* 2002, **18**:306–314.

Fetrow, J.S. *et al.*: **Genomic-scale comparison of sequence- and structure-based methods of func-**

sequence evolution. First, a lack of evolutionary constraint at one position should cause the distribution of observed amino acids at that position in the multiple sequence alignment to approach their mean abundance in all proteins, and deviances from the mean values should quantitatively represent conservation. Second, the functional coupling of two positions, even if distantly located in the structure, should mutually constrain evolution at the two positions, and this should be represented in the statistical coupling of the underlying amino-acid distributions in the multiple sequence alignment, which can then be mapped onto the protein (Figure 4-23a). For rhodopsin and for the PDZ domain family, this analysis predicted a set of coupled positions for binding-site residues (shown in red on the figure) that includes unexpected long-range interactions (Figure 4-23b). Mutational studies confirmed these predictions, demonstrating that Gstat, the statistical energy function reflecting conservation, is a good indicator of coupling in proteins. When this technique is used in combination with homology modeling, it can indicate which residues are most likely to remain in conserved positions, even at low levels of sequence identity, and it can also suggest mutagenesis experiments to verify modeled interactions.

What can be done with such models? In some cases they have proven accurate enough to be of value in structure-based drug design. They can be used to predict which amino acids may be in the catalytic site or molecular recognition site if those sites are in the same place in the modeled and experimentally determined protein structures, but they cannot be used to find new binding sites that have been added by evolution. At present, there is no well established way to interrogate an experimentally determined structure, much less a purely modeled structure, and locate such sites from first principles (although some promising new methods are described in section 4-9). Homology models cannot be used to study conformational changes induced by ligand binding, pH changes, or post-translational modification, or the structural consequences of sequence insertions and deletions. At present, computational tools to generate such changes from a starting model are not reliable.

A striking example of the limitations of homology modeling is shown by comparison of the experimentally determined crystal structures of the catalytic domains of the serine protease precursors chymotrypsinogen, trypsinogen, and plasminogen. These protein family members share a high degree of overall sequence identity (over 40%), and an attempt to model the structure of plasminogen from the structure of either of the other two should produce the correct fold. A distinctive difference between plasminogen and the other two zymogens is a complete lack of activity, whereas each of the other two precursors has some activity. This observation cannot be explained from a homology model: the arrangements of residues in the catalytic site will be similar to those of the model template. This is a fundamental limitation of homology modeling: the model is biased toward the structure of the template even in detail. The crystal structure of plasminogen shows that its inactivity is due to blockage of the substrate-binding pocket by a tryptophan residue which is conserved in the sequences of all family members but whose spatial position is different in plasminogen as a result of sequence differences elsewhere in the structure (Figure 4-24).

Homology models also usually cannot be docked together to produce good structures of protein–protein complexes; not only are the docking algorithms unreliable, but the likelihood of significant conformational changes when proteins associate makes it impossible to know whether one is docking the right structures. The same considerations mean that, unless the two homologs have the same oligomeric states, it will not be possible to predict the quaternary structure of a protein from sequence. In short, many, if not most, of the things that biologists want to do with a protein structure cannot be done with confidence using homology models alone. However, even an imperfect homology model may be of use as a guide to planning and interpreting experiments—for example, which amino acid to mutate.

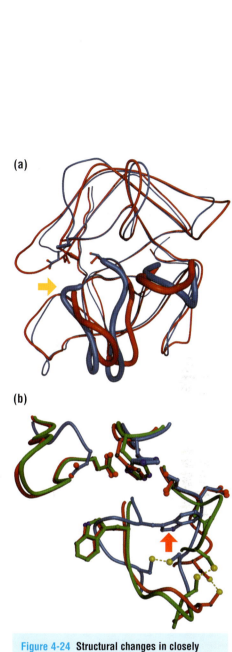

(a)

(b)

Figure 4-24 Structural changes in closely related proteins (a) The structures of plasminogen (blue) and chymotrypsinogen (red) are very similar, as befits their high sequence identity. Yet the small differences in the positions of loops have important functional consequences, as seen in (b). **(b)** Although chymotrypsinogen (red), chymotrypsin (green) and plasminogen (blue) have about the same degree of sequence identity to one another, the active sites of chymotrypsinogen and chymotrypsin differ from that of plasminogen, where a change in the conformation of the loop indicated by the yellow arrow in (a) has caused a tryptophan residue (Trp 761, red arrow), conserved in both sequences, to adopt a different conformation, where it blocks the substrate-binding pocket. (PDB 2cga, 1ab9 and 1qrz)

tion prediction: does structure provide additional insight? *Protein Sci.* 2001, **10**:1005–1014.

Irving, J.A. *et al.*: **Protein structural alignments and functional genomics.** *Proteins* 2001, **42**:378–382.

Lockless, S.W. and Ranganathan, R.: **Evolutionarily conserved pathways of energetic connectivity in protein families.** *Science* 1999, **286**:295–299.

Marti-Renom, M.A., *et al.*: **Comparative protein structure modeling of genes and genomes.** *Ann. Rev. Biophys.* 2000, **29**:291–325.

Peisach, E. *et al.*: **Crystal structure of the proenzyme domain of plasminogen.** *Biochemistry* 1999, **38**:11180–11188.

Yang, A.S. and Honig, B.: **An integrated approach to the analysis and modeling of protein sequences and structures I-III.** *J. Mol. Biol.* 2000, **301**:665–711.

Protein Data Bank website:
http://www.rcsb.org/pdb/index.html

Profile-based threading tries to predict the structure of a sequence even if no sequence homologs are known

The most important method that has been developed so far for the identification of a protein fold from sequence information alone in the absence of any apparent sequence identity to any other protein, is the method of "profile-based threading". In this method, a computer program forces the sequence to adopt every known protein fold in turn, and in each case a scoring function is calculated that measures the suitability of the sequence for that particular fold (Figure 4-25).

The function provides a quantitative measure of how well the sequence fits the fold. The method is based on the assumption that three-dimensional structures of proteins have characteristics that are at least semi-quantitatively predictable and that reflect the physical-chemical properties of strings of amino acids in sequences as well as limitations on the types of interactions allowed within a folded polypeptide chain. Does, for example, forcing the sequence to adopt particular secondary structures and intra-protein interactions place hydrophobic residues on the inside and helix-forming residues in helical segments? If so, the score will be relatively high.

Experience with profile-based threading has shown that a high score, indicating a good fit to a particular fold, can always be trusted. On the other hand, a low score only indicates that a fit was not found; it does not necessarily indicate that the sequence cannot adopt that fold. Thus, if the method fails to find any fold with a significantly high score, nothing has been learned about the sequence. Despite this limitation, profile-based threading is a powerful method that has been able to identify the general fold for many sequences. It cannot provide fine details of the structure, however, because at such low levels of sequence identity to the reference fold the local interactions and side-chain conformations will not necessarily be the same.

The Rosetta method attempts to predict protein structure from sequence without the aid of a homologous sequence or structure

Ideally, one would like to be able to compute the correct structure for any protein from sequence information alone, even in the absence of homology. Ongoing efforts to achieve this "holy grail" of structure prediction have met with mixed success. Periodically these methods are tested against proteins of known but unpublished structures in a formal competition called CASP (critical assessment of techniques for protein structure prediction). Perhaps the most promising at the moment is the Rosetta method. One of the fundamental assumptions underlying Rosetta is that the distribution of conformations sampled for a given short segment of the sequence is reasonably well approximated by the distribution of structures adopted by that sequence and closely related sequences in known protein structures. Fragment libraries for short segments of the chain are extracted from the protein structure database. At no point is knowledge of the overall native structure used to select fragments or fix segments of the structure. The conformational space defined by these fragments is then searched using a Monte Carlo procedure with an energy function that favors compact structures with paired strands and buried hydrophobic residues. A total of 1,000 independent simulations are carried out for each query sequence, and the resulting structures are clustered. One selection method was simply to choose the centers of the largest clusters as the highest-confidence models. These cluster centers are then rank-ordered according to the size of the clusters they represent, with the cluster centers representing the largest clusters being designated as the highest-confidence models. Before clustering, most structures produced by Rosetta are incorrect (that is, good

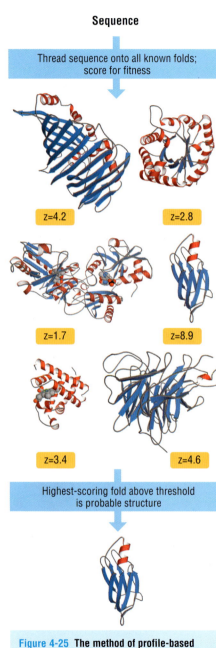

Sequence

Thread sequence onto all known folds; score for fitness

z=4.2 z=2.8

z=1.7 z=8.9

z=3.4 z=4.6

Highest-scoring fold above threshold is probable structure

Figure 4-25 The method of profile-based threading A sequence of unknown structure is forced to adopt all known protein domain folds, and scored for its suitability for each fold. The z-value relates the score for the query sequence to the average score for a set of random sequences with the same amino-acid composition and sequence length. A very high z-score indicates that the sequence almost certainly adopts that fold. Sequences can be submitted online for threading by PSIPRED (http://bioinf.cs.ucl.ac.uk/psipred/index.html).

References

Bonneau, R. et al.: **Rosetta in CASP4: Progress in** *ab initio* **protein structure prediction.** *Proteins* 2001, **45(S5)**:119–126.

Bowie, J.U. et al.: **A method to identify protein sequences that fold into a known three-dimensional structure.** *Science* 1991, **253**:164–170.

de la Cruz, X. and Thornton, J.M.: **Factors limiting the performance of prediction-based fold recognition methods.** *Protein Sci.* 1999, **8**:750–759.

Fischer, D. and Eisenberg, D.: **Protein fold recognition using sequence-derived predictions.** *Protein Sci.* 1996, **5**:947–955.

Miller, R.T. et al.: **Protein fold recognition by sequence threading: tools and assessment techniques.** *FASEB J.* 1996, **10**:171–178.

Simons, K.T. et al.: **Assembly of protein tertiary structures from fragments with similar local sequences using simulated annealing and Bayesian scoring functions.** *J. Mol. Biol.* 1997, **268**:209–225.

structures account for less than 10% of the conformations produced); for this reason, most conformations generated by Rosetta are referred to as decoys (Figure 4-26). The problem of discriminating between good and bad decoys in Rosetta populations is still under investigation. Still, in some test calculations, the best cluster center has been shown to agree fairly well with the overall fold of the protein (Figure 4-27).

Both the Rosetta method and the method of profile-based threading suffer from some of the same limitations that beset homology modeling. The issue of false positives and negatives is significant, because the failure to generate a model does not mean one cannot be generated, nor that the structure is a novel one. And the generation of a model does not mean it is right, either overall or, more usually, in detail. At best one should look to these methods, at least for the present, for rough indication of fold class and secondary structure topology. And it is important to remember that all methods of model building based on a preexisting structure, whether found by sequence homology or by threading, suffer from massive feedback and bias. The structure obtained will always look like the input structure, because the computational tools for refining the model are unable to generate the kinds of shifts in secondary structure position and local tertiary structure conformations that are likely to exist between two proteins when their overall sequence identity is low (see Figure 4-19). *Ab initio* methods like Rosetta at least do not suffer from this problem, whatever their other limitations.

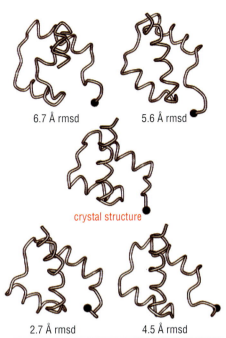

6.7 Å rmsd   5.6 Å rmsd

crystal structure

2.7 Å rmsd   4.5 Å rmsd

Figure 4-26 Some decoy structures produced by the Rosetta method The structure at the center is the target, the experimentally determined structure of a homeodomain. The other structures are generated by the Monte Carlo approach in Rosetta, using only the sequence of the protein. Although some of the structures are quite far from the true structure, others are close enough for the fold to be recognizable. Rmsd is the root mean square deviation in α-carbon positions between the computed structure and the experimentally determined structure. (Taken from Simons, K.T. *et al.*: *J. Mol. Biol.* 1997, **268**:209–225.)

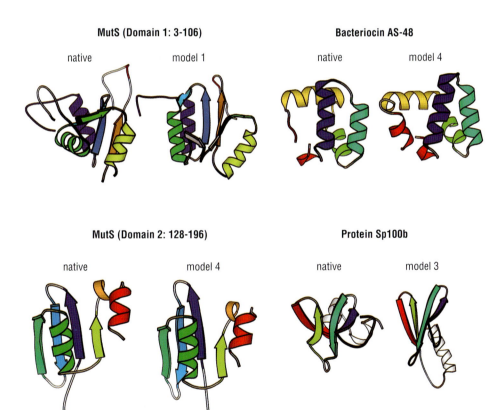

MutS (Domain 1: 3-106)

native   model 1

Bacteriocin AS-48

native   model 4

MutS (Domain 2: 128-196)

native   model 4

Protein Sp100b

native   model 3

Figure 4-27 Examples of the best-center cluster found by Rosetta for a number of different test proteins The level of agreement with the known native structure varies, but in many cases the overall fold is predicted well enough to be recognizable. Note, however, that the relative positions of the secondary structure elements are almost always shifted at least somewhat from their true values. Graphics kindly provided by Richard Bonneau and David Baker. (Adapted from Bonneau, R. *et al.*: *Proteins* 2001, **45(S5)**:119–126.)

Simons, K.T. *et al.*: **Prospects for *ab initio* protein structural genomics.** *J. Mol. Biol.* 2001, **306**: 1191–1199.

URL for threading website:
http://bioinf.cs.ucl.ac.uk/psipred/index.html

URL for CASP:
http://moult.carb.nist.gov/casp

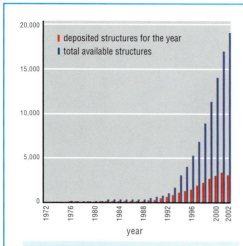

Figure 4-28 Growth in the number of structures in the protein data bank Both yearly and cumulative growth is shown. (Taken from the Protein Data Bank website: http://www.rcsb.org/pdb/holdings.html)

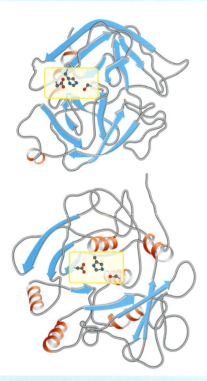

Figure 4-29 The overall folds of two members of different superfamilies of serine proteases The enzymes are chymotrypsin (top) and subtilisin (bottom). The residues in the catalytic triad are indicated for each.

Members of a structural superfamily often have related biochemical functions

In contrast to the exponential increase in sequence information, structural information, which is still chiefly obtained by X-ray crystallography and NMR, has up to now been increasing at a much lower rate (Figure 4-28). One goal of the structural genomics initiatives that have been implemented since the release of the first complete genome sequence is to increase the rate at which experimentally derived structures of the gene products are produced. The driving force behind these initiatives is the assumption that, in addition to defining the ensemble of all possible protein folds, comprehensive structural information could provide a firmer basis than sequence for functional predictions, as three-dimensional structure changes much more slowly than sequence during evolution. A good reason for optimism that these assumptions will hold true is the existence of **superfamilies** of proteins with related structures and biochemical functions.

A superfamily is loosely defined as a set of homologous proteins with similar three-dimensional structures and related, though not necessarily identical, biochemical functions. Almost all superfamilies exhibit some functional diversity, which is generated by local sequence variation and/or domain shuffling. Within enzyme superfamilies, for example, substrate diversity is common, while parts of the reaction chemistry are highly conserved. In many enzyme superfamilies, the sequence positions of catalytic residues vary from member to member, despite the fact that they have equivalent functional roles in the proteins. These variations may make the assignment of a protein to a superfamily from sequence comparison alone problematic or impossible. Although some superfamily members may be similar in sequence, it is the structural and functional relationships that place a protein in a particular superfamily. Within each superfamily, there are **families** with more closely related functions and significant (>50%) sequence identity.

Because the total number of protein folds and the total number of biochemical functions is smaller than the total number of genes in biology, if a protein can be assigned to a superfamily from sequence or structural information, at the very least the number of its possible functions can be narrowed down, and in some instances it may be possible to assign a function precisely.

The four superfamilies of serine proteases are examples of convergent evolution

Striking examples of similarities in biochemical function but quite different biological roles come from enzymes where the chemical reactions are the same but the substrates can differ considerably. There are, for instance, many hundreds of enzymes that hydrolyze peptide bonds in protein and polypeptide substrates, but they can be grouped into a small number of classes, each with its own characteristic chemical mechanism. The most numerous class comprises the serine proteases, in which the side-chain hydroxyl group of a serine residue in the active site attacks the carbonyl carbon atom of the amide bond that is to be hydrolyzed. Two other characteristic residues, a histidine and an aspartic acid (or a glutamic acid), are involved in assisting this hydrolysis, forming a catalytic triad. Serine proteases fall into several structural superfamilies, which are recognizable from their amino-acid sequences and the particular disposition of the three catalytically important residues in the active site (Figure 4-29). Each serine protease superfamily has many members but there is no obvious relationship between the superfamilies, either in sequence or structure. The three residues of the catalytic triad are

---

Definitions

family: a group of homologous proteins that share a related function. Usually these will also have closely related sequences. Members of the same enzyme family catalyze the same chemical reaction on structurally similar substrates.

superfamily: proteins with the same overall fold but with usually less than 40% sequence identity. The nature of the biochemical functions performed by proteins in the same superfamily are more divergent than those within families. For instance, members of

the same enzyme superfamily may not catalyze the same overall reaction, yet still retain a common mechanism for stabilizing chemically similar rate-limiting transition-states and intermediates, and will do so with similar active-site residues.

References

Gerlt, J.A. and Babbitt, P.C.: **Divergent evolution of enzymatic function: mechanistically diverse superfamilies and functionally distinct suprafamilies.** *Annu. Rev. Biochem.* 2001, **70**:209–246.

Krem, M.M. *et al.*: **Sequence determinants of function and evolution in serine proteases.** *Trends Cardiovasc. Med.* 2000, **10**:171–176.

Perona, J.J. and Craik, C.S.: **Evolutionary divergence of substrate specificity within the chymotrypsin-like**

found in a different order in different locations along the sequence in each superfamily: nevertheless, in the tertiary structure they come together in a similar configuration. Presumably, the existence of a similar active site is due to convergent evolution, while within each superfamily, divergent evolution has produced distinct individual proteases with very similar structures but different substrate specificity.

A given serine protease can be highly specific for a particular target amino-acid sequence— although some are relatively nonspecific—so that, in general, the substrate(s) for such a protease cannot be predicted from knowledge of the sequence or even the structure of the enzyme. However, observation of a serine protease fold combined with the right active-site residues is diagnostic for a protease. Serine proteases participate in such diverse cellular functions as blood clotting, tissue remodeling, cell-cycle control, hormone activation and protein turnover. A small number of members of some families within the serine protease superfamily have lost one or more of the catalytic residues and perform non-catalytic functions such as forming a structural matrix.

Another large enzyme superfamily with numerous different biological roles is characterized by the so-called polymerase fold, which resembles an open hand (Figure 4-30). DNA polymerases of all types from all organisms studied so far appear to share this fold, as do RNA polymerases and viral reverse transcriptases. In most cases, sequence comparisons alone do not make this functional distinction, and in some cases no sequence similarity between the families of the polymerase fold superfamily is apparent. However, structural evidence indicates that every enzyme that transcribes or replicates nucleic acid polymers has probably descended from a common ancestor.

There are many other examples of related but differing functions among members of a superfamily. It should be borne in mind, however, that in some proteins with similar function, equivalent active-site residues come from different positions in the sequence, obscuring superfamily membership until structural information is obtained.

Very closely related protein families can have completely different biochemical and biological functions

There are some well-known cases of significant differences in function at very high levels of sequence identity. One example is the crystallins, which appear to have evolved from several different enzymes. Although some crystallins retain more than 50% sequence identity to these enzymes, they function as structural proteins, not enzymes, in the eye lens.

And with increasing numbers of structures being solved for proteins of known function, the functional diversity of many other protein superfamilies has been revealed. Thornton and co-workers have assessed the functional variation within homologous enzyme superfamilies containing two or more enzymes. Combining sequence and structure information to identify relatives, the majority of superfamilies display variation in enzyme function, with 25% of the superfamilies having members of different enzyme types. For example, the α/β hydrolase superfamily has at least four different functions; the ferredoxin superfamily has at least three. For single- and multidomain enzymes, difference in biochemical function is rare above 40% sequence identity, and above 30% the overall reaction type tends to be conserved, although the identity of the substrate may not be. For more distantly related proteins, sharing less than 30% sequence identity, functional variation is significant, and below this threshold, structural data are essential for understanding the molecular basis of observed functional differences.

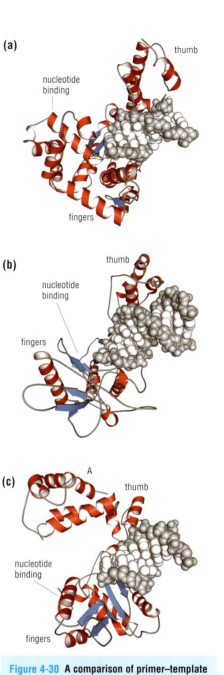

Figure 4-30 A comparison of primer–template DNA bound to three DNA polymerases
(a) Taq DNA polymerase bound to DNA. The DNA stacks against the "fingers" and is contacted across the minor groove by the "thumb" domain. **(b)** The binary complex of HIV-1 reverse transcriptase and DNA. This structure does not have a nucleotide-binding alpha helix in the fingers domain. Instead, a beta hairpin probably performs this function. **(c)** The ternary complex of rat DNA polymerase β with DNA and deoxy-ATP (not shown). Although this polymerase has an additional domain (A), the "thumb" domain similarly binds the DNA primer–template in the minor groove, while the "fingers" present a nucleotide-binding alpha helix at the primer terminus. (PDB 1tau, 2hmi and 8icp)

serine protease fold. *J. Biol. Chem.* 1997, **272**: 29987–29990.

Siezen, R.J. and Leunissen, J.A.: **Subtilases: the super-family of subtilisin-like serine proteases.** *Protein Sci.* 1997, **6**:501–523.

Steitz, T.A.: **DNA polymerases: structural diversity and common mechanisms.** *J. Biol. Chem.* 1999, **274**:17395–17398.

Todd, A.E. *et al.*: **Evolution of function in protein superfamilies, from a structural perspective.** *J. Mol.* *Biol.* 2001, **307**:1113–1143.

Evolution of function in protein superfamilies, from a structural perspective:

http://www.biochem.ucl.ac.uk/bsm/FAM-EC/

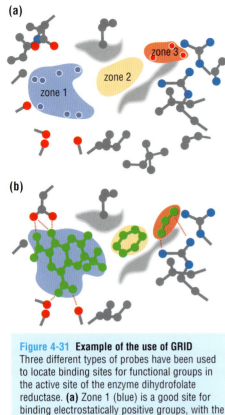

(a)

zone 3

zone 2

zone 1

(b)

Figure 4-31 Example of the use of GRID
Three different types of probes have been used to locate binding sites for functional groups in the active site of the enzyme dihydrofolate reductase. **(a)** Zone 1 (blue) is a good site for binding electrostatically positive groups, with the energy function minima from an amino probe shown in blue dots. It was also identified with a carbon probe as being a good pocket for shape complementarity. Zone 2 (yellow) is a good site for hydrophobic interaction, as illustrated by the hydrophobic molecular surface (grey shapes) in that region. Zone 3 (red) is a good binding site for electrostatically negative groups, with minima from a carboxylate probe shown by the red dots. **(b)** Overlay of three pieces of a known inhibitor of dihydrofolate reductase onto the zones of favorable interaction energy found by GRID. Figures adapted from http://thalassa.ca.sandia.gov/~dcroe/builder.html

Binding sites can sometimes be located in three-dimensional structures by purely computational means

Whether the three-dimensional structure of a new protein is determined experimentally or computationally (see sections 4-6 and 4-7)—there will always be the problem of finding those sites on the protein surface that are involved in its biochemical and cellular functions. This problem is particularly acute when the structure reveals a polypeptide chain fold that has never been seen before, and there is no obvious cofactor or other bound ligand to identify a functional site. But it can also apply to structures of proteins with known folds, as proteins often have more than one function (see section 4-13) and even a familiar protein may have acquired additional functions and functional sites in the context of a different organism.

To some extent, the characteristics of binding sites that were discussed in section 2-4 can be used to identify regions of a protein's surface that are good candidates for functional sites. These characteristics include concavity—binding sites are usually depressions rather than protrusions or flat areas, although many exceptions are known—as well as a higher than average amount of exposed hydrophobic surface area. However, such generalizations usually only narrow down the possibilities. What is needed is a method, ideally a computational method so that it can be used with homology models as well as with experimentally derived structures, that can scan the surface of a protein structure and locate those sites that have evolved to interact with small molecules or with other macromolecules. Some success has been obtained with methods that scan the surface and look for sites of specific shape. Residue conservation analysis can also be quite revealing if there are enough homologous sequences (see Figure 4-23).

Several other computational methods have been proposed and tested on experimentally determined structures. Most use a "probe" molecule and an energy function that describes the interaction of the probe with the residues on the protein surface. Binding sites are identified as regions where the computed interaction energy between the probe and the protein is favorable for binding. Two widely used methods, GRID and MCSS (multiple conformations simultaneous search), use this strategy. In GRID, the interaction of the probe group with the protein structure is computed at sample positions on a lattice throughout and around the macromolecule, giving an array of energy values. The probes, which are usually used singly, include water, the methyl group, the amine NH_2 group, the carboxylate group and the hydroxyl group, among others. Contour surfaces at various energy levels are calculated for each probe for each point on the lattice and displayed by computer graphics together with the protein structure (Figure 4-31). Contours at negative energy levels delineate regions of attraction between probe and protein that could indicate a binding site, as such contours are found at known ligand-binding clefts.

The approach taken by MCSS is similar in principle but differs in detail and can take into account the flexibility of both the probe molecule and the protein. The resulting distribution map of regions on the protein surface where functional groups show a favorable interaction energy can be used for the analysis of protein–ligand interactions and for rational drug design.

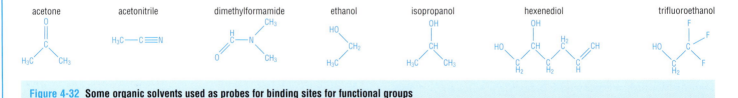

acetone   acetonitrile   dimethylformamide   ethanol   isopropanol   hexenediol   trifluoroethanol

Figure 4-32 Some organic solvents used as probes for binding sites for functional groups

References

Allen, K.N. et al.: **An experimental approach to mapping the binding surfaces of crystalline proteins.** *J. Phys. Chem.* 1996, **100**:2605–2611.

Aloy, P. et al.: **Automated structure-based prediction of functional sites in proteins: applications to assessing the validity of inheriting protein function from homology in genome annotation and to protein docking.** *J. Mol. Biol.* 2001, **311**:395–408.

Bitetti-Putzer, R. et al.: **Functional group placement in protein binding sites: a comparison of GRID and MCSS.** *J. Comput. Aided Mol. Des.* 2001, **15**:935–960.

Byerly, D.W. et al.: **Mapping the surface of *Escherichia coli* peptide deformylase by NMR with organic solvents.** *Protein Sci.* 2002, **11**:1850–1853.

Dennis, S. et al.: **Computational mapping identifies the binding sites of organic solvents on proteins.** *Proc. Natl Acad. Sci. USA* 2002, **99**:4290–4295.

English, A.C. et al.: **Experimental and computational mapping of the binding surface of a crystalline protein.** *Protein Eng.* 2001, **14**:47–59.

Goodford, P.J.: **A computational procedure for determining energetically favorable binding sites on biologically important macromolecules.** *J. Med. Chem.* 1985, **28**:849–857.

Laskowski, R.A. et al.: **Protein clefts in molecular recognition and function.** *Protein Sci.* 1996, **5**:2438–2452.

Liang, J. et al.: **Anatomy of protein pockets and cavities: measurement of binding site geometry and implications for ligand design.** *Protein Sci.*

Computational methods are, however, extremely inefficient when no information is available to limit the regions of the protein surface to be scanned. Scanning an entire protein surface with either GRID or MCSS yields hundreds of possible binding sites of roughly equivalent energy. Thus, these tools are most useful in cases where the active site is already known and one wants to determine what sorts of chemical groups might bind there.

Experimental means of locating binding sites are at present more accurate than computational methods

MSCS (multiple solvent crystal structures) is a crystallographic technique that identifies energetically favorable binding sites and orientations of small organic molecules on the surface of proteins; this experimental method can find likely functional sites on the surface of any protein that can be crystallized. The method involves soaking protein crystals in an organic solvent that mimics a functional group on a ligand: thus, ethanol will probe for hydroxymethyl-binding sites such as those that interact with a threonine side chain; dimethylformamide identifies binding sites that interact with the C=O and N–H groups of peptides, and so on (Figure 4-32). Determination of the protein structure at resolutions of the order of 2 Å in the presence of the solvent probe reveals the solvent-binding sites (Figure 4-33). If the experiment is repeated with several different probes, it is found that they cluster in only a few binding sites, regardless of their polarity. These sites are the functional sites on the surface of the protein. In contrast to the computational methods, MSCS involves direct competition between the probe and the bound water on the surface of the protein. As it is displacement of this water that drives ligand-binding events (see section 2-4), MSCS finds a much more restricted set of binding sites than do the computational methods.

High-resolution structures of crystals of the well-studied enzyme thermolysin soaked in acetone, acetonitrile, or phenol show probe molecules clustering in the main specificity pocket of the thermolysin active site and in a buried sub-site, consistent with structures of known protein–ligand complexes of thermolysin (Figure 4-34). When the experimentally determined solvent positions within the active site were compared with predictions from GRID and MCSS, both these computational methods found the same sites but gave fewer details of binding. And both GRID and MCSS predicted many other sites on the protein surface, not observed experimentally, as equally favorable for probe binding.

Related experimental methods using NMR instead of X-ray crystallography study the binding of small-molecule compounds as well as organic solvents as probes. The experimental methods are accurate but cannot be used on homology models. Computational methods are not accurate enough to discriminate among possible binding sites. What is needed is a computational analog of the experimental methods. One such new computational mapping strategy has been tested recently with promising results. Using eight different ligands for lysozyme and four for thermolysin, the computational search finds the consensus site to which all the ligands bind, whereas positions that bind only some of the ligands are ignored. The consensus sites turn out to be pockets of the enzymes' active sites, lined with partially exposed hydrophobic residues and with some polar residues toward the edge. Known substrates and inhibitors of hen egg-white lysozyme and thermolysin interact with the same side chains identified by the computational mapping, but the computational mapping did not identify the precise hydrogen bonds formed and the unique orientations of the bound substrates and inhibitors.

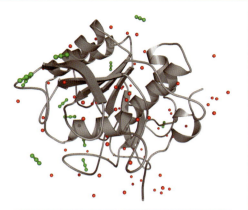

Figure 4-33 Structure of subtilisin in 100% acetonitrile Crystal structure of the serine protease subtilisin in 100% acetonitrile. The organic solvent, shown as green rods, binds at only a few sites on the protein surface, including the active site, which is approximately left of center in the figure. The red spheres are bound water molecules, which are not displaced even by this water-miscible organic solvent at 100% concentration. These bound waters should be considered an integral part of the folded structure of the protein. (PDB 1be6)

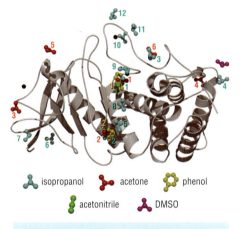

isopropanol   acetone   phenol

acetonitrile   DMSO

Figure 4-34 Ribbon representation showing the experimentally derived functionality map of thermolysin The binding sites for different organic solvent molecules were obtained by X-ray crystallography of crystals of thermolysin soaked in the solvents. The same probe molecules bound to different positions are numbered to identify their site of binding. The active-site zinc ion and the bound calcium ions are shown as grey and black spheres, respectively. Dimethyl sulfoxide (DMSO, purple) is present in the crystallization conditions of thermolysin; one molecule binds per molecule protein. Graphic kindly provided by Roderick E. Hubbard. (Adapted from English *et al.*: *Protein Eng.* 2001, **14**:47–59.)

1998, **7**:1884–1897.

Liepinsh, E. and Otting, G.: **Organic solvents identify specific ligand binding sites on protein surfaces.** *Nat. Biotechnol.* 1997, **15**:264–268.

Mattos, C. and Ringe, D.: **Locating and characterizing binding sites on proteins.** *Nat. Biotechnol.* 1996, **14**:595–599.

Mattos, C. and Ringe, D.: **Proteins in organic solvents.** *Curr. Opin. Struct. Biol.* 2001, **11**:761–764.

Miranker, A. and Karplus, M.: **Functionality maps of binding sites: a multiple copy simultaneous search method.** *Proteins* 1991, **11**:29–34.

Shuker, S.B. *et al.*: **Discovering high-affinity ligands for proteins: SAR by NMR.** *Science* 1996, **274**:1531–1534.

URL for GRID:
http://thalassa.ca.sandia.gov/~dcroe/builder.html

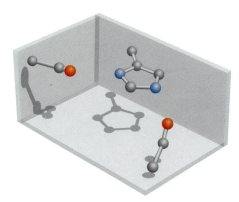

Figure 4-35 **An active-site template** The geometry of the catalytic triad of the serine proteases as used to locate similar sites in other proteins. Adapted from the rigid active-site geometries website: http://www.biochem.ucl.ac.uk/bsm/PROCAT/PROCAT.html.

Site-directed mutagenesis can identify residues involved in binding or catalysis

Locating binding sites, either experimentally or computationally (see section 4-9), does not automatically indicate which residues in those sites are responsible for ligand binding or, in the case of enzymes, catalysis. The standard experimental method for identifying these residues is alanine-scanning mutagenesis, in which candidate amino acids are replaced by alanine by site-directed mutagenesis of the gene. The effect of this side-chain excision on function—usually binding or catalysis—of the expressed mutant protein is then assayed. When combined with genetic assays for the *in vivo* phenotype of the mutated protein and information from the pH/rate profile of a catalytic reaction, for example, such experiments can reveal which side chains in a binding site may actually perform chemistry on a bound ligand.

Active-site residues in a structure can sometimes be recognized computationally by their geometry

But with the advent of genome-wide sequencing and protein-structure determination, a computational tool is needed that can identify such residues rapidly and automatically. This would be particularly useful in cases where the protein in question has no known function and the location of the active site is uncertain, because knowledge of the residues that can carry out chemistry might indicate what type of chemistry the protein actually performs.

The simplest of the computational methods searches the structure for geometrical arrangements of chemically reactive side chains that match those in the active sites of known enzymes. This rigid active-site approach has successfully identified the catalytic triad of a serine protease (Figure 4-35) in an enzyme of unknown function, but has not been used extensively to probe for other functions. Because it relies solely on geometry and not on position in the sequence, this method could find serine protease catalytic sites in any protein fold in which they occur. A more sophisticated variation uses a three-dimensional descriptor of the functional site of interest, termed a "fuzzy functional form", or FFF, to screen the structure. FFFs are based on the geometry, residue identity, and conformation of active sites using data from known crystal structures of members of a functional family and experimental biochemical data. The descriptors are made as general as possible ("fuzzy") while still being specific enough to identify the correct active sites in a database of known structures.

These fuzzy functional descriptors can identify active sites not only in experimentally determined structures, but also from predicted structures provided by *ab initio* folding algorithms (see sections 1-9, 4-7) or threading algorithms (see section 4-7). A disulfide oxidoreductase FFF has been successfully applied to find other disulfide oxidoreductases in a small structural database and, more recently, has been used to scan predicted protein structures derived from the entire *Bacillus subtilis* genome. A total of 21 candidate disulfide oxidoreductases were found, of which six turned out to be false positives. The method did not miss any of the known disulfide oxidoreductases and identified at least two potential new ones.

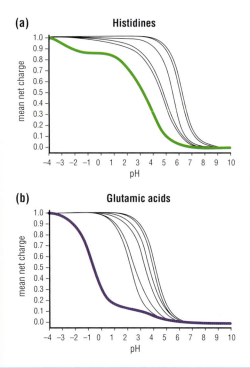

Figure 4-36 **Theoretical microscopic titration curves** Calculated curves for two types of ionizable residues—histidines and glutamic acids—in the structure of the glycolytic enzyme triosephosphate isomerase. In each case, most of the residues behave similarly, with a sharp change in charge as a function of pH, but one residue of each type (histidine 95 (curve in green) and glutamic acid 165 (curve in blue)) displays abnormal behavior.

References

Bartlett, G.J. *et al.*: **Analysis of catalytic residues in enzyme active sites.** *J. Mol. Biol.* 2002, **324**:105–121.

Di Gennaro, J.A. *et al.*: **Enhanced functional annotation of protein sequences via the use of structural descriptors.** *J. Struct. Biol.* 2001, **134**:232–245.

Ewing, T.J. *et al.*: **DOCK 4.0: search strategies for automated molecular docking of flexible molecule databases.** *J. Comput. Aided Mol. Des.* 2001, **15**:411–428.

Fetrow, J.S. *et al.*: **Structure-based functional motif identifies a potential disulfide oxidoreductase active site in the serine/threonine protein phosphatase-1 subfamily.** *FASEB J.* 1999, **13**:1866–1874.

Jones, S. and Thornton, J.M.: **Prediction of protein–protein interaction sites using patch analysis.** *J. Mol. Biol.* 1997, **272**:133–143.

Laskowski, R.A. *et al.*: **Protein clefts in molecular recognition and function.** *Protein Sci.* 1996, **5**:2438–2452.

Ondrechen, M.J. *et al.*: **THEMATICS: a simple computa-**

tional predictor of enzyme function from structure. *Proc. Natl Acad. Sci. USA* 2001, **98**:12473–12478.

Reva, B. *et al.*: **Threading with chemostructural restrictions method for predicting fold and functionally significant residues: application to dipeptidylpeptidase IV (DPP-IV).** *Proteins* 2002, **47**:180–193.

Sheinerman, F.B. *et al.*: **Electrostatic aspects of protein–protein interactions.** *Curr. Opin. Struct. Biol.* 2000, **10**:153–159.

Wallace, A.C. *et al.*: **Derivation of 3D coordinate**

The limitation with all computational methods of this type is that they can only find active-site residues that conform to known active sites. A protein with a novel function will yield no result, or worse, an incorrect identification with a known site. A more general approach has been developed that does not depend on previous knowledge of active-site geometries. This employs theoretical microscopic titration curves (THEMATICS), to identify active-site residues that are potentially involved in acid-base chemistry in proteins of known structure. Location of such residues automatically determines the position of the active site, as well as providing a clue to the biochemical function of the protein.

In THEMATICS, the mean net charge of potentially ionizable groups in each residue in the protein structure is calculated as a function of pH. The resulting family of curves for each type of residue (Figure 4-36) is then analyzed for deviations from ideal behavior. A small fraction (3–7%) of all curves for all residues differ from the others in having a flat region where the residue is partially protonated over a wide pH range. Most residues with these perturbed curves occur in active sites (Figure 4-37). The method is successful for proteins with a variety of different chemistries and structures and has a low incidence of false positives. Of course, identification of acid-base residues in active sites does not necessarily establish what the overall chemical reaction must be. Most enzymatic reactions use one or more acid-base steps but catalyze other chemistries as well. Since the pK_a values of catalytic residues are likely to be perturbed by electrostatic interactions, once they are identified, computational tools such as GRASS can be used to compute and display such interactions.

Docking programs model the binding of ligands

Even when it is possible to identify active sites and to draw some conclusions about the likely chemistry they will perform, it is still necessary to determine on what substrate(s) that chemistry will operate. At present, there is no method, experimental or computational, that will enable one to find the most likely substrate for any particular active site. Some approaches involving mass spectroscopy to identify ligands pulled out of cellular extracts by the protein in question are under development, and peptide substrates for protein kinases can often be found by screening combinatorial peptide libraries, but a computational method would be most general. One promising approach is that of the program DOCK, where the shape of the binding site on a protein is represented as a set of overlapping spheres, in which the centers of the spheres become potential locations for ligand atoms. Each ligand is divided into a small set of rigid fragments that are docked separately into the binding site, allowing a degree of flexibility at the positions that join them. The fragments are rejoined later in the calculation and an energy minimum calculated for the rejoined ligand in the receptor site. The method can find binding geometries for the ligand similar to those observed crystallographically, as well as other geometries that provide good steric fit, and has been used to find possible new compounds for drug development. In such applications, each of a set of small molecules from a structural database is individually docked to the receptor in a number of geometrically permissible orientations. The orientations are evaluated for quality of fit, with the best fits being kept for examination by molecular mechanics calculations.

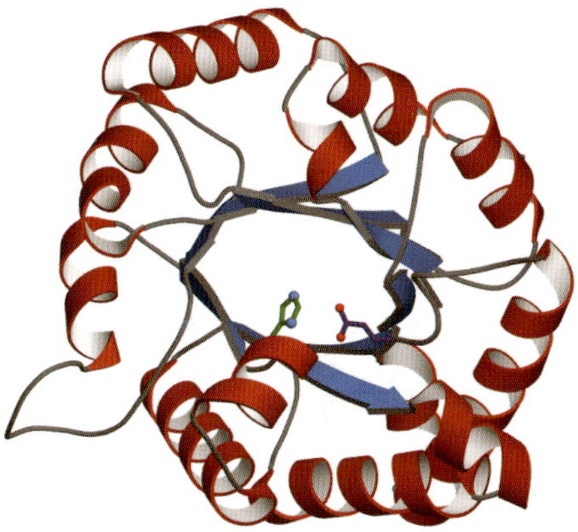

The method cannot take unknown conformational changes of the protein into account. In principle, it could be used to find candidate substrates for any active site, but in practice it is too computationally cumbersome, and all potential ligands are not contained in any database. The method also gives not one but many possible ligands from any database, and the energy function used to evaluate binding cannot discriminate among them.

Figure 4-37 Residues that show abnormal ionization behavior with changing pH define the active site The locations of the two abnormally titrating residues in Figure 4-36 are shown on the three-dimensional structure of triosephosphate isomerase. The histidine (green) and glutamic acid (blue) that are partially protonated over a wide range of pH are both located in the active site and both are important in catalysis.

templates for searching structural databases: application to Ser-His-Asp catalytic triads in the serine proteinases and lipases. *Protein Sci.* 1996, **5**:1001–1013.

Zhang, B. *et al.*: **From fold predictions to function predictions: automation of functional site conservation analysis for functional genome predictions.** *Protein Sci.* 1999, **8**:1104–1115.

Rigid active-site geometries:
http://www.biochem.ucl.ac.uk/bsm/PROCAT/PROCAT.html

GRASS, a protein surface visualization tool:
http://trantor.bioc.columbia.edu/

DOCK:
http://www.cmpharm.ucsf.edu/kuntz/dock.html

FFF:
http://www.geneformatics.com

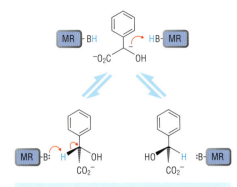

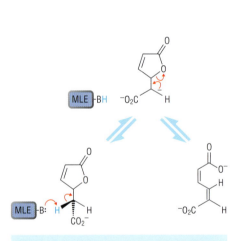

Figure 4-38 The chemical reaction catalyzed by mandelate racemase R-mandelate (left) and S-mandelate (right) can be converted into each other through the intermediate in the center. The enzyme (MR) catalyzes the reaction by removing a proton from a carbon atom adjacent to a carboxylate group and subsequently replacing it. A basic residue (B) at one side of the active site of the enzyme removes the proton and a proton is replaced by a basic residue on the other side of the active site (not shown), which can also act as an acid. The red arrows indicate the movement of electron pairs. In the reverse reaction, the two residues reverse roles.

Figure 4-39 The chemical reaction catalyzed by muconate lactonizing enzyme Although the substrates are different, the core step in the catalytic mechanism of muconate lactonizing enzyme (MLE) is similar to that of mandelate racemase.

Knowledge of a protein's structure does not necessarily make it possible to predict its biochemical or cellular functions

Perhaps the most promising case for the prediction of biochemical function from structure is when two proteins show some similarity in amino-acid sequences, share the same overall tertiary structure, and have active sites with at least some residues in common. But consideration of just such an example shows that, even with favorable parameters, function cannot always be deduced from structure.

The bacterium *Pseudomonas ovalis* lives in the soil and scavenges a wide variety of organic compounds for food. One of these is mandelate, a byproduct of decaying fruit pits (Figure 4-38). Mandelate naturally exists as two mirror-image isomers, R- and S-mandelate, and *P. ovalis* can use both as a carbon source. Two enzymes essential for the bacterium to grow on either R- or S-mandelate are mandelate racemase and muconate lactonizing enzyme.

The biochemical function of mandelate racemase (MR) is to interconvert R- and S-mandelate (Figure 4-38); because only S-mandelate is a substrate for the next enzyme in the degradative pathway, this enables *P. ovalis* to metabolize all the available mandelate instead of just half. MR is a metalloenzyme. It requires a magnesium or manganese ion for catalytic activity. Muconate lactonizing enzyme (MLE), which is further along the pathway of mandelate catabolism, transforms the *cis, cis*-muconic acid derived from mandelate into muconolactone (Figure 4-39). This is an essential step in the overall breakdown of mandelate into acetyl-CoA, a substrate for energy production via the tricarboxylic acid cycle. MLE is also a metalloenzyme. It requires a manganese ion for activity, although magnesium can be substituted.

The substrates for MR and MLE are very different molecules, and the biochemical functions of these two proteins are also different. Their amino-acid sequences are 26% identical, which falls in the "grey area" where one cannot predict for certain that two proteins will have any domains with a similar fold (see Figure 4-19). Secondary-structure prediction is also uninformative. Nevertheless, when the three-dimensional structures of MR and MLE were determined by X-ray crystallography, they showed that the overall folds were essentially identical (Figure 4-40). Both enzymes are TIM-barrel proteins (see section 1-18) with an extra, mostly antiparallel, beta-sheet domain attached. MR and MLE also bind their catalytic metal ions in the same positions in the structures.

Examination of the active sites shows that this similarity is preserved in detail (Figure 4-41). The amino acids that bind the metal ion are conserved between MR and MLE with one exception, and that is replaced by one with similar physical-chemical properties from a different position in the sequence. Thus, the way these two proteins bind their essential metal ions is structurally and functionally conserved, even though not all the residues involved are in exactly corresponding positions in the two sequences. Both active sites contain a pair of lysine residues in identical positions: in each case, one lysine acts as a catalytic base while the other serves to reduce the pK_a of the first through the proximity of its positive charge (see section 2-12). Opposite the lysine pair, however, the active sites of MR and MLE are different. MR has a second catalytic base, histidine 297, while MLE has a lysine residue, of uncertain role in catalysis, that occupies the same spatial position as histidine 297 but comes from sequence position 273.

The striking similarity between the active sites of MR and MLE, together with their virtually identical folds, implies that these enzymes are homologous, that is, that they diverged from a common ancestor. Nevertheless, they catalyze different chemical reactions on different substrates. Nor is there any conservation of binding specificity between them. Neither R- nor S-mandelate

References

Babbitt, P.C. *et al.*: **A functionally diverse enzyme superfamily that abstracts the alpha protons of carboxylic acids.** *Science* 1995, **267**:1159–1161.

Gerlt, J.A. and Babbitt, P.C.: **Can sequence determine function?** *Genome Biol.* 2000, **1**:reviews0005.1–0005.10.

Gerlt, J.A. and Babbitt, P.C.: **Divergent evolution of enzymatic function: mechanistically diverse superfamilies and functionally distinct suprafamilies.** *Annu. Rev. Biochem.* 2001, **70**:209–246.

Hasson, M.S. *et al.*: **Evolution of an enzyme active site: the structure of a new crystal form of muconate lactonizing enzyme compared with mandelate racemase and enolase.** *Proc. Natl Acad. Sci. USA* 1998, **95**:10396–10401.

Nagano, N. *et al.*: **One fold with many functions: the evolutionary relationships between TIM barrel families based on their sequences, structures and functions.** *J. Mol. Biol.* 2002, **321**:741–765.

Neidhart, D.J. *et al.*: **Mandelate racemase and muconate lactonizing enzyme are mechanistically**

distinct and structurally homologous. *Nature* 1990, **347**:692–694.

Petsko, G.A. *et al.*: **On the origin of enzymatic species.** *Trends Biochem. Sci.* 1993, **18**:372–376.

is a substrate or inhibitor of MLE, nor is *cis, cis*-muconate and muconolactone for MR. Evolutionary pressure for MR and MLE to become highly specific for their respective substrates led their substrate-binding pockets to become very different, and mutually incompatible, while the catalytic machinery remained similar. This argument implies an underlying commonality of catalytic mechanism between MR and MLE, as reflected in the conserved residues in their active sites. Both enzymes use a base—one of the pair of lysines—to abstract a hydrogen attached to a carbon atom in the substrate (see Figures 4-38 and 4-39), and in each case that carbon atom is adjacent to a carboxylate group that is coordinated to the metal ion in the active site (Figure 4-41). Presumably, the ancestral protein could carry out this chemical step on either mandelate or muconolactone or on some related molecule.

So, even if two gene products have similar sequences and share the same overall fold and some active-site residues, and even if the biochemical function of one of them is known, it is not always possible to predict the biochemical function of the other. Missing is knowledge of what molecules interact specifically with the active site of the protein of unknown function. Computational methods to determine which small molecules would bind to an active site of known shape and charge distribution do not yet exist (see section 4-9).

This particular example is not unique. MR and MLE belong to a large superfamily of TIM-barrel enzymes. All members of this superfamily have the additional beta-sheet domain, use a divalent metal ion, and have metal-binding residues in positions corresponding to those in MR and MLE. They catalyze chemical reactions as diverse as the dehydration of the sugar D-galactonate and the formation of phosphoenolpyruvate. For each of these enzymes a catalytic mechanism can be written involving base-catalyzed abstraction of a hydrogen from a carbon atom adjacent to a carboxylate group, but all of the substrates are different. A total of 21 different superfamilies of TIM-barrel enzymes have been identified on the basis of structural and functional relatedness. These 21 superfamilies include 76 difference sequence families.

Nor is the TIM barrel the only domain fold with this sort of versatility. The four-helix bundle has been found in hormones, growth factors, electron-transport proteins and enzymes. The zinc finger motif is usually found in DNA-binding proteins, where it makes sequence-specific contacts with bases in the double helix, but there are zinc finger domains that bind to RNA instead, and a number of proteins have zinc finger modules that mediate protein–protein interactions. Many other examples could be cited.

Nevertheless, there are a number of domain folds that are characteristic of a given biochemical or cellular function, and in some instances the presence of such a domain can be recognized from sequence information alone, allowing one to determine at least partial biochemical function from sequence and/or structural information. For instance, the kinase fold appears to be present almost exclusively in protein kinases; the SH2 domain appears to be used exclusively to bind phosphotyrosine-containing peptides; and in eukaryotes the seven-transmembrane helix fold appears to be used only in G-protein-coupled receptors. But many folds have so diverse a range of functions that sequence and structural information alone is unlikely to be sufficient to reveal their biochemical or cellular roles.

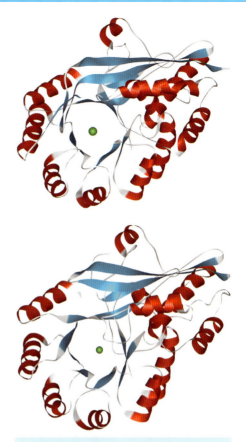

Figure 4-40 Mandelate racemase (top) and muconate lactonizing enzyme (bottom) have almost identical folds The colored spheres in the center of the structures represent the metal ions in the active sites.

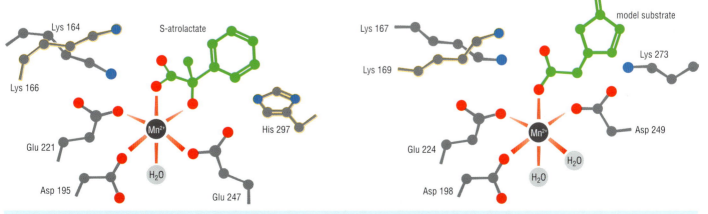

Figure 4-41 A comparison of the active sites of mandelate racemase (left) and muconate lactonizing enzyme (right) The amino acids that coordinate with the metal ion are conserved between the two enzymes, as are the catalytic residues except for histidine 297 in MR which is replaced by lysine 273 in MLE. In both cases a carboxylate group on the substrate coordinates with the metal ion. The active site of MR is shown with the inhibitor S-atrolactate bound; the MLE active site is shown with a model substrate bound. The residues shaded in yellow are the putative general acid-base catalytic residues.

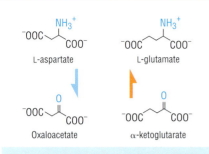

Figure 4-42 The overall reaction catalyzed by the pyridoxal phosphate-dependent enzyme L-aspartate aminotransferase

A protein's biochemical function and catalytic mechanism do not necessarily predict its three-dimensional structure

The site of biochemical function in an enzyme is characterized by a sub-site that binds the substrate and a catalytic sub-site at which the chemical reaction takes place, and these two sub-sites are usually at least partly distinct (see section 2-7). It is therefore possible to have the same arrangement of catalytic groups in combination with different arrangements of substrate-binding groups. In some cases, this produces enzymes with different specificities but which carry out the same chemistry. Such a situation is usually associated with divergent evolution from a common ancestor, in which the protein scaffold is retained but the substrate-binding sub-site is altered. It is also possible, however, for the same catalytic machinery to evolve independently on different protein scaffolds, whose substrate-binding sites may, or may not, be specific for the same substrate. This is termed convergent evolution: nature has found the same solution to the problem of catalyzing a particular reaction, but the solution has evolved in two different protein frameworks.

A prime example of convergent evolution is found among the aminotransferases. These are enzymes that "convert" one amino acid into another in a reaction known as transamination, which is central to amino-acid metabolism. In this reaction, an α-amino acid is converted to an α-keto acid, followed by conversion of a different α-keto acid to a new α-amino acid. All transaminases use the cofactor pyridoxal phosphate (PLP), derived from vitamin B_6. One of these PLP-dependent enzymes is L-aspartate aminotransferase, which converts L-aspartate to L-glutamate, via α-ketoglutarate and oxaloacetate (Figure 4-42) and is found in every living organism. Another aminotransferase, found only in bacteria, catalyzes the same reaction but is specific for the D-forms of various amino acids, including aspartate and glutamate. In bacteria, a transamination reaction involving D-glutamate, α-ketoglutarate, and pyruvate is used to produce D-alanine for synthesis of the bacterial cell wall.

Figure 4-43 The general mechanism for PLP-dependent catalysis of transamination, the interconversion of α-amino acids and α-keto acids The amino group of the amino-acid substrate displaces the side-chain amino group of the lysine residue that holds the cofactor PLP in the active site (step 1). PLP then catalyzes a rearrangement of the amino-acid substrate (step 2), followed by hydrolysis of the keto-acid portion, leaving the nitrogen of the amino acid (blue) bound to the cofactor to form the intermediate pyridoxamine phosphate (PMP) (step 3). This forward reaction is indicated by the blue arrows. To regenerate the starting form of the enzyme, a different keto acid then reverses these steps and captures the bound nitrogen, producing a new amino acid and leaving the PLP once more bound to the enzyme at the active-site lysine (orange arrows).

Both enzymes have identical catalytic mechanisms. The amino group of the amino-acid substrate displaces the side-chain amino group of the lysine residue that binds the cofactor PLP in the active site (Figure 4-43, step 1). PLP then catalyzes a rearrangement of its new bound amino acid (step 2), followed by hydrolysis of the keto-acid portion, leaving the nitrogen of the amino acid bound to the cofactor (step 3). To regenerate the starting form of the enzyme, a different keto acid then reverses these steps and captures the bound nitrogen, producing a new amino acid and leaving the PLP once more bound to the enzyme at the active-site lysine.

A number of other amino-acid side chains in the active sites of both of these aminotransferases interact specifically with the cofactor, promoting this series of transformations over the other possible chemistries that the versatile cofactor PLP can catalyze. Yet other amino acids stabilize the position of the bound substrates and confer substrate specificity. Because these two enzymes catalyze exactly the same reaction by exactly the same mechanism, one might expect their structures to be similar. On the other hand, because one of these enzymes is specific for the commonly

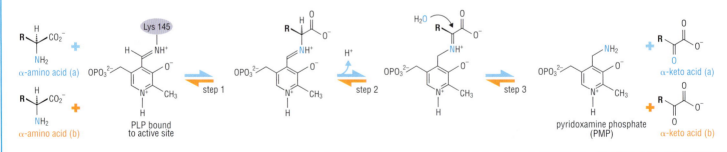

References

Doolittle, R.F.: **Convergent evolution: the need to be explicit.** *Trends Biochem. Sci.* 1994, **19**:15–18.

Kirsch, J.F. *et al.*: **Mechanism of action of aspartate aminotransferase proposed on the basis of its spatial structure.** *J. Mol. Biol.* 1984, **174**:497–525.

Smith, D.L. *et al.*: **2.8-Å-resolution crystal structure of an active-site mutant of aspartate aminotransferase from *Escherichia coli*.** *Biochemistry* 1989, **28**:8161–8167.

Sugio, S. *et al.*: **Crystal structure of a D-amino acid aminotransferase: how the protein controls stereo-selectivity.** *Biochemistry* 1995, **34**:9661–9669.

Yennawar, N. *et al.*: **The structure of human mito-chondrial branched-chain aminotransferase.** *Acta Crystallogr. D. Biol. Crystallogr.* 2001, **57**:506–515.

Yoshimura, T. *et al.*: **Stereospecificity for the hydrogen transfer and molecular evolution of pyridoxal enzymes.** *Biosci. Biotechnol. Biochem.* 1996, **60**:181-187.

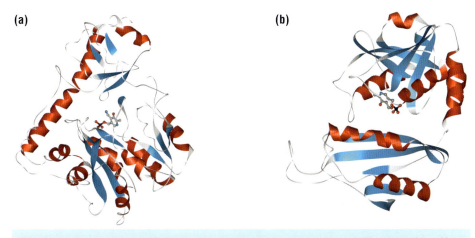

(a)　　　　　(b)

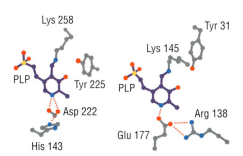

Figure 4-44 The three-dimensional structures of L-aspartate aminotransferase (left) and D-amino acid aminotransferase (right) The two proteins have completely different architectures. Not only are they different in size, they differ in their amino-acid sequence and in the folds of the protein domains. In the L-aspartate aminotransferase structure, the cofactor intermediate PMP (in a ball-and-stick representation) is shown at the active site. In the D-amino acid aminotransferase, the cofactor PLP is shown. (PDB 2aat, 1daa).

Figure 4-45 Comparison of the active sites of L-aspartate aminotransferase (left) and D-amino acid aminotransferase (right) Despite the different protein folds of these two enzymes, the active sites have converged to strikingly similar arrangements of the residues that interact with the cofactor and promote catalysis. The residues that determine which amino acid is used as a substrate and whether it will be the D- or the L-form are arranged differently in the two enzymes (not shown).

occurring L-forms of the amino acids while the other exclusively uses the rarer D-enantiomers, one might also expect that their active sites would look quite different in terms of the arrangement of side chains around the cofactor. Both these expectations are, however, far from the case.

Comparison of the amino-acid sequences of the two enzymes reveals absolutely no identity; however, as we have seen, the absence of detectable sequence identity does not necessarily mean that the protein fold will be different (see section 1-16). But in this case, comparison of the three-dimensional structures of the two enzymes shows their polypeptide-chain folds to be totally different (Figure 4-44). They clearly did not evolve from a common ancestor. When one finds two sequences and two structures that are completely different, one might on the face of it expect that they represent different mechanisms for solving the problem of catalyzing the same chemical transformation. When these two structures are examined in detail, however, the active sites are found to be strikingly similar, both in the nature of the amino acids interacting with the cofactor and their positions in space (Figure 4-45). Moreover, detailed analysis of genomic sequence data suggests that all known aminotransferases possess one or other of these polypeptide-chain folds and this same active-site configuration. It appears that this constellation of catalytic groups, in combination with the intrinsic chemistry of the PLP cofactor, is especially suited to promoting transamination, and nature has independently discovered this twice, using two different protein frameworks.

One possible explanation for the differences in three-dimensional framework is the difference in the "handedness" (or chirality) of the substrate: perhaps one fold is only suited to recognizing D-amino acids. This cannot be so, however, because there is another aminotransferase that only recognizes L-amino acids but whose sequence and polypeptide chain fold are similar to those of D-amino acid aminotransferase (Figure 4-46). These two enzymes clearly represent divergent evolution from a common ancestor. Apparently, modification of substrate specificity within the context of a given protein fold, even to the extent of reversing the handedness of the substrate, is easier than evolving a completely new catalytic mechanism.

A number of biochemical functions are carried out by enzymes that differ in their protein fold but have remarkably similar active sites. Convergent evolution to a common chemical mechanism has been observed among the serine proteases, the aminopeptidases, the NAD-dependent dehydrogenases and the sugar isomerases, to name just a few. Consequently, even if you know the biochemical function of a newly discovered protein, you cannot necessarily predict the protein fold that will carry it out. The catalytic function of enzymes can, however, sometimes be predicted by genomic analysis aimed at identification of patterns of active-site residues, and we discuss this in section 4-2.

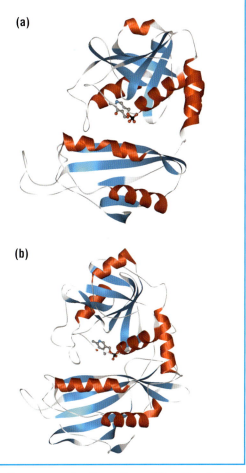

(a)

(b)

Figure 4-46 The three-dimensional structures of bacterial D-amino acid aminotransferase (top) and human mitochondrial branched-chain L-amino acid aminotransferase (bottom) These two enzymes are similar in amino-acid sequence, overall fold and active site even though their substrates are of opposite handedness. They have diverged from a common ancestor. (PDB 1daa, 1ekp).

4-13 Moonlighting: Proteins with More than One Function

In multicellular organisms, multifunctional proteins help expand the number of protein functions that can be derived from relatively small genomes

The genomes of multicellular organisms are remarkably small, in terms of number of genes (see Figure 4-10), considering the enormous increase in complexity of the organisms themselves compared with bacteria or the single-celled eukaryotes. One explanation is that the actual number of different proteins derived from a given gene can be expanded by mechanisms such as alternative splicing. An additional explanation is that, in multicellular organisms especially, a given protein may have more than one distinct biochemical and/or cellular function. The biochemical functions may include catalysis, binding, participating as a structural molecule in an assembly, or operating as a molecular switch (see sections 1-0 and 2-0). The extent of this functional diversity for any one protein is only just beginning to be appreciated.

Phosphoglucose isomerase (glucose-6-phosphate isomerase or PGI) is the second enzyme in the glycolytic pathway, a core metabolic pathway that converts glucose to pyruvate, and in which PGI converts glucose 6-phosphate to fructose 6-phosphate. The gene for PGI is thus a housekeeping gene found in nearly all organisms. Sequences of PGI from numerous organisms are well conserved, indicating that this intracellular biochemical function is catalyzed by a single type of polypeptide fold. But if one takes the protein sequence of PGI from a rabbit, say, and looks for homologous sequences in databases of other mammalian protein sequences, one finds, in addition to PGI, an identical sequence labeled neuroleukin. The protein neuroleukin was discovered as a cytokine secreted by T cells that promotes the survival of some embryonic spinal neurons and sensory nerves. It also causes B cells to mature into antibody-secreting cells. One also finds two other named protein activities with sequences identical to PGI: autocrine motility factor (AMF) and differentiation and maturation mediator (DMM). Like neuroleukin, these proteins are also secreted cytokines. AMF is produced by tumor cells and stimulates cancer-cell migration; it may be involved in cancer metastasis. DMM is isolated from culture medium in which T cells have been grown and has been shown to cause differentiation of human myeloid leukemia cells *in vitro*. Purified rabbit PGI will cause the increase in cell

Multifunctional Proteins

Protein	Function	Additional functions
Phosphoglucose isomerase	Glycolytic enzyme	Cytokine
EF-1	Elongation factor in translation	Actin-bundling protein
Cyclophilin	Peptidyl-prolyl *cis–trans* isomerase	Regulator of calcineurin
Macrophage inhibitory factor (MIF)	Activator of macrophages and T cells	Phenylpyruvate tautomerase
PutA	Proline dehydrogenase	Transcriptional repressor
Aconitase	TCA-cycle enzyme	Iron-responsive-element binding protein
Thioredoxin	Maintains SH groups in reduced state	Subunit of phage T7 DNA polymerase
Thrombin	Protease in blood clotting	Ligand for cell-surface receptor
Thymidylate synthase	Enzyme in DNA synthesis	Inhibitor of translation
FtsH	Chaperone protein in bacteria	Metalloprotease
LON	Mitochondrial protease	Chaperone protein
Methionine aminopeptidase 2	Peptidase	Protects eIF2 from phosphorylation

Figure 4-47 Some examples of multifunctional proteins with their various functions The first function column lists the biochemical function that was first identified. In most cases, this is an enzymatic activity because such activities are easily assayed. The additional functions usually depend on binding to a specific partner.

References

Cutforth, T. and Gaul, U.: **A methionine aminopeptidase and putative regulator of translation initiation is required for cell growth and patterning in Drosophila.** *Mech. Dev.* 1999, **82**:23–28.

Datta, B.: **MAPs and POEP of the roads from prokaryotic to eukaryotic kingdoms.** *Biochimie* 2000, **82**:95–107.

Griffith, E.C. *et al.*: **Methionine aminopeptidase (type 2) is the common target for angiogenesis inhibitors AGM-1470 and ovalicin.** *Chem. Biol.* 1997, **4**:461–471.

Haga, A. *et al.*: **Phosphohexose isomerase/autocrine motility factor/neuroleukin/maturation factor is a multifunctional phosphoprotein.** *Biochim. Biophys. Acta* 2000, **1480**:235–244.

Jeffery, C.J.: **Moonlighting proteins.** *Trends Biochem. Sci.* 1999, **24**:8–11.

Jeffery, C.J. *et al.*: **Crystal structure of rabbit phosphoglucose isomerase, a glycolytic enzyme that moonlights as neuroleukin, autocrine motility factor, and differentiation mediator.** *Biochemistry* 2000, **39**:955–964.

Liu, S. *et al.*: **Structure of human methionine aminopeptidase-2 complexed with fumagillin.** *Science* 1998, **282**:1324–1327.

Lubetsky, J.B. *et al.*: **The tautomerase active site of macrophage migration inhibitory factor is a potential target for discovery of novel anti-inflammatory agents.** *J. Biol. Chem.* 2002, **277**:24976–24982.

Sun, Y.J. *et al.*: **The crystal structure of a multifunctional**

motility seen with AMF and the dosage-dependent differentiation of human leukemia cells seen with DMM; conversely, both AMF and DMM have PGI activity. PGI, neuroleukin, AMF and DMM are the same protein, encoded by the same gene.

To carry out its cytokine and growth-factor functions it is likely that PGI/neuroleukin/AMF/ DMM binds to at least one type of cell-surface receptor on a variety of target cells, and a receptor corresponding to AMF activity has been cloned from fibrosarcoma cells. Inhibitors of the PGI reaction block some, but not all, of the cytokine functions of the protein, indicating that the sites on the protein surface responsible for these different activities are at least partly distinct. Remarkably, PGI from a bacterium has been reported to have activity in the AMF assay.

The PGI reaction is extremely similar to that catalyzed by the glycolytic enzyme triosephosphate isomerase, and the catalytic mechanisms of the two reactions are identical. Nevertheless, the three-dimensional structure of PGI is completely different from that of TIM. Nothing in PGI's sequence or structure, however, suggests its additional cytokine functions; there are no obvious domains with structural similarity to any known cytokine and no sequence motifs suggestive of known growth factors or signal transduction molecules. One concludes that the extracellular cytokine functions of this protein in higher eukaryotes are at least partially independent of each other, and have evolved without gross modification of the ancestral fold.

Now that numerous genome sequences are available, each annotated according to the literature of previous studies of that organism, many other examples of multiple functions for the same protein are being discovered (Figure 4-47). One is methionine aminopeptidase type 2 (MetAP2), which catalyzes the removal of the amino-terminal methionine from the growing polypeptide chain of many proteins in eukaryotes. MetAP2 is the target for the anti-angiogenesis drugs ovalicin and fumagillin, which act by inhibiting this catalytic activity. DNA sequence analysis of MetAP2 genes reveals, as expected, sequence homologies with MetAP2 genes from other organisms but also with various eukaryotic homologs of a rat protein known originally as p67. This intracellular protein protects the alpha-subunit of eukaryotic initiation factor 2 (eIF2) from phosphorylation by its kinases. This activity of p67 is observed in different stress-related situations such as heme deficiency in reticulocytes (immature red blood cells), serum starvation and heat shock in mammalian cells, vaccinia virus infection of mammalian cells, baculovirus infection of insect cells, mitosis, apoptosis, and even possibly during normal cell growth. MetAP2 and p67 are identical proteins encoded by the same gene. Inhibitors of MetAP2 activity do not inhibit the translational cofactor activity of MetAP2, suggesting that the two functions are independent. Some mutations in the MetAP2 gene in *Drosophila* result in loss of ventral tissue in the compound eye as well as extra wing veins, whereas others impair tissue growth. However, it is not clear whether these phenotypes are due to loss of MetAP2's catalytic activity, translational cofactor activity, or both. Another example is the cytokine macrophage inhibitory factor, MIF (Figure 4-48), which also has enzymatic activity.

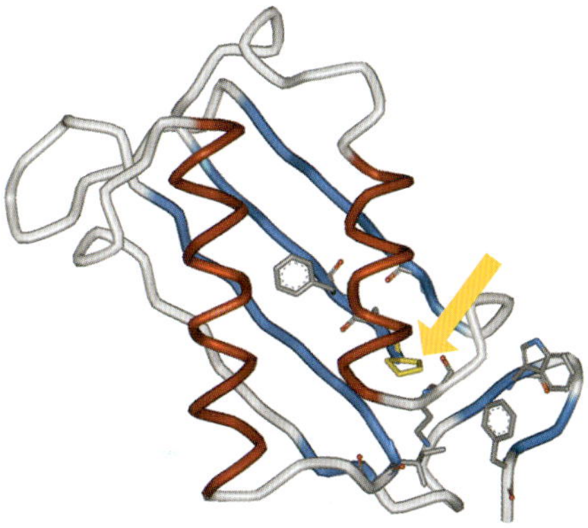

The term "moonlighting" has been coined to describe the performance of more than one job by the same protein. From the point of view of any one experiment, each job may appear to be the main activity and the other(s) to be the sideline. Consequently, in multicellular organisms especially, knowledge of one function of a gene product does not necessarily mean that all its functions have been determined. This fact has profound consequences for gene knockout experiments (by, for example, antisense RNA, RNA interference (RNAi), or gene disruption) as a means of determining function. The phenotype of a knockout animal or cell may be the result of the loss of all the different functions that a protein can carry out, or may differ in different tissues or under different conditions in which various functions are dominant.

Figure 4-48 The three-dimensional structure of the monomer of macrophage inhibitory factor, MIF The protein is an important proinflammatory cytokine that activates T cells and macrophages. It is also an enzyme that catalyzes the tautomerization of phenylpyruvic acid. The residues involved in substrate binding and catalysis are shown. The proline associated with the active site is indicated by the yellow arrow. Because the active site overlaps with the binding site for receptors of the cytokine function of MIF, inhibitors of its enzymatic activity are also potential antiinflammatory drugs. (PDB 1ljt)

protein: phosphoglucose isomerase/ autocrine motility factor/neuroleukin. *Proc. Natl Acad. Sci. USA* 1999, **96**:5412–5417.

Swope, M.D. and Lolis, E.: **Macrophage migration inhibitory factor: cytokine, hormone, or enzyme?** *Rev. Physiol. Biochem. Pharmacol.* 1999, **139**:1–32.

Watanabe, H. *et al.*: **Tumor cell autocrine motility factor is the neuroleukin/phosphohexose isomerase polypeptide.** *Cancer Res.* 1996, **56**:2960–2963.

Xu, W. *et al.*: **The differentiation and maturation**

mediator for human myeloid leukemia cells shares homology with neuroleukin or phosphoglucose isomerase. *Blood* 1996, **87**:4502–4506.

Some amino-acid sequences can assume different secondary structures in different structural contexts

The concept that the secondary structure of a protein is essentially determined locally by the amino-acid sequence is at the heart of most methods of secondary structure prediction; it also underlies some of the computational approaches to predicting tertiary structure directly from sequence. Although this concept appears to be valid for many sequences, as the database of protein structures has grown, a number of exceptions have been found. Some stretches of sequence up to seven residues in length have been identified that adopt an alpha-helical conformation in the context of one protein fold but form a beta strand when embedded in the sequence of a protein with a different overall fold. These sequences have been dubbed **chameleon sequences** for their tendency to change their appearance with their surroundings. One survey of all known protein structures up to 1997 found three such sequences seven residues long (Figure 4-49), 38 such sequences six residues long, and 940 chameleon sequences five residues long. Some were buried and some were on the surface; their sequences varied considerably but there tended to be a preponderance of alanines, leucines and valines and a dearth of charged and aromatic residues.

We have already seen that some segments in certain proteins can change their conformation from, for example, an alpha helix to a loop in response to the binding of a small molecule or another protein or to a change in pH. For example, when elongation factor Tu switches from its GTP-bound form to its GDP-bound form, a portion of the switch helix unravels, breaking an interaction between two domains (see section 3-9).

The ability of amino-acid sequences to convert from an alpha-helical to a beta-strand conformation has received extensive attention recently, as this structural change may induce many proteins to self-assemble into so-called amyloid fibrils and cause fatal diseases (see section 4-15). A number of sequences that are not natural chameleons can become such by a single point mutation, suggesting a possible mechanism whereby such diseases may be initiated. One example is the bacterial protein Fis, a DNA-binding protein that is implicated in the regulation of DNA replication and recombination as well as in transcriptional regulation. A peptide segment in Fis can be converted from a beta strand to an alpha helix by a single-site mutation, proline 26 to alanine. Proline 26 in Fis occurs at the point where a flexible extended beta-hairpin arm leaves the core structure (Figure 4-50a). Thus it can be classified as a "hinge proline" located at the carboxy-terminal end of one beta strand and the amino-terminal cap of the following alpha helix. The replacement of proline 26 with alanine extends the alpha helix for two additional turns in one of the dimeric subunits of Fis; therefore, the structure of the peptide from residues 22 to 26 is converted from a beta strand to an alpha helix by this one mutation (Figure 4-50b). Interestingly, this peptide in the second monomer subunit retains its beta-strand conformation in the crystal structure of Fis, suggesting that the alpha-helical and beta-sheet conformation are very similar in energy for this sequence and that only small local changes in environment are needed to cause it to flip from one form to the other.

While the conversion of a beta strand to an alpha helix in Fis is caused by a mutation, and has no implications for normal function, some proteins contain natural chameleon sequences that may be important to their function. One example is a DNA-binding transcriptional regulator from yeast, the MATα2 protein, which helps determine two differentiated cell types (mating types) in growing yeast cells by repressing genes whose expression is required for one of the two types. MATα2 binds to DNA in association with a second protein, MCM1, so that one copy of MATα2 binds on each side of MCM1 (Figure 4-51a). In the crystal structure of this complex

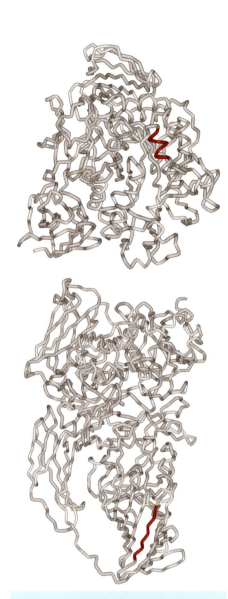

Figure 4-49 Chameleon sequences The protein backbones of the enzymes cyclodextrin glycosyltransferase (PDB 1cgu) (top) and beta-galactosidase (PDB 1bgl) (bottom), each of which contains the chameleon sequence LITTAHA (shown in red), corresponding to residues 121–127 in the sequence of cyclodextrin glycosyltransferase and residues 835–841 in beta-galactosidase. In the former structure, the sequence forms two turns of alpha helix; in the latter, it is a beta strand.

Definitions

chameleon sequence: a sequence that exists in different conformations in different environments.

Figure 4-50 Chameleon sequence in the DNA-binding protein Fis (a) The structure of the dimer of the sequence-specific DNA-binding protein Fis shows a predominantly alpha-helical fold with two strands of antiparallel beta sheet, β_1 and β_2 (red), at the amino terminus. (PDB 1f36) **(b)** The replacement of proline 26 at the end of the second beta strand with an alanine converts this beta strand into two additional turns of the alpha helix that follows.

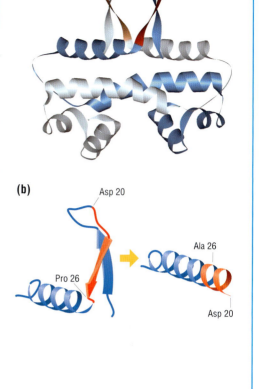

bound to DNA, an eight-amino-acid sequence adopts an alpha-helical conformation in one of two copies of the MATα2 monomer and a beta-strand conformation in the other (Figure 4-51b). Although there is no direct evidence that both forms exist in biology, such an alternative fold could have functional consequences. In most sites the sequences recognized by the MATα2 monomers are identical. However, there are separations of two to three base pairs between the MCM1- and MATα2-binding sites in the natural yeast promoters to which this transcription factor binds, and the different conformations may permit such variations to be tolerated. MATα2 can also form a complex with another transcriptional modulator, MATa1, and in this context, the change in conformation may again allow MATα2 to accommodate differences in the spacing of the sites on DNA. The ability of parts of MATα2 to change conformation in different contexts could help this protein to bind to a number of sites on the genome.

To probe the context dependence of the structures of short polypeptide sequences, an 11-amino-acid chameleon sequence has been designed that folds as an alpha helix when in one position but as a beta sheet when in another position of the primary sequence of the immunoglobulin-binding domain of protein G. This protein from *Staphylococcus aureus* binds to the Fc region of IgG antibodies and is thought to protect the bacteria from these antibodies by blocking their interactions with complement and Fc receptors. Both proteins, chameleon-alpha and chameleon-beta, are folded into structures similar to native protein G except for the small region of the chameleon sequence.

These examples illustrate the general principle that the secondary structures of short peptide segments can often depend more on the tertiary structural context in which they are placed than on their intrinsic secondary structure propensities. The balance between inherent tendency and the effect of environment will be different for different sequences. If the free energies of a peptide in its alpha-helical and beta-sheet conformations are similar, then the energies of interaction between the peptide and the environment could be enough to tip the balance in favor of one or the other.

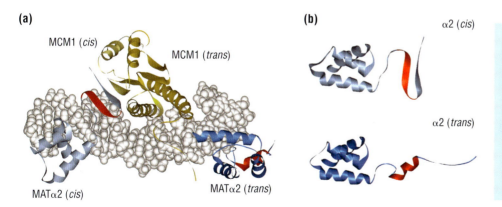

Figure 4-51 Chameleon sequence in the DNA-binding protein MATα2 from yeast (a) The structure of the complex of MATα2 (blue and red) with its transcriptional co-regulator MCM1 (yellow) bound to a target site in DNA. At such a site, two monomers of MATα2 bind to two (usually identical) DNA sequences on either side of two monomers of MCM1. **(b)** An eight-amino-acid sequence (red) adopts a beta-strand conformation in one MATα2 molecule (the *cis* monomer; light blue) and an alpha-helical conformation in the other (the *trans* monomer; dark blue). (PDB 1mnm)

References

Mezei, M.: **Chameleon sequences in the PDB.** *Protein Eng.* 1998, **11**:411–414.

Minor, D.L. Jr. and Kim, P.S.: **Context-dependent secondary structure formation of a designed protein sequence.** *Nature* 1996, **380**:730–734.

Smith, C.A. *et al.*: **An RNA-binding chameleon.** *Mol. Cell* 2000, **6**:1067–1076.

Sudarsanam, S.: **Structural diversity of sequentially identical subsequences of proteins: identical octapeptides can have different conformations.** *Proteins* 1998, **30**:228–231.

Tan, S. and Richmond T.J.: **Crystal structure of the yeast MATalpha2/MCM1/DNA ternary complex.** *Nature* 1998, **391**:660–666.

Yang, W.Z. *et al.*: **Conversion of a beta-strand to an alpha-helix induced by a single-site mutation observed in the crystal structure of Fis mutant Pro26Ala.** *Protein Sci.* 1998, **7**:1875–1883.

4-15 Prions, Amyloids and Serpins: Metastable Protein Folds

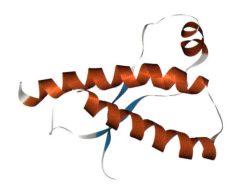

Figure 4-52 The prion protein This figure shows the soluble form of Syrian hamster prion protein PrPSc, which was generated by removing the amino terminus. This protein consists of residues 90–231, incorporating the region thought to be involved in the structural change. The protein was generated by refolding to produce the structure of the cellular form of the protein. The precise structure of the disease-causing form is not yet known, but is known to have much more beta sheet than the cellular form.

A single sequence can adopt more than one stable structure

The existence of chameleon sequences may reflect a general principle: that not all sequences fold into one unique structure. Some structures may be **metastable**—able to change into one or more different stable structures. But are complete protein sequences of such plasticity found in nature? It appears that the answer is yes although rarely, and that such cases are often, at least so far, associated with severe mammalian disease. The best characterized of these changeable structures is the prion associated with Creutzfeldt-Jakob disease (CJD) in humans and scrapie and bovine spongiform encephalopathy (BSE) in sheep and cattle (Figure 4-52). Prions are infectious proteins whose misfolded form is identical in sequence to the normal cellular form of the same protein. The two forms have, however, quite different conformations and physical properties. The infectious form has a propensity to form aggregates, possesses secondary structure content that is rich in beta sheets, is partially resistant to proteolysis, and is insoluble in nonionic detergents. In contrast, the cellular form contains little beta structure, is sensitive to protease digestion, and is soluble in nonionic detergent. Contact with the infectious form causes the cellular counterpart to undergo pronounced conformational changes that lead, ultimately, to the formation of cytotoxic protein aggregates consisting almost entirely of the infectious conformation. Although the molecular events that lead to this profound conformational change are poorly understood, there is substantial evidence that the infectious form acts as a template directing the structural rearrangement of the normal form into the infectious one. Studies of synthetic peptides derived from the prion sequence indicate that a stretch of up to 55 residues in the middle of the protein has the propensity to adopt both alpha-helical and beta-sheet conformations. Presumably, the infectious form arises spontaneously in a small number of molecules as a result of this inherent plasticity.

A similar mechanism may underlie protein aggregate formation in a group of about 20 diseases called amyloidoses, which include Alzheimer's, Parkinson's, and type II diabetes. Each disease is associated with a particular protein, and extracellular aggregates of these proteins are thought to be the direct or indirect origin of the pathological conditions associated with the disease. Strikingly, the so-called amyloid fibrils characteristic of these diseases arise from well known proteins, including lysozyme and transthyretin, that have well defined, stable, non-identical folds but produce fibrous protein aggregates of identical, largely beta-sheet, structure (Figure 4-53).

Recent studies suggest that the ability to undergo a refolding leading to amyloid formation is not unique to these proteins, but can be observed in many other proteins under laboratory conditions such as low pH. One conclusion from such findings is that prions and other proteins that cause disease by this mechanism may differ from the vast array of "normal" cellular proteins only in having sequences that can undergo such refolding spontaneously under physiological conditions. Clearly, if this is the case, it is possible that a single point mutation may convert a harmless protein into one that can refold spontaneously; such mutations have been found associated with some of the amyloidoses.

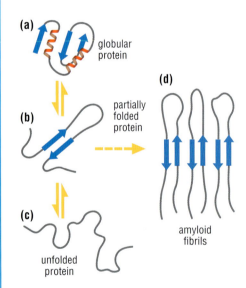

Figure 4-53 A possible mechanism for the formation of amyloid fibrils by a globular protein The correctly folded protein **(a)** is secreted from the cell. Under certain conditions, or because it contains a mutation, the protein unfolds partially **(b)** or completely **(c)**; the unfolded forms can also refold partially or completely. The partially unfolded form is prone to aggregation, which results in the formation of fibrils **(d)** and other aggregates that accumulate in the extracellular space.

Definitions

metastable: only partially stable under the given conditions. In the case of protein structures, a metastable fold exists in equilibrium with other conformations or with the unfolded state.

References

Cohen, B.I. *et al.*: **Origins of structural diversity within sequentially identical hexapeptides.** *Protein Sci.* 1993, **2**:2134–2145.

Dalal, S. and Regan, L.: **Understanding the sequence determinants of conformational switching using protein design.** *Protein Sci.* 2000, **9**:1651–1659.

Dobson, C.M.: **Protein misfolding, evolution and disease.** *Trends Biochem. Sci.* 1999, **24**:329–332.

Engh, R.A. *et al.*: **Divining the serpin inhibition mechanism: a suicide substrate "springe"?** *Trends Biotechnol.* 1995, **13**:503–510.

Kabsch, W. and Sander, C.: **On the use of sequence homologies to predict protein structure: identical pentapeptides can have completely different conformations.** *Proc. Natl Acad. Sci. USA* 1984, **81**:1075–1078.

Ko, Y.H. and Pedersen, P.L.: **Cystic fibrosis: a brief look at some highlights of a decade of research focused on elucidating and correcting the molecular basis of**

Structural plasticity can also be part of the normal function of a protein. Often, when this is the case, ligand binding or specific proteolytic modification is needed to drive one folded form into the other. Large, ligand-induced domain rearrangements such as that found in elongation factor Tu (see section 3-9) can be considered examples of different overall protein folds induced by the state of assembly. Perhaps the best example of a large structural rearrangement caused by limited proteolysis is found in the family of protein protease inhibitors called the serpins. Some protein protease inhibitors function as rigid substrate mimics; the serpins differ fundamentally in that the loop that recognizes and initially binds the protease active site is flexible and is cleaved by the protease. The serpin remains bound to the enzyme but the cleavage triggers a refolding of the cleaved structure that makes it more stable: one segment of the cleaved loop becomes the central strand of an existing beta sheet in the center of the serpin, converting it from a mixed beta sheet to a more stable antiparallel form (Figure 4-54); in at least one case, the downstream segment also becomes the edge strand of another beta sheet. If the cleaved serpin is released it cannot reassociate with the protease because this refolding has made the recognition strand unavailable for binding.

Mutations in serpins leading to misfolding and aggregation have also been found in some human diseases. For example, the so-called Z-variant of the serpin alpha$_1$-antitrypsin (in which glutamic acid 342 is mutated to lysine) is retained within hepatocytes as inclusion bodies; this is associated with neonatal hepatitis and cirrhosis. The inclusion bodies form because the mutation perturbs the conformation of the protein, facilitating a sequential interaction between the recognition loop of one molecule and beta-sheet A of a second; this could be thought of as a pathological case of domain swapping (see section 2-4).

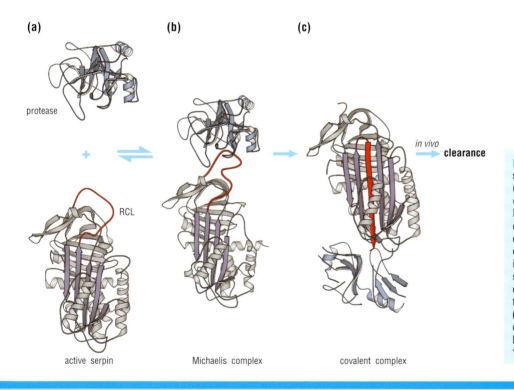

(a) protease

RCL

active serpin

(b) Michaelis complex

(c) *in vivo* **clearance**

covalent complex

Figure 4-54 Structural transformation in a serine protease inhibitor on binding protease (a) The exposed reactive center loop (RCL; red) of the serpin alpha$_1$-antitrypsin (grey) is shown binding to the target protease (blue) to form an enzyme–substrate complex, the Michaelis complex **(b)**. The RCL of the serpin is then cleaved by the target protease, leading to the insertion of the unconstrained RCL into the serpin beta sheet and the formation of a covalent complex that is trapped by release of the newly formed amino terminus of the serpin **(c)**. This complex is then targeted for clearance. Adapted from Ye, S. and Goldsmith, E.J.: *Curr. Opin. Struct. Biol.* 2001, **11**:740–745.

the disease. *J. Bioenerg. Biomembr.* 2001, **33**:513–521.

Leclerc, E. *et al.*: **Immobilized prion protein undergoes spontaneous rearrangement to a conformation having features in common with the infectious form.** *EMBO J.* 2001, **20**:1547–1554.

Mahadeva, R. *et al.*: **6-mer peptide selectively anneals to a pathogenic serpin conformation and blocks polymerization. Implications for the prevention of Z alpha(1)-antitrypsin-related cirrhosis.** *J. Biol. Chem.* 2002, **277**:6771–6774.

Ye, S. and Goldsmith, E.J.: **Serpins and other covalent protease inhibitors.** *Curr. Opin. Struct. Biol.* 2001, **11**:740–745.

Zahn, R. *et al.*: **NMR solution structure of the human prion protein.** *Proc. Natl Acad. Sci. USA* 2000, **97**:145–150.

Determining biochemical function from sequence and structure becomes more accurate as more family members are identified

The identification of the function of the *E. coli* protein f587, known originally only as an uncharacterized open reading frame in the *E. coli* genome sequence, is an illustration of the importance of lateral thinking. In this case, the sequence and structural information was eventually interpreted in the light of the known genetics and physiology of the bacterium.

The story starts with the related enzymes mandelate racemase (MR) and muconate lactonizing enzyme (MLE) (see section 4-11). The enzyme enolase, which catalyzes the conversion of 2-phospho-D-glycerate to phosphoenolpyruvate, was subsequently found to have a similar degree of sequence identity to MR and MLE (26%) and a similar distribution of conserved residues. It has the same polypeptide chain fold and many of the conserved residues map to the active-site region of the fold. Together, these three enzymes form part of the so-called enolase superfamily.

With three related sequences and structures in hand, it became apparent that not every part of the active site was preserved (Figure 4-55). All three enzymes require at least one divalent metal ion for activity, and the carboxylate ligands to these metal ions are present in all three proteins; in enolase however, glutamic acid and aspartic acid are substituted for one another. The remainder of the catalytic machinery is even more divergent in enolase. Yet, the residues are conserved in terms of their broad chemical role although not in terms of their identity. Glutamic acid 211 in enolase could in principle act as a general acid-base group, just like lysine 166 in MR and lysine 169 in MLE. On the other side of the substrate-binding pocket, lysine 345 in enolase occupies the same position as lysine 273 in MLE; the corresponding position in the structure of MR is occupied by histidine 297, which might have a different role. All three enzymes use their bound metal ions in the same way—to activate a C–H bond adjacent to a carboxylate group for abstraction of the hydrogen by a base on the enzyme.

MR has two substrates with the C–H bond in different positions (see Figure 4-38) and its function is to interconvert them; thus, two different acid-base groups are needed. In the case of enolase and MLE, only one proton needs to be abstracted and so only a single base is required. Thus, both lysine 166 and histidine 297 in MR function as acid-base groups. In MLE and enolase, however, the abstraction of a single proton is carried out only by lysine 169 and lysine 345, respectively. This conserved base-catalyzed, metal-promoted proton-transfer step is the common function linking all three enzymes.

Alignments based on conservation of residues that carry out the same active-site chemistry can identify more family members than sequence comparisons alone

If chemistry is the conserved feature, rather than the absolute identity and position in the sequence of the groups that carry out the chemistry, then a more sophisticated approach to finding homologous sequences would be to search for patterns of residues that can perform the same chemistry regardless of their specific amino-acid identities. Such a search, using a specialized computer program, identified dozens of potential members of the enolase/MR/MLE superfamily, most of which could not have been detected by conventional sequence comparison.

One of these predicted homologs, open reading frame f587 in the *E. coli* genome, coded for a protein of unknown function. Alignment of the f587 sequence with those of the other three proteins on the basis of the conservation of active-site chemical function showed that f587 contains the requisite metal-ion ligands and conservation of an active-site histidine—histidine 285—which aligns with histidine 297 of MR. From the position of this base, the prediction would be that the substrate for f587, whatever it might be, would be a carboxylate-containing molecule with a proton on an adjacent carbon that has the R-configuration, as does R-mandelate.

In well studied model organisms, information from genetics and cell biology can help identify the substrate of an "unknown" enzyme and the actual reaction catalyzed

The remaining problems—what is the substrate and overall reaction of f587—are insoluble from sequence and structural information alone. For f587, however, additional information was available. *E. coli*, like most bacteria, tends to organize its genes into operons encoding a set

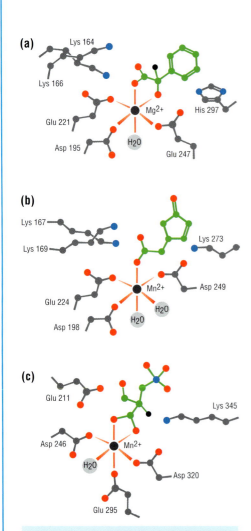

(a) Lys 164, Lys 166, Glu 221, Asp 195, Mg^{2+}, H_2O, His 297, Glu 247

(b) Lys 167, Lys 169, Glu 224, Asp 198, Mn^{2+}, H_2O, H_2O, Lys 273, Asp 249

(c) Glu 211, Asp 246, H_2O, Mn^{2+}, Lys 345, Asp 320, Glu 295

Figure 4-55 Active sites of MR, MLE, and enolase Schematic diagrams of the arrangements of the active-site residues of **(a)** mandelate racemase (MR), **(b)** muconate lactonizing enzyme (MLE) and **(c)** enolase. The types of amino acid that coordinate the divalent metal ion are conserved between the three enzymes. The other catalytic residues, however, are conserved neither in exact position nor in chemical type; nevertheless, these various residues can carry out similar chemistry. In each reaction, the carboxy group of the substrate forms a ligand to the metal ion in the active site, facilitating abstraction of a proton. The enolic intermediate resulting from this common core step is stabilized by interactions with other electrophilic groups in the active site. These groups differ among the three enzymes. The rest of the substrate-binding pocket differs considerably among the three enzymes. (PDB 1mns, 1muc and 1one)

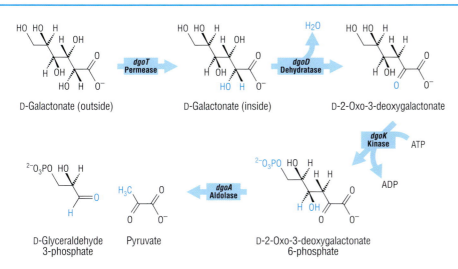

D-Galactonate (outside) → *dgoT* Permease → D-Galactonate (inside) → *dgoD* Dehydratase (H₂O) → D-2-Oxo-3-deoxygalactonate

D-2-Oxo-3-deoxygalactonate → *dgoK* Kinase (ATP → ADP) → D-2-Oxo-3-deoxygalactonate 6-phosphate → *dgoA* Aldolase → D-Glyceraldehyde 3-phosphate + Pyruvate

Figure 4-56 The pathway for the utilization of galactonate in *E. coli* D-galactonate is transported into the cell, dehydrated, phosphorylated, and then cleaved to produce glyceraldehyde 3-phosphate and pyruvate. This pathway allows the bacterium to grow on galactonate as a sole carbon source. The names of the genes and the enzyme activity they represent are given in the blue arrows. F587 has now been identified as the gene *dgoD*, encoding galactonate dehydratase.

of proteins that act in the same pathway. The genome sequence suggested that f587 is the fourth in a series of five open reading frames that could constitute a single operon. The other open reading frames were known to encode proteins involved in galactonate metabolism. They appear to be part of the pathway in which galactonate is imported into the cell by galactonate permease and then degraded stepwise into glyceraldehyde 3-phosphate and pyruvate. Given the proposed functions of the other proteins, the only function needed to complete this pathway was galactonate dehydratase (GalD) (Figure 4-56).

Dehydration of galactonate could be catalyzed by abstracting the 2-R proton via a catalytic base with assistance of a metal ion to activate the C–H bond. From its sequence, f587 appeared to contain all the necessary elements in the right stereochemical configuration to catalyze just such a reaction. To test the hypothesis that f587 encodes GalD, a cell-free extract of *E. coli* transformed with a plasmid overexpressing f587 was assayed for GalD activity. An alpha-keto acid was produced from galactonate, but not from other similar sugars. Subsequent purification of the f587 gene product confirmed it as GalD. Finally, the crystal structure of GalD was determined with an analog of galactonate bound. The overall polypeptide chain fold (Figure 4-57) is the same as that of MR, MLE, and enolase and the active site also resembles those of the other family members (Figure 4-58).

The case of galactonate dehydratase shows that sequence comparisons can identify overall protein structure class and locate the active-site residues, even in cases of very low sequence identity. They can also provide specific information about bound ligands—in this case, a divalent metal ion—and about the residues that bind them. Comparison of a protein of unknown function with just one other homologous sequence will usually not identify the other active-site residues or establish any commonality of function unless the overall sequence identity is very high. Comparisons of multiple sequences are much more informative, however, and will often be able to detect the functional groups that carry out the common core chemical step. When the active-site structures of at least some of the proteins being compared are known, the arrangement of these groups may also reveal the stereochemistry of the reaction being catalyzed. But what sequence and structure usually cannot do alone is to identify the substrate(s) of the reaction and the overall chemistry. Identification of chemistry is more reliable than identification of specificity. Proceeding from sequence to consequence in cases of very low sequence identity requires other sources of information, as illustrated by this case study.

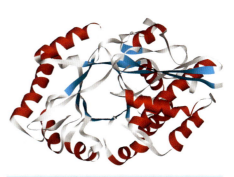

Figure 4-57 **Structure of galactonate dehydratase** The fold is the same as those of MR, MLE, and enolase (see Figure 4-40 for MR and MLE).

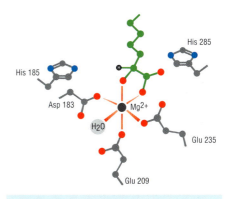

His 285
His 185
Asp 183
Mg²⁺
H₂O
Glu 235
Glu 209

Figure 4-58 Schematic diagram of a model of the active site of galactonate dehydratase with substrate bound The metal-ion coordination and the disposition of the catalytic base histidine 285 and the electrophilic groups that interact with the substrate are similar to those found in MR, MLE and enolase, even though the overall reactions they catalyze are completely different.

References

Babbitt, P.C. and Gerlt, J.A.: **Understanding enzyme superfamilies. Chemistry as the fundamental determinant in the evolution of new catalytic activities.** *J. Biol. Chem.* 1997, **272**:30591–30594.

Babbitt, P.C. *et al.*: **The enolase superfamily: a general strategy for enzyme-catalyzed abstraction of the alpha-protons of carboxylic acids.** *Biochemistry* 1996, **35**:16489–16501.

Babbitt, P.C. *et al.*: **A functionally diverse enzyme superfamily that abstracts the alpha protons of carboxylic acids.** *Science* 1995, **267**:1159–1161.

Function cannot always be determined from sequence, even with the aid of structural information and chemical intuition

About 30% of the 6,282 genes in the genome of the budding yeast *Saccharomyces cerevisiae* code for proteins whose function is completely unknown. One of these is gene *YBL036c*, whose sequence indicates that it encodes a protein of 257 amino acids. A comparison of this sequence against all genomic DNA sequences in the databases as of 1 June 2002 indicates about 200 other proteins whose amino-acid sequences show a greater-than-chance similarity to that of YBL036c. The putative homologs come from every kingdom of life—and none has a known function. At the time of writing, this is still true: although there are now clues to the function of the gene in yeast, the last chapter in this story is still to be written. What follows is an account of the avenues explored and where they lead.

As homologs of YBL036c are ubiquitous, the protein is more likely to function in some fundamental cellular process than to be involved in, for example, some aspect of cell–cell communication that would be confined to multicellular organisms. Thus yeast, a single-celled eukaryote whose complete genome sequence is known and whose metabolic processes can easily be studied by genetic methods, should be an ideal model organism in which to uncover the functions of this family of proteins.

YBL036c was selected as one of the first gene products to be studied in a project aimed at determining structures for yeast proteins which, because of the absence of clear sequence similarity to other proteins, seemed likely to have novel folds. In fact when the three-dimensional structure was determined by X-ray crystallography it proved to have a familiar fold: the triosephosphate isomerase alpha/beta barrel (Figure 4-59). This is yet another clear illustration of the fact that a protein fold can be encoded by very divergent sequences. Structural comparison between this three-dimensional fold and all other folds in the structural database shows the greatest similarity with the large domain of the bacterial enzyme alanine racemase. Like alanine racemase, the structure of YBL036c also revealed a covalently bound pyridoxal phosphate cofactor, which accounts for the yellow color of the purified protein.

Comparison of the active sites of alanine racemase and YBL036c revealed both similarities and important differences. The essential lysine residue required of all pyridoxal-phosphate-dependent enzymes is present, covalently linked to the cofactor (Figure 4-60). A second interaction between the protein and the cofactor that is diagnostic for the chemical function of bacterial alanine racemase, an arginine interacting with the pyridine nitrogen of the cofactor, is also present in YBL036c. However, there were several significant differences. Alanine racemase is an obligatory dimer: residues from both subunits contribute to each other's active sites. In addition, alanine racemase has a second domain, which also contributes residues to the active site. YBL036c is a monomer and lacks the second domain entirely. Consequently, a number of residues found in the active site of alanine racemase are not present in the active site of YBL036c, raising the question of whether the biochemical function of alanine racemase has been preserved.

At this point, sequence and structure can tell us nothing further about function. Additional experimental approaches are needed (see Figure 4-1). The purified protein was first assayed to see if it had any alanine racemase activity, but it did not. More general methods of determining function must be tried. (Several of these are briefly described in section 4-4.)

Location of a protein within the cell is often informative about the cellular, if not biochemical, function. Sometimes, clues to location can be found in the sequence: for example, a carboxy-terminal KDEL sequence codes for retention of a protein in the endoplasmic reticulum. Other

(a)

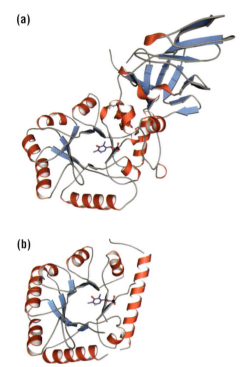

(b)

Figure 4-59 The three-dimensional structures of bacterial alanine racemase and yeast YBL036c The structure of the large domain of alanine racemase (a) is similar to the overall structure of YBL036c (b). The yeast protein lacks the largely antiparallel beta-sheet domain of the racemase; however, the active sites, indicated by the presence of the bound pyridoxal phosphate cofactor (shown in ball-and-stick form), are located in the same place in both proteins. (PDB 1sft and 1ct5)

References

Brent, R. and Finley, R.L. Jr.: **Understanding gene and allele function with two-hybrid methods.** *Annu. Rev. Genet.* 1997; **31**:663–704.

Eswaramoorthy, S. *et al.*: **Structure of a yeast hypothetical protein selected by a structural genomics approach.** *Acta Crystallogr. D. Biol. Crystallogr.* 2003; **59**:127–135.

Shaw, J.P. *et al.*: **Determination of the structure of alanine racemase from *Bacillus stearothermophilus***

at 1.9-Å resolution. *Biochemistry* 1997, **36**:1329–1342.

Templin, M. F. *et al.*: **Protein microarray technology.** *Trends Biotechnol.* 2002, **20**:160–166.

Tucker, C.L. *et al.*: **Towards an understanding of complex protein networks.** *Trends Cell Biol.* 2001, **11**:102–106.

van Roessel, P. and Brand, A.H.: **Imaging into the future: visualizing gene expression and protein interactions with fluorescent proteins.** *Nat. Cell Biol.* 2002, **4**:E15–E20.

von Mering, C. *et al.*: **Comparative assessment of large-scale data sets of protein–protein interactions.** *Nature* 2002, **417**:399–403.

Web site:
http://genome-www.stanford.edu/Saccharomyces/

sequence motifs specify transport into the nucleus, secretion from the cell, and so forth. YBL036c has no such motifs in its sequence, so more direct methods of determining localization must be used. The most common of these is the fusion of the protein of interest with a protein that can be visualized in the cell by antibody staining or intrinsic fluorescence. Fusion to green fluorescent protein (GFP), originally isolated from jellyfish, is a widely used strategy. Efforts are underway to apply these methods systematically to all the gene products in the yeast genome. By the GFP method, YBL036c was found distributed throughout the cell.

In eukaryotic cells in particular, the function of every protein is likely to depend in some manner on interaction with one or more other proteins. Demonstrating a physical interaction between two proteins can thus provide a clue to cellular or biochemical function if the function of one of them is known. Several different approaches to discovering such interactions have been developed. These include co-immunoprecipitation or affinity chromotography from cell extracts followed by mass spectroscopy to identify the interacting partner(s), and cell-based methods such as the yeast two-hybrid screen (see section 4-4). Application of the two-hybrid method to YBL036c detected one interacting partner, another protein of unknown function.

Proteins function in regulatory networks inside the cell; their expression patterns change with changes in external and internal conditions and proteins that perform similar functions often display similar patterns of expression. Thus, clues to the function of a gene product can come from analysis of the pattern of its expression under different conditions and comparison with the patterns of other proteins of known function. Two widely used methods for studying expression are two-dimensional gel electrophoresis, which measures protein levels directly, and DNA microarrays, which measure levels of mRNA. Both can be applied to whole genomes or subsets of the genome. Microarray analyses of yeast gene expression have been carried out by many different laboratories under hundreds of different conditions; most contain information about YBL036c. In general, YBL036c is expressed in all stages of the cell cycle and in all growth conditions tested. Its expression is broadly the same as that of a number of genes that code for proteins involved in amino-acid metabolism. Its expression is upregulated slightly in a variety of stress conditions.

If a gene is not essential for the survival of an organism, deleting it from the genome can often give rise to a phenotype suggestive of function. Microarray analysis of such a deletion strain should show changes in the expression of genes whose function is in some manner coupled with that of the gene that has been deleted. YBL036c is not an essential gene in yeast: the deletion strain is viable and shows no growth defect under a variety of conditions. However, there is a subtle phenotype when the yeast cells form spores. Instead of being dispersed, the spores clump together. Electron microscopy of the spore shows that the ascus containing the spores and the spores themselves have an abnormal wall structure. Since sporulation requires remodeling of cell-wall structures, this phenotype implies that YBL036c is involved in this process. Microarray analysis of the deletion strain supports this conclusion: genes involved in cell-wall biosynthesis show changes in expression levels when YBL036c is absent.

Although these genome-wide methods have suggested a cellular process in which YBL036c participates, a great deal of information is still needed to fully describe the workings of this gene product within the cell. The active-site architecture suggests that the protein may be an amino-acid racemase, but the substrate has yet to be identified. If YBL036c produces a D-amino acid, as such activity would indicate, the role of this product in cell-wall structure remains to be determined. As animal cells do not have cell walls, the cellular function of YBL036c homologs in those organisms might be somewhat different. It is clear from this example that the task of determining the function(s) of a gene is one that does not end with a single organism.

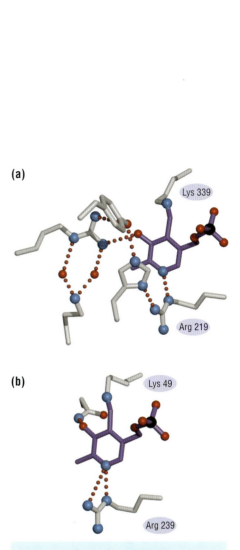

(a)

(b)

Figure 4-60 Comparison of the active sites of bacterial alanine racemase and YBL036c **(a)** Alanine racemase; **(b)** YBL036c. Although many of the interactions between the pyridoxal phosphate cofactor (shown in purple) and protein side chains are different in the two active sites, two interactions are preserved: a covalent linkage to a lysine residue, and the interaction of a nitrogen atom in the pyridine ring of the cofactor with an arginine residue. The lysine interaction is diagnostic of all pyridoxal-phosphate-dependent enzymes; the interaction with arginine is diagnostic for pyridoxal-phosphate-dependent amino-acid racemases.

5

Structure Determination

Most of the protein structures described and discussed in this book have been determined either by X-ray crystallography or by nuclear magnetic resonance (NMR) spectroscopy. Although these techniques both depend on data derived from physical techniques for probing structure, their interpretation is not unambiguous and entails assumptions and approximations often depending upon knowledge of the protein from other sources, including its biology. This chapter briefly describes how structures are determined by X-ray crystallography and NMR, how the data are interpreted, and what contributes to the accuracy of a structure determination.

(a)

(b)

(c)

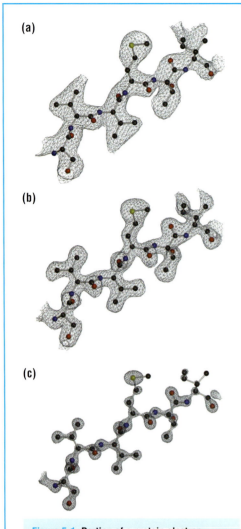

Figure 5-1 Portion of a protein electron density map at three different resolutions
The peptide corresponding to the electron density map (serine, valine, valine, methionine, threonine, isoleucine) is superimposed. **(a)** At 3-Å resolution the fold of the polypeptide chain can be seen and the approximate positions of side chains can be determined. Interatomic distances can only be measured to ± about 0.5 Å. **(b)** At 2-Å resolution side chains are well delineated and the peptide carbonyls of the backbone are discernible, allowing the chain to be oriented with precision. Interatomic distances can be measured to a precision of about 0.2 Å. Approximately three times as many data are required for this resolution as were used at 3-Å resolution. **(c)** At 1-Å resolution atoms are visible and resolved. Interatomic distances can be measured to a precision of a few hundredths of an Ångstrom. Almost 30 times more data are required for this resolution as were used at 3-Å resolution. In favorable cases, the position of hydrogen atoms can be inferred at this resolution. Note that at high resolution (c) there is no electron density for the methyl group of the methionine side chain because it is disordered. At lower resolutions (a and b), it appears that such density is present, but this is actually the density of the heavier sulfur atom. Since at low resolution electron density is more diffuse, the entire thio-ether portion of the side chain can be fitted into the sulfur density. Kindly provided by Aaron Moulin.

Experimentally determined protein structures are the result of the interpretation of different types of data

Much of the content of this book depends on a detailed understanding of the structure of proteins at the atomic level. The atomic structures of biological macromolecules can be determined by several techniques: although a few have been determined by electron microscopy, the vast majority have been obtained by either single crystal *X-ray diffraction*, generally known as *X-ray crystallography*, or by *nuclear magnetic resonance spectroscopy*, generally known as NMR. How this is done, in both cases, is briefly explained in the next section. Here, we describe the kind of information these techniques produce, and how it can be interpreted.

Both X-ray crystallography and NMR produce information on the relative positions of the atoms of the molecule: these are termed **atomic coordinates**. Because X-rays are scattered from the electron cloud of the atoms, whereas NMR measures the interactions of atomic nuclei, X-ray crystallography provides the positions of all non-hydrogen atoms whereas NMR provides the positions of all atoms including hydrogen. However, the coordinate sets provided in this way represent the interpretation of the primary data obtained in each case. The precision and accuracy with which coordinates can be derived from these data depend on several factors that are different for the two methods. The objective end-product of a crystallographic structure determination is an **electron density map** (Figure 5-1), which is essentially a contour plot indicating those regions in the crystal where the electrons in the molecule are to be found. Human beings must interpret this electron density map in terms of an atomic model, aided by semi-automatic computational procedures. The objective end-product of an NMR structure determination is usually a set of distances between atomic nuclei that define both bonded and non-bonded close contacts in the molecule. These must be interpreted in terms of a molecular structure, a process that is aided by automated methods.

Because interpretation in each case requires assumptions and approximations, macromolecular structures can have errors. In most cases, these errors are small, and only affect a small part of the structure. However, such errors can be quite large: cases of completely incorrect structures, though rare, have been reported.

The choice of technique depends on several factors, including the molecular weight, solubility and ease of crystallization of the protein in question. For proteins and protein complexes with molecular weights above 50–100 kDa, X-ray crystallography is the method of choice. For smaller proteins or protein complexes, either method may be usable but X-ray diffraction will usually provide more precise structural information than NMR provided the species crystallizes easily and is well ordered in the crystal. For molecules that are hard to crystallize and can be dissolved at reasonably high concentrations without aggregation, NMR is the method of choice. In many cases however, both techniques can be used and provide complementary information, since crystallographic images are static but NMR can be used to study the flexibility of proteins and their dynamics over a wide range of time scales.

Both the accuracy and the precision of a structure can vary

It is essential to appreciate the distinction between the accuracy of an experimentally determined structure and its precision. The latter is much easier to assess than the former. We define the precision of a structure determination in terms of the reproducibility of its atomic coordinates. If the data allow the equilibrium position of each atom to be determined precisely, then the

Definitions

atomic coordinate: the position in three-dimensional space of an atom in a molecule relative to all other atoms in the molecule.

electron density map: a contour plot showing the distribution of electrons around the atoms of a molecule.

resolution: the level of detail that can be derived from a given process.

same structure determined independently elsewhere should yield atomic coordinates that agree very closely with the first set. Precise coordinates are usually reproducible to within a few tenths of an Ångstrom.

Accuracy refers to whether or not the structure is correct, but there are two ways to define correctness. A structure may be right but irrelevant: the protein may be in a conformation that is determined by the experimental conditions rather than its biological context. Such cases are rare, but they have occurred. It is also possible to interpret the experimental data incorrectly, yielding a structure that is not and could not possibly be correct, either in whole (rarely) or in part. Often it is possible to determine that a structure has been built incorrectly on the basis of the known rules of how proteins fold and their properties when they are folded. It is not possible to determine the correctness of minor details unless there is a biological experiment that can probe the structure at that position. For example, the involvement of a residue in catalysis predicted by the structure can be tested by mutating it to one that cannot perform the required catalytic function. Ultimately, the accuracy of a structure can best be assessed by the answer to a simple question: is the structure consistent with the body of biochemical and biological information about the protein?

Sometimes, part of a structure may be invisible to the experimental method used. This can arise because that part of the molecule is disordered and therefore does not contribute to the reflected X-rays in a crystallographic experiment or, in an NMR experiment, atoms may not be close enough to interact strongly.

The information content of a structure is determined by its resolution

Precision and accuracy of a structure are related: the more precise a structure determination is, the less likely it is to have gross errors. What links precision with accuracy is the concept of **resolution**.

Crystallographers express the resolution of a structure in terms of a distance: if a structure has been determined at 2-Å resolution then any atoms separated by more than about this distance will appear as separate maxima in the electron density contour plot, as we explain in the next section, and their positions can be obtained directly to high precision. Atoms closer together than the resolution limit will appear as a fused electron density feature, and their exact positions must be inferred from the shape of the electron density and knowledge of the chemical structure of amino acids or nucleotides (Figure 5-1). Since the average C–C single bond distance is 1.5 Å, the precision with which the individual atoms of, say, an inhibitor bound to an enzyme can be located will be much greater at 1.5-Å resolution than in a structure determined at, say, 3-Å resolution.

NMR spectroscopists express resolution somewhat differently. NMR structures are determined as ensembles of similar models (Figure 5-2): we explain in the next section how these are derived. However, since more than one closely related model will fit the data, the effective resolution of an NMR structure is given by the extent of the differences between these models, expressed as a root-mean-square deviation (RMSD) of their atomic coordinates. The smaller the deviation, the more precise (and, presumably accurate) the NMR structure is believed to be, and therefore the higher its effective resolution.

In the case of both X-ray and NMR structures, what sets the resolution limit is the intrinsic order of the protein plus the amount of data that the experimenter is able to measure. The greater the amount of data, the higher the resolution, and the higher the resolution, the more precise and accurate the information that can be extracted from the structure.

In favorable cases, the resolution of a macromolecular structure determination by crystallography is such that the relative positions of all non-hydrogen atoms are known to a precision of a few tenths of an Ångstrom.

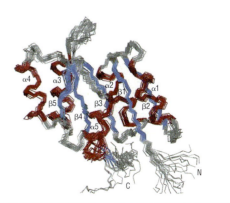

Figure 5-2 NMR structure ensemble The figure shows the superposition of the set of models derived from the internuclear distances measured for this protein in solution. Note that different portions of the structure are determined with different precision. Blue represents beta strands, red represents alpha helices, grey represents loops. This figure should not be taken to indicate the flexibility of segments of the protein: different regions may be poorly defined because there are insufficient data to constrain the structure, and not because the structures are mobile.

Protein crystallography involves summing the scattered X-ray waves from a macromolecular crystal

The steps in solving a protein crystal structure at high resolution are diagrammed in Figure 5-3. First, the protein must be crystallized. This is often the rate-limiting step in straightforward structure determinations, especially for membrane proteins. Then, the X-ray diffraction pattern from the crystal must be recorded. When X-rays strike a macromolecular crystal, the atoms in the molecules produce scattered X-ray waves which combine to give a complex diffraction pattern consisting of waves of different amplitudes. What is measured experimentally are the amplitudes and positions of the scattered X-ray waves from the crystal. The structure can be reconstructed by summing these waves, but each one must be in the correct registration with respect to every other wave, that is, the origin of each wave must be determined so that they sum to give some image instead of a sea of noise. This is called the **phase problem**. Phase values must be assigned to all of the recorded data; this can sometimes be done computationally, but is usually done experimentally by labeling the protein with one or more heavy atoms whose position in the crystal can be determined independently. The phased waves are then summed in three dimensions to generate an image of the electron density distribution of the molecule in the crystal. This can be done semi-automatically or by hand on a computer graphics system. A chemical model of part of the molecule is docked into the shape of each part of the electron part of the electron density (as shown in Figure 5-3). This fitting provides the first picture of the structure of the protein. The overall model is improved by an iterative process called refinement whereby the positions of the atoms in the model are tweaked until the calculated diffraction pattern from the model agrees as well as can be with the experimentally measured diffraction pattern from the actual protein. There is no practical limit to the size of the protein or protein complex whose structure can be determined by X-ray crystallography.

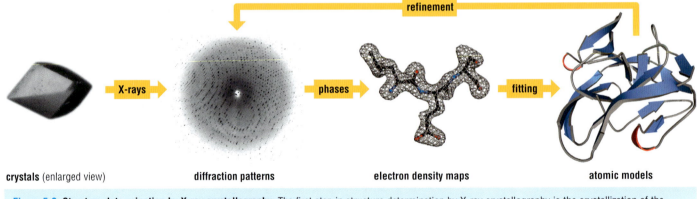

crystals (enlarged view)   **diffraction patterns**   **electron density maps**   **atomic models**

Figure 5-3 Structure determination by X-ray crystallography The first step in structure determination by X-ray crystallography is the crystallization of the protein. The source of the X-rays is often a synchrotron and in this case the typical size for a crystal for data collection may be 0.3 × 0.3 × 0.1 mm. The crystals are bombarded with X-rays which are scattered from the planes of the crystal lattice and are captured as a diffraction pattern on a detector such as film or an electronic device. From this pattern, and with the use of reference—or phase—information from labeled atoms in the crystal, electron density maps (shown here with the corresponding peptide superimposed) are computed for different parts of the crystal. A model of the protein is constructed from the electron density maps and the diffraction pattern for the modeled protein is calculated and compared with the actual diffraction pattern. The model is then adjusted—or refined— to reduce the difference between its calculated diffraction pattern and the pattern obtained from the crystal, until the correspondence between model and reality is as good as possible. The quality of the structure determination is measured as the percentage difference between the calculated and the actual pattern.

Definitions

phase problem: in the measurement of data from an X-ray crystallographic experiment only the amplitude of the wave is determined. To compute a structure, the phase must also be known. Since it cannot be determined directly, it must be determined indirectly or by some other experiment.

NMR spectroscopy involves determining internuclear distances by measuring perturbations between assigned resonances from atoms in the protein in solution

Unlike crystallography, structure determination by NMR (Figure 5-4) is carried out on proteins in solution, but the protein must be soluble without aggregating at concentrations close to those of a protein in a crystal lattice. NMR structure determination requires two types of data. The first is measurement of nuclear magnetic resonances from protons and isotopically labeled carbons and nitrogens in the molecule. Different nuclei in a protein absorb electromagnetic energy (resonance) at different frequencies because their local electromagnetic environments differ due to the three-dimensional structure of the protein. These resonances must be assigned to atoms in specific amino acids in the protein sequence, a process that requires several specific types of experiments. The second set of data consists of internuclear distances that are inferred from perturbing the resonance of different atoms and observing which resonances respond; only atoms within 5 Å of each other show this effect, and its magnitude varies with the distance between them. The set of approximate internuclear distances is then used to compute a structure model consistent with the data. Since the distances are imprecise, many closely related models may be consistent with the observations, so NMR structures are usually reported as ensembles of atomic coordinates. In practice, if a structure is determined by both NMR and X-ray diffraction, it is usually found that the average of the NMR ensemble closely resembles the crystal structure. As a general rule, there are practical limitations to the determination of the structure of a complete protein by NMR: a molecular weight of about 50 kDa is considered a very large protein for NMR. In special cases, domains or portions of much larger proteins or complexes can be studied.

Unlike X-ray diffraction, which presents a static picture (an average in time and space) of the structure of a protein, NMR has the capability of measuring certain dynamic properties of proteins over a wide range of time scales.

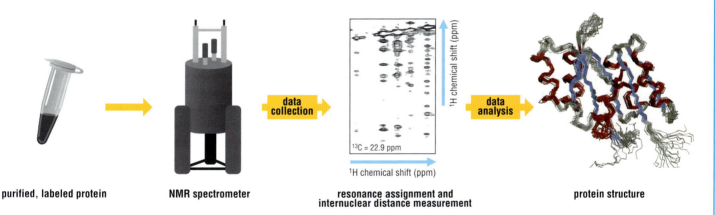

| purified, labeled protein | NMR spectrometer | resonance assignment and internuclear distance measurement | protein structure |

Figure 5-4 Structure determination by NMR For protein structure determination by NMR, a labeled protein is dissolved at very high concentration and placed in a magnetic field, which causes the spin of the hydrogen atoms to align along the field. Radio frequency pulses are then applied to the sample, perturbing the nuclei of the atoms which when they relax back to their original state emit radio frequency radiation whose properties are determined by the environment of the atom in the protein. This emitted radiation is recorded in the NMR spectrometer for pulses of differing types and durations (for simplicity, only one such record is shown here), and compared with a reference signal to give a measure known as the chemical shift. The relative positions of the atoms in the molecule are calculated from these data to give a series of models of the protein which can account for these data. The quality of the structure determination is measured as the difference between the different models.

References

Drenth, J.: *Principles of Protein X-Ray Crystallography* 2nd ed. (Springer-Verlag, New York, 1999).

Evans, J.N.S.: *Biomolecular NMR Spectroscopy* (Oxford University Press, Oxford, 1995).

Markley, J.L. *et al.*: **Macromolecular structure determination by NMR spectroscopy.** *Methods Biochem. Anal.* 2003, **44**:89–113.

Rhodes, G.: *Crystallography Made Crystal Clear: A Guide*

to *Users of Macromolecular Models* 2nd ed. (Academic Press, New York and London, 1999).

Schmidt, A. and Lamzin, V.S.: ***Veni, vidi, vici*—atomic resolution unravelling the mysteries of protein function.** *Curr. Opin. Struct. Biol.* 2002, **12**:698–703.

Gorenstein, N.: *Nuclear Magnetic Resonance (NMR)* (Biophysics Textbooks Online): http://www.biophysics.org/btol/NMR.html

The quality of a finished structure depends largely on the amount of data collected

Both X-ray and NMR structure determination have statistical criteria for the quality of the atomic model produced. Crystallographers usually speak of R-factors, which represent the percentage disagreement between the observed diffraction pattern and that calculated from the final model. R-factors of around 20% or less are considered indicative of well determined structures that are expected to contain relatively few errors. NMR spectroscopists usually report overall root mean square deviation (RMSD) between the atoms in secondary structure elements in all coordinate sets in the ensemble of structures consistent with the experimental data. In practice, RMSDs of 0.7 Å are considered good, indicating a structure determination of high precision. RMSDs of around 1 Å are considered acceptable.

There is no substitute for high resolution. It makes the structure determination easier and more reliable. The closer one gets to true atomic resolution in X-ray crystallography (better than 1.5 Å), the less ambiguity one has in positioning every atom. Atomic resolution allows one to detect mistakes in the biochemically determined or genomically derived amino-acid sequence, to correct preliminary incorrect chain connectivity, and to identify unexpected chemical features in the molecule. Incorrect crystal structures are almost never reported from high-resolution data. Most of the mistakes in protein crystallography have been made because a medium-resolution structure has been misinterpreted or overinterpreted. In NMR, the general rule is the more internuclear distances measured the better. Most of the mistakes that have been made in NMR structure determination have resulted from either incorrect assignment of a set of resonances to a particular part of a protein or the failure to measure enough internuclear distances.

Different conventions for representing the structures of proteins are useful for different purposes

Atomic coordinate sets make for boring reading, so protein structures are presented visually. There are a number of different ways to render a protein structure, depending on the information that one wishes to convey. The fold of the polypeptide chain can be depicted as a wire model that follows the path of the backbone (Figure 5-5a), which is useful for example in comparisons of two conformations of the same molecule (see Figure 1-80); or it can be depicted as a ribbon diagram in which alpha helices and beta sheets are graphically stylized (Figure 5-5b): this not only makes the overall fold easily recognizable, but makes particular secondary structure elements or loops that may have particular functional significance easily recognizable. Detail of the structure at the atomic level can be rendered by means of a ball-and-stick model (Figure 5-5c) in which the balls are colored or sized by type of atom and covalent bonds are represented by perspective sticks; such drawings are to scale so relative bonded and non-bonded distances can be assessed, which is important for evaluating interactions. Atoms as volumes can be represented by space-filling drawings in which each atom is given a sphere scaled to its van der Waals radius (Figure 5-5d): this representation is particularly useful for assessing the fit of a ligand to a binding site (see Figure 2-8b). Finally, to emphasize the protein surface that is created by the space-filling nature of atoms, a surface topography image can be produced (Figure 5-5e). Such an image can be colored according to different local properties such as the electrostatic potential at different points in the molecule. In this book, all of these different methods of visualizing structures are used.

References

Holyoak, T. *et al.*: **The 2.4 Å crystal structure of the processing protease Kex2 in complex with an Ala–Lys–Arg boronic acid inhibitor.** *Biochemistry*, in the press.

Martz, E.: **Protein Explorer: easy yet powerful macromolecular visualization.** *Trends Biochem. Sci.* 2002, **27**:107–109.

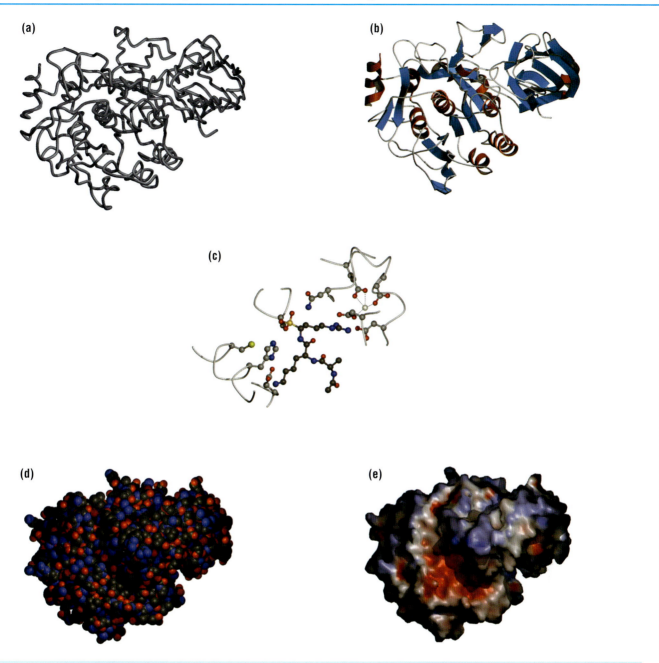

Figure 5-5 Different ways of presenting a protein structure The structure of the catalytic and P domains of the protease Kex2 is shown in different styles. **(a)** "Wire" diagram showing the path of the polypeptide backbone. **(b)** Ribbon diagram highlighting the secondary structure elements. Beta strands are depicted as arrows with the arrow head being at the carboxyl terminus. Alpha helices are drawn as coiled ribbons. **(c)** Ball-and-stick model of a small part of the protein structure, showing details of amino-acid interactions. **(d)** Space-filling representation in which every non-hydrogen atom is shown as a sphere of its van der Waals radius. **(e)** Surface representation (sometimes called a GRASP image after the program that computes it) in which the topography of the protein surface is shown and the electrostatic characteristics of the surface are highlighted in color (red for negative, blue for positive). Kindly provided by Todd Holyoak.

Glossary

acid: a molecule or chemical group that donates a proton, either to water or to some other base. (2-12)

acid-base catalysis: catalysis in which a proton is transferred in going to or from the transition state. When the acid or base that abstracts or donates the proton is derived directly from water (H⁺ or OH⁻) this is called specific acid-base catalysis. When the acid or base is not H⁺ or OH⁻, it is called general acid-base catalysis. Nearly all enzymatic acid-base catalysis is general acid-base catalysis. (2-12)

activation energy: the energy required to bring a species in a chemical reaction from the **ground state** to a state of higher free energy, in which it can transform spontaneously to another low-energy species. (2-6)

activation-energy barrier: the higher-energy region between two consecutive chemical species in a reaction. (2-6)

activation loop: a stretch of polypeptide chain that changes conformation when a kinase is activated by phosphorylation and/or protein binding. This segment may or may not be the one containing the residue that is phosphorylated to activate the kinase. Usually, in the inactive state, the activation loop blocks access to the active site. (3-13)

activation segment: see **activation loop**.

active site: asymmetric pocket on or near the surface of a macromolecule that promotes chemical catalysis when the appropriate **ligand** (substrate) binds. (2-1)

affinity: the tightness of a protein–ligand complex. (2-4)

alignment: procedure of comparing two or more sequences by looking for a series of characteristics (residue identity, similarity, and so on) that match up in both and maximize conservation, in order to assess overall similarity. (4-1)

allosteric activator: a ligand that binds to a protein and induces a conformational change that increases the protein's activity. (3-5)

allosteric inhibitor: a ligand that binds to a protein and induces a conformational change that decreases the protein's activity. (3-5)

allostery: the property of being able to exist in two structural states of differing activity. The equilibrium between these states is modulated by ligand binding. (3-5)

alpha/beta barrel: a parallel beta barrel formed usually of eight strands, each connected to the next by an alpha-helical segment. Also known as a **TIM barrel**. (1-18)

alpha/beta domain: a protein domain composed of beta strands connected by alpha helices. (1-17)

alpha+beta domain: a protein domain containing separate alpha-helical and beta-sheet regions. (1-17)

alpha/beta twist: a twisted parallel beta sheet with a saddle shape. Helices are found on one side of the sheet for the first half and the other side for the second half. (1-18)

alpha domain: a protein domain composed entirely of alpha helices. (1-17)

alpha helix: a coiled conformation, resembling a right-handed spiral staircase, for a stretch of consecutive amino acids in which the backbone –N–H group of every residue n donates a hydrogen bond to the C=O group of every residue n+4. (1-5)

alternative splicing: the production of different versions of the final protein sequence from a gene sequence by the removal during RNA processing of portions of the RNA containing or affecting coding sequences. (1-2)

amide bond: a chemical bond formed when a carboxylic acid condenses with an amino group with the expulsion of a water molecule. (1-3)

amphipathic: having both polar and nonpolar character and therefore a tendency to form interfaces between **hydrophobic** and **hydrophilic** molecules. (1-1)

amphipathic alpha helix: an alpha helix with a hydrophilic side and a hydrophobic side. (1-6)

anisotropic: behaving differently in different directions; dependent on geometry and direction. (2-4)

antiparallel beta sheet: a beta sheet, often formed from contiguous regions of the polypeptide chain, in which each strand runs in the opposite direction from its immediate neighbors. (1-7)

atomic coordinate: the position in three-dimensional space of an atom in a molecule relative to all other atoms in the molecule. (5-1)

autophosphorylation: phosphorylation of a protein kinase by itself. Autophosphorylation may occur when the active site of the protein molecule to be phosphorylated catalyzes this reaction (*cis* autophosphorylation) or when another molecule of the same kinase provides the active site that carries out the chemistry (*trans* autophosphorylation). Autophosphorylation *in trans* often occurs when kinase molecules dimerize, a process that can be driven by ligand binding as in the receptor tyrosine kinases. (3-13)

backbone: the regularly repeating part of a polymer. In proteins it consists of the amide –N–H, alpha carbon –C–H and the carbonyl –C=O groups of each amino acid residue in the **polypeptide** chain. Residues are linked to each other by means of **peptide bonds**. (1-0, 1-3)

base: (in a nucleic acid) the aromatic group attached to the sugar of a **nucleotide**. (1-2)

base: (in chemistry) a molecule or chemical group that accepts a proton, either from water or from some other acid. (2-12)

beta barrel: a beta sheet in which the last strand is hydrogen bonded to the first strand, forming a closed cylinder. (1-7)

beta domain: a protein domain containing only beta sheet. (1-17)

beta sandwich: a structure formed of two antiparallel beta sheets packed face to face. (1-17)

beta sheet: a **secondary structure** element formed by backbone hydrogen bonding between segments of extended polypeptide chain. (1-5)

beta turn: a tight turn that reverses the direction of the polypeptide chain, stabilized by one or more backbone hydrogen bonds. Changes in chain direction can also occur by loops, which are peptide chain segments with no regular conformations. (1-5)

bifunctional: having two distinct biochemical functions in one gene product. Bifunctional enzymes catalyze two distinct chemical reactions. (2-15)

BLAST: a family of programs for searching protein and DNA databases for sequence similarities by optimizing a specific similarity measure between the sequences being compared. (4-2)

catalyst: a substance that accelerates the rate of a reaction without itself being permanently altered. (2-6)

catalytic triad: a set of three amino acids that are hydrogen bonded together and cooperate in catalysis. (2-14)

cavity: a completely enclosed hole in the interior of a protein. Cavities may contain one or more disordered water molecules but some are believed to be completely empty. (2-3)

chameleon sequence: a sequence that exists in different conformations in different environments. (4-14)

chaperone: a protein that aids in the folding of another protein by preventing the unwanted association of the unfolded or partially folded forms of that protein with itself or with others. (1-9)

chromatin: the complex of DNA and protein that comprises eukaryotic nuclear chromosomes. The DNA is wound around the outside of highly conserved histone proteins, and decorated with other DNA-binding proteins. (3-20)

co-activator: a regulatory molecule that binds to a gene activator protein and assists its binding to DNA. (3-5)

codon: three consecutive **nucleotides** in a strand of DNA or RNA that represent either a particular amino acid or a signal to stop translating the transcript of the gene. The formula for translating the codons is given by the genetic code. (1-2)

coenzyme: a cofactor that is an organic or organometallic molecule and that assists catalysis. (2-13)

cofactor: a small, non-protein molecule or ion that is bound in the functional site of a protein and assists in ligand binding or catalysis or both. Some cofactors are bound covalently, others are not. (1-13, 2-13)

coiled coil: a protein or a region of a protein formed by a dimerization interaction between two alpha helices in which hydrophobic side chains on one face of each helix interdigitate with those on the other. (1-19)

competitive inhibitor: a species that competes with substrate for binding to the active site of an enzyme and thus inhibits catalytic activity. (3-0)

conservative substitution: replacement of one amino acid by another that has similar chemical and/or physical properties. (1-2)

conserved: identical in all sequences or structures compared. (4-1)

convergent evolution: evolution of structures not related by ancestry to a common function that is reflected in a common structure. (1-16, 4-5)

cooperative binding: interaction between two sites on a protein such that the binding of a ligand to the first one affects the properties—usually binding or catalytic—of the second one. (3-4)

cooperativity: interaction between two sites on a protein such that something that happens to the first one affects the properties of the second one. (3-4)

coordinate covalent bond: a bond formed when a lone pair of electrons from an atom in a ligand is donated to a vacant orbital on a metal ion. (1-13)

co-repressor: a regulatory molecule that binds to a gene repressor protein and assists its binding to DNA. (3-5)

cross-linked domain: a small protein domain with little or no secondary structure and stabilized by disulphide bridges or metal ions. (1-17)

decarboxylation: removal of carbon dioxide from a molecule. (2-11)

degenerate: having more than one **codon** for an amino acid. (1-2)

denaturant: a chemical capable of unfolding a protein in solution at ordinary temperatures. (1-12)

denatured state: the partially or completely unfolded form of a biological macromolecule in which it is incapable of carrying out its biochemical and biological functions. (1-12)

dimer: an assembly of two identical (homo-) or different (hetero-) subunits. In a protein, the subunits are individual folded polypeptide chains. (1-19)

dipole moment: an imaginary vector between two separated charges that may be full or partial. Molecules or functional groups having a dipole moment are said to be polar. (1-3)

disulfide bridge: a covalent bond formed when the reduced –S–H groups of two cysteine residues react with one another to make an oxidized –S–S– linkage. (1-4)

divergent evolution: evolution from a common ancestor. (4-5)

DNA microarray: an ordered array of nucleic acid molecules, either cDNA fragments or synthetic oligonucleotides, where each position in the array represents a single gene. (4-4)

domain: a compact unit of protein structure that is usually capable of folding stably as an independent entity in solution. Domains do not need to comprise a contiguous segment of peptide chain, although this is often the case. (1-14)

domain fold: the particular topographical arrangement of secondary structural elements that characterizes a single domain. Examples are an antiparallel arrangement of four helices in a four-helix bundle, or an open twisted beta sandwich with a particular sequence that binds nucleotides. (1-15)

domain swapping: the replacement of a structural element of one subunit of an oligomer by the same structural element of the other subunit, and vice versa. The structural element may be a secondary structure element or a whole domain. (2-4)

dominant-negative: dominant loss of function due to a single mutant copy of a gene. This can occur when the mutant subunit is able to oligomerize with normal subunits to form a non-functional protein, thereby producing a loss-of-function phenotype even in the presence of a normal copy of the gene. (1-20)

effector: a species that binds to a protein and modifies its activity. Effectors may be as small as a proton or as large as a membrane and may act by covalent binding, noncovalent binding, or covalent modification. (3-0)

effector ligand: a ligand that induces a change in the properties of a protein. (3-4)

electron density map: a contour plot showing the distribution of electrons around the atoms of a molecule. (5-1)

electrostatic interactions: noncovalent interaction between atoms or groups of atoms due to attraction of opposite charges. (1-4)

enthalpy: a form of energy, equivalent to work, that can be released or absorbed as heat at constant pressure. (1-12)

entropy: a measure of the disorder or randomness in a molecule or system. (1-12)

equilibrium: the state at which the rate of the forward reaction and the rate of the reverse reaction in a chemical transformation are equal. At equilibrium, the relative concentrations of reactants and products no longer change, although the reaction continues to occur. (2-6)

equilibrium constant: the ratio of the product of the concentrations of reaction products to the product of the concentrations of reaction reactants. For a reaction of the general form A + B = C + D, the equilibrium constant K_{eq} is [C][D]/[A][B], where [X] is the concentration of X, usually in moles per liter. This definition is a simplification that neglects effects at high concentrations. (2-6)

E-value: the probability that an **alignment** score as good as the one found between two sequences would be found in a comparison between two random sequences; that is, the probability that such a match would occur by chance. (4-1)

evolutionary distance: the number of observed changes in nucleotides or amino acids between two related sequences. (4-1)

exon: coding segment of a gene. The coding DNA of many eukaryotic genes is interrupted by segments of non-coding DNA (**introns**). (1-2)

extein: the sequences flanking an intein and which are religated after **intein** excision to form the functional protein. (3-17)

family: a group of homologous proteins that share a related function. Usually these will also have closely related sequences. Members of the same enzyme family catalyze the same chemical reaction on structurally similar substrates. (4-8)

four-helix bundle: a structure of four antiparallel alpha helices. Parallel bundles are possible but rare. (1-17)

free energy: a function, designed to produce a criterion for spontaneous change, that combines the **entropy** and **enthalpy** of a molecule or system. Free energy decreases for a spontaneous process, and is unchanged at equilibrium. (1-12)

functional motif: sequence or structural **motif** that is always associated with a particular biochemical function. (1-16, 4-2)

gated binding: binding that is controlled by the opening and closing of a physical obstacle to substrate or inhibitor access in the protein. (2-7)

gene knockout: inactivation of the function of a specific gene in a cell or organism, usually by recombination with a marker sequence but sometimes by antisense DNA, RNA interference, or by antibody binding to the gene product. The phenotype resulting from the knockout can often provide clues to the function of the gene. (4-4)

genetic code: the relationship between each of the 64 possible three-letter combinations of A, U, G and C (which stand for the RNA bases adenine, uracil, guanine and cytosine, respectively) and the 20 naturally occurring amino acids that make up proteins. U is the RNA equivalent of T (thymine) in DNA. (1-2)

genomics: the study of the DNA sequence and gene content of whole genomes. (4-0)

globin fold: a predominantly alpha-helical arrangement observed in certain heme-containing proteins. (1-17)

glycosylation: the post-translational covalent addition of sugar molecules to asparagine, serine or threonine residues on a protein molecule. Glycosylation can add a single sugar or a chain of sugars at any given site and is usually enzymatically catalyzed. (1-13, 3-18)

glycosylphosphatidylinositol anchor: a complex structure involving both lipids and carbohydrate molecules that is reversibly attached to some proteins to target them to the cell membrane. (3-19)

G protein: a member of a large class of proteins with GTPase activity that act as molecular switches in many different cellular pathways, controlling processes such as sensory perception, intracellular transport, protein synthesis and cell growth and differentiation. They undergo a large conformational change when a bound GTP is hydrolyzed to GDP. (3-6)

Greek-key motif: an arrangement of antiparallel strands in which the first three strands are adjacent but the fourth strand is adjacent to the first, with a long connecting loop. (1-17)

ground state: a species with low free energy; usually, the non-activated state of any substance. (2-6)

ground-state destabilization: raising the free energy, (relative to some reference state), of the ground state, usually referring to the bound substrate in the active site before any chemical change has occurred. Geometric or electronic strain are two ways of destabilizing the ground state. (2-8)

GTPase-activating protein (GAP): a protein that accelerates the intrinsic GTPase activity of switch GTPases. (3-7)

guanine-nucleotide-binding protein: see **G protein**.

guanine-nucleotide exchange factor (GEF): a protein that facilitates exchange of GDP for GTP in switch GTPases. (3-7)

hairpin turn: another name for **beta turn**.

helical parameters: set of numerical values that define the geometry of a helix. These include the number of residues per turn, the translational rise per residue, and the main-chain torsional angles. (1-6)

helix dipole: the macrodipole that is thought to be formed by the cumulative effect of the individual peptide dipoles in an alpha helix. The positive end of the dipole is at the beginning (amino terminus) of the helix; the negative end is at the carboxyl terminus of the helical rod. (1-6)

heptad repeat: a sequence in which hydrophobic residues occur every seven amino acids, a pattern that is reliably indicative of a **coiled-coil** interaction between two alpha helices in which the hydrophobic side chains of each helix interdigitate with those of the other. (1-19)

heterotetramer: an assembly of four subunits of more than one kind of polypeptide chain. (1-19)

heterotrimeric G protein: a GTPase switch protein composed of three different subunits, an α subunit with GTPase activity, and associated β and γ subunits, found associated with the cytoplasmic tails of G-protein-coupled receptors, where it acts to relay signals from the receptor to downstream targets. Exchange of bound GDP for GTP on the α subunit causes dissociation of the heterotrimer into a free α subunit and a βγ heterodimer; hydrolysis of the bound GTP causes reassociation of the subunits. (3-8)

hexamer: an assembly of six identical or different subunits. In a protein the subunits are individual folded polypeptide chains. (1-19)

Hidden Markov Model: a probabilistic model of a sequence **alignment**. (4-1)

homologous: describes genes or proteins related by divergent evolution from a common ancestor. Homologous proteins, or homologs, will generally have similar sequences, structures and biochemical functions, although the sequence and/or functional similarity may be difficult to recognize. (4-1)

homology: the similarity seen between two gene or protein sequences that are both derived by evolution from a common ancestral sequence. (4-1)

homology modeling: a computational method for modeling the structure of a protein based on its sequence similarity to one or more proteins of known structure. (4-6)

homotrimer: an assembly of three identical subunits: in a protein, these are individual folded polypeptide chains. (1-19)

hydride ion: a hydrogen atom with an extra electron. (2-9)

hydrogen bond: a noncovalent interaction between the **donor atom**, which is bound to a positively polarized hydrogen atom, and the acceptor atom, which is negatively polarized. Though not covalent, the hydrogen bond holds the donor and **acceptor atom** close together. (1-4)

hydrolysis: breaking a covalent bond by addition of a molecule of water. (1-3)

hydrophilic: tending to interact with water. Hydrophilic molecules are polar or charged and, as a consequence, are very soluble in water. In polymers, hydrophilic **side chains** tend to associate with other hydrophilic side chains, or with water molecules, usually by means of hydrogen bonds. (1-1)

hydrophobic: tending to avoid water. Hydrophobic molecules are nonpolar and uncharged and, as a consequence, are relatively insoluble in water. In polymers, hydrophobic **side chains** tend to associate with each other to minimize their contact with water or polar side chains. (1-1)

hydrophobic effect: the tendency of nonpolar groups in water to self-associate and thereby minimize their contact surface area with the polar solvent. (1-9)

induced fit: originally, the change in the structure of an enzyme, induced by binding of the substrate, that brings the catalytic groups into proper alignment. Now generalized to the idea that specific ligands can induce the protein conformation that results in optimal binding interactions. (1-22, 2-2)

intein: a protein intron (intervening sequence). An internal portion of a protein sequence that is post-translationally excised in an autocatalytic reaction while the flanking regions are spliced together, making an additional protein product. (3-17)

interaction domain: a protein that recognizes another protein, usually via a specific recognition motif. (3-1)

intermediate: a species that forms transiently along the path from substrate to product. (2-6)

intron: an intervening sequence in a gene that does not correspond to any portion of the final protein sequence and is spliced out of the RNA transcript before translation. (1-2)

jelly roll fold: a beta sandwich built from two sheets with topologies resembling a Greek key design. The sheets pack almost at right-angles to each other. (1-17)

K_d: the dissociation constant for the binding of a ligand to a macromolecule. Typical values range from 10^{-3} M

to 10^{-10} M. The lower the K_d, the tighter the ligand binds. (1-13)

ligand: small molecule or macromolecule that recognizes and binds to a specific site on a macromolecule. (2-1)

ligand-binding site: site on the surface of a protein at which another molecule binds. (2-1)

limited proteolysis: specific cleavage by a protease of a limited number of the peptide bonds in a protein substrate. The fragments thus produced may remain associated or may dissociate. (1-13)

lipid anchor: lipid attached to a protein that inserts into a membrane, thereby anchoring the protein to the bilayer. (3-2)

lipidation: covalent attachment of a fatty-acid group to a protein. (3-19)

lipid bilayer: the structure of cellular membranes, formed when two sheets of lipid molecules pack against each other with their hydrophobic tails forming the interior of the sandwich and their polar headgroups covering the outside. (1-6)

local alignment: alignment of only a part of a sequence with a part of another. (4-2)

mesophilic: favoring moderate temperatures. Mesophilic organisms normally cannot tolerate extremes of heat or cold. Mesophilic enzymes typically denature at moderate temperatures (over 40 °C or so). (1-12)

messenger RNA (mRNA): the RNA molecule transcribed from a gene sequence after removal of **introns** and editing. (1-2)

metastable: only partially stable under the given conditions. In the case of protein structures, a metastable fold exists in equilibrium with other conformations or with the unfolded state. (4-15)

methylation: modification, usually of a nitrogen or oxygen atom of an amino-acid side chain, by addition of a methyl group. Some bases on DNA and RNA can also be methylated. (3-20)

mixed beta sheet: beta sheet containing both parallel and antiparallel strands. (1-7)

monomer: a single subunit: in a protein, this is a folded peptide chain. (1-19)

motif: characteristic sequence or structure that in the case of a **structural motif** may comprise a whole domain or protein but usually consists of a small local arrangement of secondary structure elements which then coalesce to form domains. **Sequence motifs**, which are recognizable amino-acid sequences found in different proteins, usually indicate biochemical function. Structural motifs are less commonly associated with specific biochemical functions. (1-16)

multifunctional: having a number of distinct biochemical functions in one gene product. (2-15)

multiple sequence alignment: alignment of more than two sequences to maximize their overall mutual identity or similarity. (4-1)

myristoylation: irreversible attachment of a myristoyl group to a protein via an amide linkage. (3-19)

N-acetylation: covalent addition of an acetyl group from acetyl-CoA to a nitrogen atom at either the amino-terminus of a polypeptide chain or in a lysine side-chain. The reaction is catalyzed by N-acetyltransferase. (1-13, 3-20)

native state: the stably folded and functional form of a biological macromolecule. (1-9)

negative cooperativity: binding of one molecule of a ligand to a protein makes it more difficult for a second molecule of that ligand to bind at another site. (3-4)

nitrosylation: modification of the –SH group of a cysteine residue by addition of nitric oxide produced by nitric oxide synthase. (3-20)

northern blot: technique for detecting and identifying individual RNAs by hybridization to specific nucleic acid probes, after separation of a complex mixture of mRNAs by electrophoresis and blotting onto a nylon membrane. (4-4)

nucleophile: a group that is electron-rich, such as an alkoxide ion (–O$^-$), and can donate electrons to an electron-deficient center. (2-14)

nucleotide: the basic repeating unit of a nucleic acid polymer. It consists of a **base** (A, U [in RNA, T in DNA], G or C), a sugar (ribose in RNA, deoxyribose in DNA) and a phosphate group. (1-2)

nucleotide-binding fold: an open parallel beta sheet with connecting alpha helices that is usually used to bind NADH or NADPH. It contains a characteristic sequence motif that is involved in binding the cofactor. Also known as the Rossmann fold. (1-18)

oligomer: an assembly of more than one subunit: in a protein, the subunits are individual folded polypeptide chains. (1-19)

oxidation: the loss of electrons from an atom or molecule. (2-10)

oxyanion hole: a binding site for an alkoxide in an enzyme active site. The "hole" is a pocket that fits the –O$^-$ group precisely, and has two hydrogen-bond-donating groups that stabilize the oxyanion with –O$^-$···H–X hydrogen bonds. (2-14)

packing motif: an arrangement of secondary structure elements defined by the number and types of such elements and the angles between them. The term motif is used in structural biology in a number of contexts and thus can be confusing. (1-10)

pairwise alignment: alignment of two sequences. (4-1)

palmitoylation: reversible attachment of a palmitoyl group to a protein via a thioester linkage. (3-19)

parallel beta sheet: a beta sheet, formed from non-contiguous regions of the polypeptide chain, in which every strand runs in the same direction. (1-7)

partner swapping: exchange of one protein for another in multiprotein complexes. (2-4)

pentamer: an assembly of five identical or different subunits: in a protein, these are individual folded polypeptide chains. (1-19)

peptide bond: another name for **amide bond**, a chemical bond formed when a carboxylic acid condenses with an amino group with the expulsion of a water molecule. The term peptide bond is used only when both groups come from amino acids. (1-3)

percent identity: the percentage of columns in an **alignment** of two sequences that contain identical amino acids. Columns that include gaps are not counted. (4-1)

phase problem: in the measurement of data from an X-ray crystallographic experiment only the amplitude of the wave is determined. To compute a structure, the phase must also be known. Since it cannot be determined directly, it must be determined indirectly or by some other experiment. (5-2)

phi torsion angle: see **torsion angle**.

phosphate-binding loop: see **P-loop**.

phosphorylation: covalent addition of a phosphate group, usually to one or more amino-acid side chains on a protein, catalyzed by protein kinases. (1-13)

phylogenetic tree: a branching diagram, usually based on the evolutionary distances between sequences, that illustrates the evolutionary history of a protein family or superfamily, or the relationships between different species of organism. (4-1)

pK$_a$ value: strictly defined as the negative logarithm of the equilibrium constant for the acid-base equation. For ranges of pK$_a$ between 0 and 14, it can be thought of as the pH of an aqueous solution at which a proton-donating group is half protonated and half deprotonated. pK$_a$ is a measure of the proton affinity of a group: the lower the pK$_a$, the more weakly the proton is held. (2-12)

pleated sheet: another name for **beta sheet**.

P-loop: a conserved loop in GTPase- and ATPase-based nucleotide switch proteins that binds to phosphate groups in the bound nucleotide. (3-6)

polypeptide: a polymer of amino acids joined together by **peptide bonds**. (1-3)

positive cooperativity: binding of one molecule of a ligand to a protein makes it easier for a second molecule of that ligand to bind at another site. (3-4)

prenylation: irreversible attachment of either a farnesyl or geranylgeranyl group to a protein via thioether linkage. (3-19)

primary structure: the amino-acid sequence of a polypeptide chain. (1-2)

primary transcript: the RNA molecule directly transcribed from a gene, before removal of introns or other editing. (1-2)

profile: a table or matrix of information that characterizes a protein family or superfamily. It is typically composed of sequence variation or identity with respect to a reference sequence, expressed as a function of each position in the amino-acid sequence of a protein. It can be generalized to include structural information. Three-dimensional profiles express the three-dimensional structure of a protein as a table which represents the local environment and conformation of each residue. (4-2)

propinquity factor: another term for **proximity factor**.

proteasome: a multiprotein complex that degrades ubiquitinated proteins into short peptides. (3-11)

protein kinase: enzyme that transfers a phosphate group from ATP to the OH group of serines, threonines and tyrosines of target proteins. Kinases that phosphorylate carboxylates and histidines also occur as part of **two-component systems** in prokaryotes, fungi and plants, but not in animals. (3-2)

protein phosphatase: enzyme that specifically removes phosphate groups from phosphorylated serines, threonines or tyrosines on proteins. (3-12)

proteolytic cascade: a sequential series of protein cleavages by proteases, each cleavage activating the next protease in the cascade. (3-16)

protomer: the asymmetric repeating unit (or units) from which an oligomeric protein is built up. (1-21)

proximity factor: the concept that a reaction will be facilitated if the reacting species are brought close together in an orientation appropriate for chemistry to occur. (2-8)

pseudosymmetric: having approximate but not exact symmetry. A protein with two non-identical subunits of very similar three-dimensional structure is a pseudosymmetric dimer. (1-21)

psi torsion angle: see **torsion angle**.

quaternary structure: the subunit structure of a protein. (1-19)

Ramachandran plot: a two-dimensional plot of the values of the backbone torsion angles phi and psi, with allowed regions indicated for conformations where there is no steric interference. Ramachandran plots are used as a diagnosis for accurate structures: when the phi and psi torsion angles of an experimentally determined protein structure are plotted on such a diagram, the observed values should fall predominantly in the allowed regions. (1-5)

reaction sub-site: that part of the active site where chemistry occurs. (2-7)

redox reactions: reactions in which oxidation and reduction occur. (2-10)

reducing environment: a chemical environment in which the reduced states of chemical groups are favored. In a reducing environment, free –S–H groups are favored over –S–S– bridges. The interior of most cells is a highly reducing environment. (1-4)

reduction: the gain of electrons by an atom or molecule. (2-10)

residue: the basic building block of a polymer; the fragment that is released when the bonds that hold the polymer segments together are broken. In proteins, the residues are the amino acids. (1-1)

resolution: the level of detail that can be derived from a given process. (5-1)

resonance: delocalization of bonding electrons over more than one chemical bond in a molecule. Resonance greatly increases the stability of a molecule. It can be represented, conceptually, as if the properties of the molecule were an average of several structures in which the chemical bonds differ. (1-3)

reverse turn: another name for **beta turn**.

RGS protein: regulator of G-protein signaling protein; protein that binds to the free GTP-bound α subunit of a **heterotrimeric G protein** and stimulates its GTPase activity. (3-8)

RNA editing: enzymatic modification of the mRNA base sequence. It may involve changes in the bases or the insertion of entirely new stretches of bases. RNA editing produces a protein sequence that does not correspond precisely to the sequence of amino acids that would be predicted from the gene sequence by the genetic code. (1-2)

RNA interference (RNAi): abolition of the expression of a gene by a small (~22 base pair) double-stranded RNA. (4-4)

S-acylation: reversible attachment of a fatty-acid group to a protein via a thioester linkage; **palmitoylation** is an example of S-acylation. (3-19)

salt bridge: a **hydrogen bond** in which both donor and acceptor atoms are fully charged. The bonding energy of a salt bridge is significantly higher than that of a hydrogen bond in which only one participating atom is fully charged or in which both are partially charged. (1-4)

scaffold protein: a protein that serves as a platform onto which other proteins assemble to form functional complexes. (2-5)

secondary structure: folded segments of a polypeptide chain with repeating, characteristic phi, psi backbone torsion angles, that are stabilized by a regular pattern of hydrogen bonds between the peptide –N–H and C=O groups of different residues. (1-5)

side chain: a chemical group in a polymer that protrudes from the repeating backbone. In proteins, the side chain, which is bonded to the alpha carbon of the backbone, gives each of the 20 amino acids its particular chemical identity. Glycine has no side chain, and the end of the side chain of proline is fused to the nitrogen of the backbone, creating a closed ring. (1-1)

single-nucleotide polymorphism (SNP): a mutation of a single base in a codon, usually one that does not affect the identity of the amino acid that it encodes. (1-2)

specificity sub-site: that part of the active site where recognition of the ligand takes place. (2-7)

stop codon: a codon that signals the end of the coding sequence and usually terminates **translation**. (1-2)

stress-response proteins: proteins whose synthesis is induced when cells are subjected to environmental stress, such as heat. (3-11)

structural domain: a compact part of the overall structure of a protein that is sufficiently independent of the rest of the molecule to suggest that it could fold stably on its own. (2-3)

substrate: the molecule that is transformed in a reaction. (2-6)

sumoylation: modification of the side chain of a lysine residue by addition of a small ubiquitin-like protein (SUMO). The covalent attachment is an amide bond between the carboxy-terminal carboxylate of SUMO and the NH$_2$ on the lysine side chain of the targeted protein. (3-20)

superfamily: proteins with the same overall fold but with usually less than 40% sequence identity. The nature of the biochemical functions performed by proteins in the same superfamily are more divergent than those within families. For instance, members of the same enzyme superfamily may not catalyze the same overall reaction, yet still retain a common mechanism for stabilizing chemically similar rate-limiting transition-states and intermediates, and will do so with similar active site residues. (4-8)

switch I region: a conserved sequence motif in GTPase- and ATPase-based nucleotide switch proteins that, with the **switch II region**, binds the terminal gamma-phosphate in the triphosphate form of the bound nucleotide and undergoes a marked conformational change when the nucleotide is hydrolyzed. (3-6)

switch II region: a conserved sequence motif in GTPase- and ATPase-based nucleotide switch proteins that, with the **switch I region**, binds the terminal gamma-phosphate in the triphosphate form of the bound nucleotide and undergoes a marked conformational change when the nucleotide is hydrolyzed. (3-6)

temperature-sensitive: losing structure and/or function at temperatures above physiological or room temperature. A temperature-sensitive mutation is a change in the amino-acid sequence of a protein that causes the protein to inactivate or fail to fold properly at such temperatures. (1-12)

temperature-sensitive mutants: organisms containing a genetic mutation that makes the resulting protein sensitive to slightly elevated temperatures. The temperature at which the mutant protein unfolds is called the restrictive temperature. The term is also used for the protein itself. (3-11)

tertiary structure: the folded conformation of a protein, formed by the condensation of the various secondary elements, stabilized by a large number of weak interactions. (1-10)

tetramer: an assembly of four identical or different subunits. (1-19)

thermophilic: favoring high temperatures. A thermophilic organism is one that requires high temperatures (above approximately 50 °C) for survival. A thermophilic enzyme is one that functions optimally and is stable at temperatures at which **mesophilic** proteins denature. (1-12)

TIM barrel: another name for the alpha/beta barrel fold. (1-18)

torsion angle: the angle between two groups on either side of a rotatable chemical bond. If the bond is the C_α–N bond of a peptide backbone the torsion angle is called **phi**. If the bond is the C_α–C backbone bond, the angle is called **psi**. (1-3)

transcription: the synthesis of RNA from the coding strand of DNA by DNA-dependent RNA polymerase. (1-2)

transition state: the species of highest free energy either in a reaction or a step of a reaction; the highest region on the **activation-energy barrier**. (2-6)

translation: the transfer of genetic information from the sequence of **codons** on **mRNA** into a sequence of amino acids and the synthesis on the ribosome of the corresponding polypeptide chain. (1-2)

trimer: an assembly of three identical or different subunits. (1-19)

two-component systems: signal transduction systems found in bacteria and some eukaryotes involving a membrane-bound histidine kinase and a cytoplasmic response regulator protein that is activated by phosphorylation. (3-15)

ubiquitin: a small protein that when attached to other proteins (**ubiquitination**), targets them for degradation to the **proteasome**. Sometimes ubiquitin tagging targets a protein to other fates such as endocytosis. (3-11)

ubiquitination: the attachment of ubiquitin to a protein. (3-11)

up-and-down structural motif: a simple fold in which beta strands in an antiparallel sheet are all adjacent in sequence and connectivity. (1-17)

van der Waals interaction: a weak attractive force between two atoms or groups of atoms, arising from the fluctuations in electron distribution around the nuclei. Van der Waals forces are stronger between less electronegative atoms such as those found in hydrophobic groups. (1-4)

yeast two-hybrid: a method for finding proteins that interact with another protein, based on activation of a reporter gene in yeast. (4-4)

zinc finger: a small, irregular domain stabilized by binding of a zinc ion. Zinc fingers usually are found in eukaryotic DNA-binding proteins. They contain signature metal-ion binding sequence motifs. (1-18)

References

Abel, K., Yoder, M.D., Hilgenfeld, R. and Jurnak, F.: **An alpha to beta conformational switch in EF-Tu.** *Structure* 1996, **4**:1153–1159. (3-9)

Aitken, A.: **Protein consensus sequence motifs.** *Mol. Biotechnol.* 1999, **12**:241–253. (1-16, 4-2)

Alberts, B., Johnson, A., Lewis, J., Raff, M., Roberts, K. and Walter, P.: *Molecular Biology of the Cell* 4th ed. (Garland, New York, 2002). (1-0, 1-2, 2-0)

Al-Lazikani, B., Jung, J., Xiang, Z. and Honig, B.: **Protein structure prediction.** *Curr. Opin. Chem. Biol.* 2001, **5**:51–56. (4-6)

Allen, K.N., Bellamacina, C.R., Ding, X., Jeffrey, C.J., Mattos, C.J., Petsko, G.A. and Ringe, D.: **An experimental approach to mapping the binding surfaces of crystalline proteins.** *J. Phys. Chem.* 1996, **100**:2605–2611. (4-9)

Almo, S.C., Smith, D.L., Danishefsky, A.T. and Ringe, D.: **The structural basis for the altered substrate specificity of the R292D active site mutant of aspartate aminotransferase from *E. coli*.** *Protein Eng.* 1994, **7**:405–412. (2-7)

Aloy, P., Querol, E., Aviles, F.X. and Sternberg, M.J.: **Automated structure-based prediction of functional sites in proteins: applications to assessing the validity of inheriting protein function from homology in genome annotation and to protein docking.** *J. Mol. Biol.* 2001, **311**:395–408. (4-9)

Altschul, S.F., Madden, T.L., Schaffer, A.A., Zhang, J., Zhang, Z., Miller, W. and Lipman, D.J.: **Gapped BLAST and PSI-BLAST: a new generation of protein database search programs.** *Nucleic Acids Res.* 1997, **25**:3389–3402. (4-2)

Anston, A.A., Dodson, E.J. and Dodson, G.G.: **Circular assemblies.** *Curr. Opin. Struct. Biol.* 1996, **6**:142–150. (1-19)

Argyle, E.: **A similarity ring for amino acids based on their evolutionary substitution rates.** *Orig. Life* 1980, **10**:357–360. (1-2)

Arrondo, J.L. and Goni, F.M.: **Structure and dynamics of membrane proteins as studied by infrared spectroscopy.** *Prog. Biophys. Mol. Biol.* 1999, **72**:367–405. (1-22)

Authier, F., Metioui, M., Fabrega, S., Kouach, M. and Briand, G.: **Endosomal proteolysis of internalized insulin at the C-terminal region of the B chain by cathepsin D.** *J. Biol. Chem.* 2002, **277**:9437–9446. (3-3)

Authier, F., Posner, B.I. and Bergeron, J.J.: **Endosomal proteolysis of internalized proteins.** *FEBS Lett.* 1996, **389**:55–60. (3-3)

Babbitt, P.C. and Gerlt, J.A.: **Understanding enzyme superfamilies. Chemistry as the fundamental determinant in the evolution of new catalytic activities.** *J. Biol. Chem.* 1997, **272**:30591–30594. (4-16)

Babbitt, P.C., Hasson, M.S., Wedekind, J.E., Palmer, D.R., Barrett, W.C., Reed, G.H., Rayment, I., Ringe, D., Kenyon, G.L. and Gerlt, J.A.: **The enolase superfamily: a general strategy for enzyme-catalyzed abstraction of the alpha-protons of carboxylic acids.** *Biochemistry* 1996, **35**:16489–16501. (4-16)

Babbitt, P.C., Mrachko, G.T., Hasson, M.S., Huisman, G.W., Kolter, R., Ringe, D., Petsko, G.A., Kenyon, G.L. and Gerlt, J.A.: **A functionally diverse enzyme superfamily that abstracts the alpha protons of carboxylic acids.** *Science* 1995, **267**:1159–1161. (4-11, 4-16)

Badger, J., Minor, I., Oliveira, M.A., Smith, T.J. and

Rossmann, M.G.: **Structural analysis of antiviral agents that interact with the capsid of human rhinoviruses.** *Proteins* 1989, **6**:1–19. (2-1)

Baker, D. and Sali, A.: **Protein structure prediction and structural genomics.** *Science* 2001, **294**:93–96. (4-6)

Ban, N., Nissen, P., Hansen, J., Moore, P.B. and Steitz, T.A.: **The complete atomic structure of the large ribosomal subunit at 2.4 Å resolution.** *Science* 2000, **289**:905–920. (2-5)

Barford, D., Hu, S.H. and Johnson, L.N.: **Structural mechanism for glycogen phosphorylase control by phosphorylation and AMP.** *J. Mol. Biol.* 1991, **218**:233–260. (3-12)

Barlow, D.J. and Thornton, J.M.: **Helix geometry in proteins.** *J. Mol. Biol.* 1988, **201**:601–619. (1-10)

Bartlett, G.J., Porter, C.T., Borkakoti, N. and Thornton, J.M.: **Analysis of catalytic residues in enzyme active sites.** *J. Mol. Biol.* 2002, **324**:105–121. (4-10)

Berisio, R., Vitagliano, L., Mazzarella, L. and Zagari, A.: **Recent progress on collagen triple helix structure, stability and assembly.** *Curr. Pharm. Des.* 2002, **9**:107–116. (2-5)

Beissinger, M., Sticht, H., Sutter, M., Ejchart, A., Haehnel, W., and Rosch, P.: **Solution structure of cytochrome c6 from the thermophilic cyanobacterium *Synechococcus elongatus*.** *EMBO J.* 1998, **2**:27–36. (2-4)

Bellamacina, C.R.: **The nicotinamide dinucleotide binding motif: a comparison of nucleotide binding proteins.** *FASEB J.* 1996, **10**:1257–1269. (1-18)

Benaroudj, N., Tarcsa, E., Cascio, P. and Goldberg, A.L.: **The unfolding of substrates and ubiquitin-independent protein degradation by proteasomes.** *Biochimie* 2001, **83**:311–318. (3-11)

Bennett, M.J., Choe, S. and Eisenberg, D.: **Domain swapping: entangling alliances between proteins.** *Proc. Natl Acad. Sci. USA* 1994, **91**:3127–3131. (2-4)

Bernstein, B.E. and Hol, W.G.: **Crystal structures and products bound to the phosphoglycerate kinase active site reveal the catalytic mechanism.** *Biochemistry* 1998, **37**:4429–4436. (2-9)

Bernstein, B.E. Michels, P.A. and Hol, W.G.: **Synergistic effects of substrate-induced conformational changes in phosphoglycerate kinase activation.** *Nature* 1997, **385**:275–278. (2-9)

Bitetti-Putzer, R., Joseph-McCarthy, D., Hogle, J. M. and Karplus, M.: **Functional group placement in protein binding sites: a comparison of GRID and MCSS.** *J. Comput Aided Mol Des.* 2001, **15**:935–960. (4-9)

Bond, C.J., Wong, K.B., Clarke, J., Fersht, A.R. and Daggett, V.: **Characterization of residual structure in the thermally denatured state of barnase by simulation and experiment: description of the folding pathway.** *Proc. Natl Acad. Sci. USA* 1997, **94**:13409–13413. (1-9)

Bonneau, R., Tsai, J., Ruczinski, I., Chivian, D., Rohl, C., Strauss, C.E. and Baker, D.: **Rosetta in CASP4: Progress in *ab initio* protein structure prediction.** *Proteins* 2001, **45(S5)**:119–126. (4-7)

Bork, P., Holm, L. and Sander, C.: **The immunoglobulin fold. Structural classification, sequence patterns and common core.** *J. Mol. Biol.* 1994, **242**:309–320. (1-17)

Bosshard, H.R., Durr, E., Hitz, T. and Jelesarov, I.: **Energetics of coiled coil folding: the nature of the transition state.** *Biochemistry.* 2001, **40**:3544–3552. (1-19)

Bourne, Y., Arvai, A.S., Bernstein, S.L., Watson, M.H., Reed,

S.I., Endicott, J.E., Noble, M.E., Johnson, L.N. and Tainer, J.A.: **Crystal structure of the cell cycle-regulatory protein suc1 reveals a beta-hinge conformational switch.** *Proc. Natl Acad. Sci. USA* 1995, **92**:10232–10236. (2-4)

Bowie, J.U., Luthy, R. and Eisenberg, D.A.: **A method to identify protein sequences that fold into a known three-dimensional structure.** *Science* 1991, **253**:164–170. (4-7)

Branden, C. and Tooze, J.: *Introduction to Protein Structure* 2nd ed. (Garland, New York, 1999). (1-15, 1-17, 1-18)

Brazhnik, P., de la Fuente, A. and Mendes, P.: **Gene networks: how to put the function in genomics.** *Trends Biotechnol.* 2002, **20**:467–472. (4-0)

Brenner, S.: **Theoretical biology in the third millennium.** *Philos. Trans. R. Soc. Lond. B. Biol. Sci.* 1999, **354**:1963–1965. (4-3)

Brent, R. and Finley, R.L. Jr.: **Understanding gene and allele function with two-hybrid methods.** *Annu. Rev. Genet.* 1997, **31**:663–704. (4-17)

Brizuela, L., Richardson, A., Marsischky, G., and Labaer, J.: **The FLEXGene repository: exploiting the fruits of the genome projects by creating a needed resource to face the challenges of the post-genomic era.** *Arch. Med. Res.* 2002, **33**:318–324. (4-3)

Brown, N.R., Noble, M.E., Endicott, J.A. and Johnson, L.N.: **The structural basis for specificity of substrate and recruitment peptides for cyclin-dependent kinases.** *Nat. Cell Biol.* 1999, **1**:438–443. (3-14)

Brown, N.R., Noble, M.E., Lawrie, A.M., Morris, M.C., Tunnah, P., Divita, G., Johnson, L.N. and Endicott, J.A.: **Effects of phosphorylation of threonine 160 on cyclin-dependent kinase 2 structure and activity.** *J. Biol. Chem.* 1999, **274**:8746–8756. (3-14)

Bruckner, K., Perez, L., Clausen, H. and Cohen, S.: **Glycosyltransferase activity of Fringe modulates Notch-Delta interactions.** *Nature* 2000, **406**:411–415. (3-18)

Bruice, T.C. and Benkovic, S.J.: **Chemical basis for enzyme catalysis.** *Biochemistry* 2000, **39**:6267–6274. (2-6)

Bullock, T.L., Clarkson, W.D., Kent, H.M. and Stewart, M.: **The 1.6 angstroms resolution crystal structure of nuclear transport factor 2 (ntf2).** *J. Mol. Biol.* 1996, **260**:422–431. (4-5)

Bunn, H.F. and Forget, B.G.: *Hemoglobin: Molecular, Genetic and Clinical Aspects* (Saunders, New York, 1986). (1-20)

Burchett, S.A.: **Regulators of G protein signaling: a bestiary of modular protein binding domains.** *J. Neurochem.* 2000, **75**:1335–1351. (1-14)

Burley, S.K. and Petsko, G.A.: **Weakly polar interactions in proteins.** *Adv. Prot. Chem.* 1988, **39**:125–189. (1-4)

Butikofer, P., Malherbe, T., Boschung, M. and Roditi, I.: **GPI-anchored proteins: now you see 'em, now you don't.** *FASEB J.* 2001, **15**:545–548. (3-19)

Byerly, D.W., McElroy, C.A. and Foster, M.P.: **Mapping the surface of *Escherichia coli* peptide deformylase by NMR with organic solvents.** *Protein Science* 2002, **11**:1850–1853. (4-9)

Campbell, I.D. and Downing, A.K.: **Building protein structure and function from modular units.** *Trends Biotechnol.* 1994, **12**:168–172. (1-14)

Capaldi, R.A. and Aggeler, R.: **Mechanism of the F(1)F(0)-type ATP synthase, a biological rotary motor.** *Trends Biochem. Sci.* 2002, **27**:154–160. (3-10)

Cardozo, T., Batalov, S. and Abagyan, R.: **Estimating local backbone structural deviation in homology models.** *Comput. Chem.* 2000, **24**:13–31. (4-6)

Casey, P.J.: **Lipid modifications** in *Encyclopedic Reference of Molecular Pharmacology.* Offermanns, S. and Rosenthal, W. eds (Springer-Verlag, Heidelberg, 2003) in the press. (3-19)

Casey, P.J.: **Protein lipidation in cell signaling.** *Science* 1995, **268**:221–225. (3-19)

Chan, T.-F., Carvalho, J., Riles, L. and Zheng, X.F.: **A chemical genomics approach toward understanding the global functions of the target of rapamycin protein (TOR).** *Proc. Natl Acad. Sci. USA* 2000, **97**:13227–13232. (4-0)

Chatterjee, S. and Mayor, S.: **The GPI-anchor and protein sorting.** *Cell. Mol. Life Sci.* 2001, **58**:1969–1987. (3-19)

Chevalier, B.S. and Stoddard, B.L.: **Homing endonucleases: structural and functional insight into the catalysts of intron/intein mobility.** *Nucleic Acids Res.* 2001, **29**:3757–3774. (3-17)

Chook, Y.M., Ke, H., and Lipscomb, W.N.: **Crystal structures of the monofunctional chorismate mutase from *Bacillus subtilis* and its complex with a transition state analog.** *Proc. Natl Acad. Sci. USA* 1993, **90**:8600–8603. (2-8)

Chothia, C.: **Asymmetry in protein structure.** *Ciba Foundation Symp.* 1991, **162**:36–49. (1-18)

Chou, P.Y. and Fasman, G.D.: **Prediction of protein conformation.** *Biochemistry* 1974, **13**:222–245. (1-8)

Cline, M., Hughey, R. and Karplus, K.: **Predicting reliable regions in protein sequence alignments.** *Bioinformatics* 2002, **18**:306–314. (4-6)

Cohen, C. and Parry, D.A.: **Alpha-helical coiled coils and bundles: how to design an alpha-helical protein.** *Proteins* 1990, **7**:1–15. (1-19)

Cohen, B.I., Presnell, S.R. and Cohen, F.E.: **Origins of structural diversity within sequentially identical hexapeptides.** *Protein Sci.* 1993, **2**:2134–2145. (4-15)

Colas, P. and Brent, R.: **The impact of two-hybrid and related methods on biotechnology.** *Trends Biotechnol.* 1998, **16**:355–363. (4-4)

Cook, A., Lowe, E.D., Chrysina, E.D., Skamnaki, V.T., Oikonomakos, N.G. and Johnson, L.N.: **Structural studies on phospho-CDK2/cyclin A bound to nitrate, a transition state analogue: implications for the protein kinase mechanism.** *Biochemistry* 2002, **41**:7301–7311. (3-14)

Cozzone, A.J.: **Regulation of acetate metabolism by protein phosphorylation in enteric bacteria.** *Annu. Rev. Microbiol.* 1998, **52**:127–164. (3-12)

Cramer, W.A., Engelman, D.M., Von Heijne, G. and Rees, D.C.: **Forces involved in the assembly and stabilization of membrane proteins.** *FASEB J.* 1992, **6**:3397–3402. (1-9)

Creighton, T.E.: *Proteins: Structure and Molecular Properties* 2nd ed. (Freeman, New York, 1993). (1-1, 1-19)

Cronin, C.N. and Kirsch, J.F.: **Role of arginine-292 in the substrate specificity of aspartate aminotransferase as examined by site-directed mutagenesis.** *Biochemistry.* 1988, **27**:4572–4579. (2-7)

Cutforth, T. and Gaul, U.: **A methionine aminopeptidase and putative regulator of translation initiation is required for cell growth and patterning in *Drosophila*.** *Mech Dev.* 1999, **82**:23–28. (4-13)

Cyert, M.S.: **Regulation of nuclear localization during signaling.** *J. Biol. Chem.* 2001, **276**:20805–20808. (3-2)

Dalal, S. and Regan, L.: **Understanding the sequence determinants of conformational switching using protein design.** *Protein Sci.* 2000, **9**:1651–1659. (4-15)

Daggett, V.: **Long timescale simulations.** *Curr. Opin. Struct. Biol.* 2000, **10**:160–164. (1-22)

Darnell, J.E. Jr.: **STATs and gene regulation.** *Science* 1997, **277**:1630–1635. (2-4)

Datta, B.: **MAPs and POEP of the roads from prokaryotic to eukaryotic kingdoms.** *Biochimie* 2000, **82**:95–107. (4-13)

Dayhoff, M.O, Barker, W.C. and Hunt, L.C.: **Establishing homologies in protein sequences.** *Methods Enzymol.* 1983, **91**:524–545. (1-2)

Dean, A.M. and Koshland, D.E. Jr.: **Electrostatic and steric contributions to regulation at the active site of isocitrate dehydrogenase.** *Science* 1990, **249**:1044–1046. (3-12)

Deane, C.M., Allen, F.H., Taylor, R. and Blundell, T.L.: **Carbonyl-carbonyl interactions stabilize the partially allowed Ramachandran conformations of asparagine and aspartic acid.** *Protein Eng.* 1999, **12**:1025–1028. (1-5)

de la Cruz, X. and Thornton, J.M.: **Factors limiting the performance of prediction-based fold recognition methods.** *Protein Sci.* 1999, **8**:750–759. (1-16, 4-7)

Deleage, G., Blanchet, C. and Geourjou, C.: **Protein structure prediction. Implications for the biologist.** *Biochimie* 1997, **79**:681–686. (1-8)

Dengler, U., Siddiqui, A.S. and Barton, G.J.: **Protein structural domains: analysis of the 3Dee domains database.** *Proteins* 2001, **42**:332–344. (1-14)

Dennis, S., Kortvelyesi, T. and Vajda, S.: **Computational mapping identifies the binding sites of organic solvents on proteins.** *Proc. Natl Acad. Sci. USA* 2002, **99**:4290–4295. (4-9)

Desai, A. and Mitchison, T.J.: **Microtubule polymerization dynamics.** *Annu. Rev. Cell Dev. Biol.* 1997, **13**:83–117. (2-0)

Dhillon, A.S. and Kolch, W.: **Untying the regulation of the Raf-1 kinase.** *Arch. Biochem. Biophys.* 2002, **404**:3–9. (3-2)

Di Gennaro, J.A., Siew, N., Hoffman, B.T., Zhang, L., Skolnick, J., Neilson, L.I. and Fetrow, J.S.: **Enhanced functional annotation of protein sequences via the use of structural descriptors.** *J. Struct. Biol.* 2001, **134**:232–245. (4-10)

Dill, K.A. and Bromberg, S.: *Molecular Driving Forces: Statistical Thermodynamics in Chemistry and Biology* (Garland Science, New York and London 2003). (1-12)

Ding, X., Rasmussen, B.F., Petsko, G.A. and Ringe, D.: **Direct structural observation of an acyl-enzyme intermediate in the hydrolysis of an ester substrate by elastase.** *Biochemistry* 1994, **33**:9285–9293. (2-2)

Dinner, A.R., Sali, A., Smith, L.J., Dobson, C.M. and Karplus, M.: **Understanding protein folding via free-energy surfaces from theory and experiment.** *Trends Biochem. Sci.* 2000, **25**:331–339. (1-9)

Dobson, C.M.: **Protein misfolding, evolution and disease.** *Trends Biochem Sci.* 1999, **24**:329–332. (4-15)

Domingues, F.S., Lackner, P., Andreeva, A. and Sippl, M.J.: **Structure-based evaluation of sequence comparison and fold recognition alignment accuracy.** *J. Mol. Biol.* 2000, **297**:1003–1013. (4-3)

Doolittle R.F.: **Convergent evolution: the need to be explicit.** *Trends Biochem. Sci.* 1994, **19**:15–18. (4-12)

Dorn, G.W. 2nd and Mochly-Rosen, D.: **Intracellular transport mechanisms of signal transducers.** *Annu. Rev. Physiol.* 2002, **64**:407–429. (3-2)

Doyle, D.A., Morais Cabral, J., Pfuetzner, R.A., Kuo, A., Gulbis, J.M., Cohen, S.L., Chait, B.T. and MacKinnon, R.: **The structure of the potassium channel: molecular basis of K+ conduction and selectivity.** *Science* 1998, **280**:69–77. (1-11)

Drenth, J.: *Principles of Protein X-Ray Crystallography* 2nd ed. (Springer-Verlag, New York, 1999). (5-2)

Dunitz, J.D.: **Win some, lose some: enthalpy-entropy compensation in weak intermolecular interactions.** *Chem. Biol.* 1995, **2**:709–712. (1-4)

Eaton, W.A., Munoz, V., Hagen, S.J., Jas, G.S., Lapidus, L.J., Henry, E.R. and Hofrichter, J.: **Fast kinetics and mechanisms in protein folding.** *Annu. Rev. Biophys. Biomol. Struct.* 2000, **29**:327–259. (1-9)

Eilers, M., Shekar, S.C., Shieh, T., Smith, S.O. and Fleming, P.J.: **Internal packing of helical membrane proteins.** *Proc. Natl Acad. Sci. USA* 2000, **97**:5796–5801. (1-10)

Elia, A.E., Cantley, L.C. and Yaffe, M.B.: **Proteomic screen finds pSer/pThr-binding domain localizing Plk1 to mitotic substrates.** *Science* 2003, **299**:1228–1231. (3-1)

Elion, E.A.: **The Ste5p scaffold.** *J. Cell Sci.* 2001, **114**:3967–3978. (2-5)

Elofsson, A., Fischer, D., Rice, D.W., Le Grand, S.M. and Eisenberg, D.: **A study of combined structure/sequence profiles.** *Fold. Des.* 1996, **1**:451–461. (4-2)

Engh, R.A., Huber, R., Bode, W. and Schulze, A.J.: **Divining the serpin inhibition mechanism: a suicide substrate "springe"?** *Trends Biotechnol.* 1995, **13**:503–510. (4-15)

English, A.C., Groom, C.R. and Hubbard, R.E.: **Experimental and computational mapping of the binding surface of a crystalline protein.** *Protein Eng.* 2001, **14**:47–59. (4-9)

Eswaramoorthy, S., Gerchman, S., Graziano, V., Kycia, H., Studier, F.W. and Swaminathan, S.: **Structure of a yeast hypothetical protein selected by a structural genomics approach.** *Acta Crystallogr. D. Biol. Crystallogr.* 2003, **59**:127–135. (4-17)

Evans J.N.S.: *Biomolecular NMR Spectroscopy* (Oxford University Press, Oxford, 1995). (5-2)

Ewing, T.J., Makino, S., Skillman A.G. and Kuntz, I.D.: **DOCK 4.0: search strategies for automated molecular docking of flexible molecule databases.** *J. Comput. Aided Mol. Des.* 2001, **15**:411–428. (4-10)

Falquet, L., Pagni, M., Bucher, P., Hulo, N., Sigrist, C.J., Hofmann, K. and Bairoch, A.: **The PROSITE database, its status in 2002.** *Nucleic Acids Res.* 2002, **30**:235–238. (4-2)

Fan, J.S. and Zhang, M.: **Signaling complex organization by PDZ domain proteins.** *Neurosignals* 2002, **11**:315–321. (3-1)

Ferguson, A.D., Hofmann, E., Coulton, J.W., Diederichs, K. and Welte, W.: **Siderophore-mediated iron transport: crystal structure of FhuA with bound lipopolysaccharide.** *Science* 1998, **282**:2215–2220. (1-11)

Ferreira, S.T. and De Felice, F.G.: **Protein dynamics, folding and misfolding: from basic physical chemistry to human conformational diseases.** *FEBS Lett.* 2001, **498**:129–134. (1-12)

Fersht, A.: *Structure and Mechanism in Protein Science. A Guide to Enzyme Catalysis and Protein Folding.* (Freeman, New York, 1999). (1-9)

Fersht, A.R.: **The hydrogen bond in molecular recognition.** *Trends Biochem. Sci.* 1987, **12**:301–304. (1-4)

Fetrow, J.S., Siew, N., Di Gennaro, J.A., Martinez-Yamout, M., Dyson, H.J. and Skolnick, J.: **Genomic-scale comparison of sequence- and structure-based methods of function prediction: does structure provide additional insight?** *Protein Sci.* 2001, **10**:1005–1014. (4-6)

Fetrow, J.S., Siew, N. and Skolnick, J.: **Structure-based functional motif identifies a potential disulfide oxidoreductase active site in the serine/threonine protein phosphatase-1 subfamily.** *FASEB J.* 1999, **13**:1866–1874. (4-10)

Fischer, D. and Eisenberg, D.: **Protein fold recognition using sequence-derived predictions.** *Protein Sci.* 1996, **5**:947–955. (4-7)

Garrington, T.P. and Johnson, G.L.: **Organization and regulation of mitogen-activated protein kinase signaling pathways.** *Curr. Opin. Cell Biol.* 1999, **2**:211–218. (3-2)

Gasch, A.P., Huang, M., Metzner, S., Botstein, D., Elledge, S.J. and Brown, P.O.: **Genomic expression programs in the response of yeast cells to environmental changes.** *Mol. Biol. Cell* 2000, **11**:4241–4257. (4-4)

Gaucher, E.A., Gu, X., Miyamoto, M.M. and Benner, S.A.: **Predicting functional divergence in protein evolution by site-specific rate shifts.** *Trends Biochem. Sci.* 2002, **27**:315–321. (4-2)

Gerlt, J.A. and Babbitt, P.C.: **Can sequence determine function?** *Genome Biol.* 2000, **1**: reviews 0005.1–0005.10. (4-11)

Gerlt, J.A. and Babbitt, P.C.: **Divergent evolution of enzymatic function: mechanistically diverse superfamilies and functionally distinct suprafamilies.** *Annu. Rev. Biochem.* 2001, **70**:209–246. (4-8, 4-11)

Gerstein, M. and Honig, B.: **Sequences and topology.** *Curr. Opin. Struct. Biol.* 2001, **11**:327–329. (4-1)

Geyer, M. and Wittinghofer A.: **GEFs, GAPs, GDIs and effectors: taking a closer (3D) look at the regulation of Ras-related GTP-binding proteins.** *Curr. Opin. Struct. Biol.* 1997, **7**:786–792. (3-7)

Goldberg, A.L., Elledge, S.J. and Harper, J.W.: **The cellular chamber of doom.** *Sci. Am.* 2001, **284**:68–73. (3-11)

Goldstein, L.S.: **Molecular motors: from one motor many tails to one motor many tales.** *Trends Cell Biol.* 2001, **11**:477–482. (3-10)

Gonzalez, L. Jr., Woolfson, D.N. and Alber, T.: **Buried polar residues and structural specificity in the GCN4 leucine zipper.** *Nat. Struct. Biol.* 1996, **3**:1011–1018. (1-19)

Goodford, P.J.: **A computational procedure for determining energetically favorable binding sites on biologically important macromolecules.** *J. Med. Chem.* 1985, **28**:849–857. (4-9)

Goodsell, D.S.: **Inside a living cell.** *Trends Biochem. Sci.* 1991, **16**:203–206. (3-0)

Goodsell, D.S.: **Structural symmetry and protein function.** *Annu. Rev. Biophys. Biomol. Struct.* 2000,

29:105–153. (1-21)

Greasley, S.E., Horton, P., Ramcharan, J., Beardsley, G.P., Benkovic, S.J. and Wilson, I.A.: **Crystal structure of a bifunctional transformylase and cyclohydrolase in purine biosynthesis.** *Nat. Struct. Biol.* 2001, **8**:402–406. (2-15)

Greenfield, N.J. and Fasman, G.D.: **Computed circular dichroism spectra for the evaluation of protein conformation.** *Biochemistry* 1969, **8**:4108–4116. (1-12)

Gribskov, M., McLachlan, A.D. and Eisenberg, D.: **Profile analysis: detection of distantly related proteins.** *Proc. Natl Acad. Sci. USA* 1987, **84**:4355–4358. (4-2)

Griffith, E.C., Su, Z., Turk, B.E., Chen, S., Chang, Y.H., Wu, Z., Biemann, K. and Liu, J.O.: **Methionine aminopeptidase (type 2) is the common target for angiogenesis inhibitors AGM-1470 and ovalicin.** *Chem Biol.* 1997, **4**:461–471. (4-13)

Guttmacher A.E. and Collins, F.S.: **Genomic medicine— a primer.** *N. Engl. J. Med.* 2002, **347**:1512–1520. (4-0)

Haga, A., Niinaka, Y. and Raz, A.: **Phosphohexose isomerase/autocrine motility factor/neuroleukin/maturation factor is a multifunctional phosphoprotein.** *Biochim. Biophys. Acta.* 2000, **1480**:235–244. (4-13)

Hall, T.M., Porter, J.A., Young, K.E., Koonin, E.V., Beachy, P.A. and Leahy, D.J.: **Crystal structure of a Hedgehog autoprocessing domain: homology between Hedgehog and self-splicing proteins.** *Cell* 1997, **91**:85–97. (3-17)

Hamad, N.M., Elconin, J.H., Karnoub, A.E., Bai, W., Rich, J.N., Abraham, R.T., Der, C.J. and Counter, C.M.: **Distinct requirements for Ras oncogenesis in human versus mouse cells.** *Genes Dev.* 2002, **16**:2045–2057. (3-7)

Hamm, H.E. and Gilchrist, A.: **Heterotrimeric G proteins.** *Curr. Opin. Cell Biol.* 1996, **8**:189–196. (3-8)

Hammes, G.G.: **Multiple conformational changes in enzyme catalysis.** *Biochemistry* 2002, **41**:8221–8228. (2-2)

Han, G.W., Lee, J.Y., Song, H.K., Chang, C., Min, K., Moon, J., Shin, D.H., Kopka, M.L., Sawaya, M.R., Yuan, H.S., Kim, T.D., Choe, J., Lim, D., Moon, H.J. and Suh, S.W.: **Structural basis of non-specific lipid binding in maize lipid-transfer protein complexes revealed by high-resolution X-ray crystallography.** *J. Mol. Biol.* 2001, **308**:263–278. (2-4)

Harrington, D.J., Adachi, K. and Royer, W.E. Jr.: **The high resolution structure of deoxyhemoglobin S.** *J. Mol. Biol.* 1997, **272**:398–407. (1-20)

Hasson, M.S., Muscate, A., McLeish, M., Polovnikova, L.S., Gerlt, J.A., Kenyon. G.L., Petsko, G.A. and Ringe, D.: **The crystal structure of benzoylformate decarboxylase at 1.6Å resolution: diversity of catalytic residues in thiamine diphosphate dependent enzymes.** *Biochemistry* 1998, **37**:9918–9930. (4-5)

Hasson, M.S., Schlichting, I., Moulai, J., Taylor, K., Kenyon, G.L., Babbitt, P.C., Gerlt, J.A. Petsko, G.A. and Ringe, D.: **Evolution of an enzyme active site: the structure of a new crystal form of muconate lactonizing enzyme compared with mandelate racemase and enolase.** *Proc. Natl Acad. Sci. USA* 1998, **95**:10396–10401. (4-11)

Hawkins, A.R. and Lamb, H.K.: **The molecular biology of multidomain proteins. Selected examples.** *Eur. J. Biochem.* 1995, **232**:7–18. (1-14)

Hegyi, H. and Bork, P.: **On the classification and evolution of protein molecules.** *J. Prot. Chem.* 1997, **16**:545–551. (1-14)

Hegyi, H. and Gerstein, M.: **Annotation transfer for genomics: measuring functional divergence in multi-domain proteins.** *Genome Res.* 2001, **11**:1632–1640. (4-3)

Heller, H., Schaefer, M. and Schulten, K.: **Molecular dynamics simulation of a bilayer of 200 lipids in the gel and in the liquid crystal phases.** *J. Phys. Chem.* 1993, **97**:8343–8360. (1-11)

Herskowitz, I.: **Functional inactivation of genes by dominant negative mutations.** *Nature* 1987, **329**:219–222. (1-20)

Highbarger, L.A., Gerlt, J.A., and Kenyon, G.L.: **Mechanism of the reaction catalyzed by acetoacetate decarboxylase. Importance of lysine 116 in determining the pK_a of active-site lysine.** *Biochemistry* 1996, **35**:41–46. (2-1)

Hochstrasser, M.: **SP-RING for SUMO: New functions bloom for a ubiquitin-like protein.** *Cell* 2001, **107**:5–8. (3-20)

Hodgson, D.A. and Thomas, C.M. (eds): **Signals, switches, regulons, and cascades: control of bacterial gene expression.** *61st Symposium of the Society for General Microbiology* (Cambridge University Press, Cambridge, 2002). (3-0)

Hogg, T., Mesters, J.R. and Hilgenfeld, R.: **Inhibitory mechanisms of antibiotics targeting elongation factor Tu.** *Curr. Protein Pept. Sci.* 2002, **3**:121–131. (3-9)

Hol, W.G.: **The role of the alpha helix dipole in protein function and structure.** *Prog. Biophys. Mol. Biol.* 1985, **45**:149–195. (1-6)

Holmes, K.C. and Geeves, M.A.: **The structural basis of muscle contraction.** *Philos. Trans. R. Soc. Lond. B. Biol. Sci.* 2000, **355**:419–431. (3-10)

Holyoak, T., Wilson, M.A., Fenn, T.D., Petsko, G.A., Fuller, R.S. and Ringe, D.: **The 2.4 Å crystal structure of the processing protease Kex2 in complex with an Ala–Lys–Arg boronic acid inhibitor.** *Biochemistry* in the press. (5-3)

Hook, V.Y., Azaryan, A.V., Hwang, S.R. and Tezapsidis, N.: **Proteases and the emerging role of protease inhibitors in prohormone processing.** *FASEB J.* 1994, **8**:1269–1278. (3-16)

Houdusse, A., Szent-Gyorgyi, A.G. and Cohen, C.: **Three conformational states of scallop myosin S1.** *Proc. Natl Acad. Sci. USA.* 2000, **97**:11238–11243. (2-5)

Houry, W.A., Frishman, D., Eckerskorn, C., Lottspeich, F. and Hartl, F.U.: **Identification of *in vivo* substrates of the chaperonin GroEL.** *Nature* 1999, **402**:147–154. (4-0)

Huang, X., Holden, H.M. and Raushel, F.M.: **Channeling of substrates and intermediates in enzyme-catalyzed reactions.** *Annu. Rev. Biochem.* 2001, **70**:149–180. (2-16)

Hunter, T.: **Signaling—2000 and beyond.** *Cell* 2000, **100**:113–127. (3-12)

Hurley, J.H., Dean, A.M., Koshland, D.E. Jr. and Stroud, R.M.: **Catalytic mechanism of NADP(+)-dependent isocitrate dehydrogenase: implications from the structures of magnesium-isocitrate and NADP+ complexes.** *Biochemistry* 1991, **30**:8671–8678. (2-15)

Hurley, J.H., Dean, A.M., Sohl, J.L., Koshland, D.E. Jr. and Stroud, R.M.: **Regulation of an enzyme by phosphorylation at the active site.** *Science* 1990, **249**:1012–1016. (3-12)

Huse, M. and Kuriyan, J.: **The conformational plasticity of protein kinases**. *Cell* 2002, **109**:275–282. (3-13)

Hyde, C.C., Ahmed, S.A., Padlan, E.A., Miles, E.W. and Davies, D.R.: **Three-dimensional structure of the tryptophan synthase alpha 2 beta 2 multienzyme complex from *Salmonella typhimurium***. *J. Biol. Chem.* 1988, **263**:17857–17871. (2-16)

Imada, K., Sato, M., Tanaka N., Katsube, Y., Matsuura, Y. and Oshima, T.: **Three-dimensional structure of a highly thermostable enzyme, 3-isopropylmalate dehydrogenase of *Thermus thermophilus* at 2.2 Å resolution**. *J. Mol. Biol.* 1991, **222**:725–738. (2-3)

Imperiali, B. and O'Connor, S.E.: **Effect of *N*-linked glycosylation on glycopeptide and glycoprotein structure**. *Curr. Opin. Chem. Biol.* 1999, **3**:643–649. (3-18)

Irving, J.A., Whisstock, J.C., and Lesk, A.M.: **Protein structural alignments and functional genomics**. *Proteins* 2001, **42**:378–382. (4-5)

Ishima, R. and Torchia, D.A.: **Protein dynamics from NMR**. *Nat. Struct. Biol.* 2000, **7**:740–743. (1-22)

Jaenicke, R.: **Stability and stabilization of globular proteins in solution**. *J. Biotechnol.* 2000, **79**:193–203. (1-12)

James, C.L. and Viola, R.E.: **Production and characterization of bifunctional enzymes. Substrate channeling in the aspartate pathway**. *Biochemistry* 2002, **41**:3726–3731. (2-16)

Jansen, R. and Gerstein, M.: **Analysis of the yeast transcriptome with structural and functional categories: characterizing highly expressed proteins**. *Nucleic Acids Res.* 2000, **28**:1481–1488. (1-0)

Jeffery, C.J.: **Moonlighting proteins**. *Trends Biochem. Sci.* 1999, **24**:8–11. (4-13)

Jeffery, C.J., Bahnson, B.J., Chien, W., Ringe, D. and Petsko, G.A.: **Crystal structure of rabbit phosphoglucose isomerase, a glycolytic enzyme that moonlights as neuroleukin, autocrine motility factor, and differentiation mediator**. *Biochemistry* 2000, **39**:955–964. (4-13)

Jencks, W.P.: **Binding energy, specificity, and enzymic catalysis: the Circe effect**. *Adv. Enzymol. Relat. Areas Mol. Biol.* 1975, **43**:219–410. (2-8)

Jencks, W.P.: *Catalysis and Enzymology* (Dover Publications, New York, 1987). (2-6)

Jensen, R.B. and Shapiro, L.: **Proteins on the move: dynamic protein localization in prokaryotes**. *Trends Cell Biol.* 2000, **10**:483–488. (3-0)

Jenuwein, T. and Allis, C.D.: **Translating the histone code**. *Science*, **293**:1074–1080. (3-20)

Jeong, J. and McMahon, A.P.: **Cholesterol modification of Hedgehog family proteins**. *J. Clin. Invest.* 2002, **110**:591–596. (3-17)

Johnson, L.N. and O'Reilly, M.: **Control by phosphorylation**. *Curr. Opin. Struct. Biol.* 1996, **6**:762–769. (3-12)

Jones, D.T. and Swindells, M.B.: **Getting the most from PSI-BLAST**. *Trends Biochem. Sci* 2002, **27**:161–164. (4-2)

Jones, D.T., Taylor, W.R. and Thornton, J.M.: **The rapid generation of mutation data matrices from protein sequences**. *Comput. Appl. Biosci.* 1992, **8**:275–282. (1-2)

Jones, S. and Thornton, J.M.: **Principles of protein–protein interactions** *Proc. Natl Acad. Sci. USA* 1996, **93**:13–20. (1-19)

Jones, S. and Thornton, J.M.: **Prediction of protein–protein interaction sites using patch analysis.**

J. Mol. Biol. 1997, **272**:133–143. (4-10)

Jouanguy, E., Lamhamedi-Cherradi, S., Lammas, D., Dorman, S.E., Fondaneche, M.C., Dupuis, S., Doffinger, R., Altare, F., Girdlestone, J., Emile, J.F., Ducoulombier, H., Edgar, D., Clarke, J., Oxelius, V.A., Brai, M., Novelli, V., Heyne, K., Fischer, A., Holland, S.M., Kumararatne, D.S., Schreiber, R.D. and Casanova, J.L.: **A human IFNGR1 small deletion hotspot associated with dominant susceptibility to mycobacterial infection**. *Nat. Genet.* 1999, **21**:370–378. (1-20)

Kabsch, W. and Sander, C.: **On the use of sequence homologies to predict protein structure: identical pentapeptides can have completely different conformations**. *Proc. Natl Acad. Sci. USA* 1984, **81**:1075–1078. (4-15)

Kalafatis, M., Egan, J.O., van 't Veer, C., Cawthern, K.M. and Mann, K.G.: **The regulation of clotting factors**. *Crit. Rev. Eukaryot. Gene. Expr.* 1997, **7**:241–280. (3-16)

Kallal, L. and Benovic, J.L.: **Using green fluorescent proteins to study G-protein-coupled receptor localization and trafficking**. *Trends Pharmacol Sci.* 2000, **21**:175–180. (4-4)

Kantrowitz, E.R. and Lipscomb, W.N.: *Escherichia coli* **aspartate transcarbamylase: the relation between structure and function**. *Science* 1988, **241**:669–674. (3-5)

Karpusas, M., Branchaud, B. and Remington, S.J.: **Proposed mechanism for the condensation reaction of citrate synthase: 1.9Å structure of the ternary complex with oxaloacetate and carboxymethyl coenzyme A**. *Biochemistry* 1990, **29**:2213–2219. (2-9)

Kauzmann, W.: **Some factors in the interpretation of protein denaturation**. *Adv. Protein Chem.* 1959, **14**:1–63. (1-12)

Kawabata, T., Fukuchi, S., Homma, K., Ota, M., Araki, J., Ito, T., Ichiyoshi, N. and Nishikawa, K.: **GTOP: a database of protein structures predicted from genome sequences**. *Nucleic Acids Res.* 2002, **30**:294–298. (4-2)

Karplus, M. and Petsko, G.A.: **Molecular dynamics simulations in biology**. *Nature* 1990, **347**:631–639. (1-22)

Kerrebrock, A.W., Moore, D.P., Wu, J.S. and Orr-Weaver, T.L.: **Mei-S332, a *Drosophila* protein required for sister-chromatid cohesion, can localize to meiotic centromere regions**. *Cell* 1995, **83**:247–256. (4-4)

Kim, S.W., Cha, S.S., Cho, H.S., Kim, J.S., Ha, N.C., Cho, M.J., Joo, S., Kim, K.K., Choi, K.Y. and Oh, B.H.: **High-resolution crystal structures of delta5-3-ketosteroid isomerase with and without a reaction intermediate analogue**. *Biochemistry* 1997, **36**: 14030–14036. (4-5)

Kirsch, J.F., Eichele, G., Ford, G.C., Vincent, M.G., Jansonius, J.N., Gehring, H. and Christen, P.: **Mechanism of action of aspartate aminotransferase proposed on the basis of its spatial structure**. *J. Mol. Biol.* 1984, **174**:497–525. (4-12)

Kisselev, A.F. and Goldberg, A.L.: **Proteasome inhibitors: from research tools to drug candidates**. *Chem. Biol.* 2001, **8**:739–758. (3-11)

Klabunde, T., Sharma, S., Telenti, A., Jacobs, W.R. Jr. and Sacchettini, J.C.: **Crystal structure of GyrA intein from *Mycobacterium xenopi* reveals structural basis of protein splicing**. *Nat. Struct. Biol.* 1998, **5**:31–36. (3-17)

Knighton, D.R., Kan, C.C., Howland, E., Janson, C.A., Hostomska, Z., Welsh, K.M. and Matthews, D.A.: **Structure and kinetic channelling in bifunctional dihydrofolate reductase-thyidylate synthase**. *Nat. Struct. Biol.* 1994, **1**:186–194. (2-15)

Ko, Y.H. and Pedersen, P.L.: **Cystic fibrosis: a brief look at some highlights of a decade of research focused on elucidating and correcting the molecular basis of the disease**. *J. Bioenerg. Biomembr.* 2001, **33**:513–521. (4-15)

Koebnik, R., Locher, K.P. and Van Gelder, P.: **Structure and function of bacterial outer membrane proteins: barrels in a nutshell**. *Mol. Microbiol.* 2000, **37**:239–253. (1-11)

Koonin E.V., Wolf, Y.I. and Karev, G.P.: **The structure of the protein universe and genome evolution**. *Nature* 2002, **420**:218–223. (4-0)

Kornitzer, D. and Ciechanover, A.: **Modes of regulation of ubiquitin-mediated protein degradation**. *J. Cell Physiol.* 2000, **182**:1–11. (3-0)

Koshland, D.E. Jr. and Hamadani, K.: **Proteomics and models for enzyme cooperativity**. *J. Biol. Chem.* 2002, **277**:46841–46844. (3-4)

Koshland, D.E. Jr., Nemethy, G. and Filmer, D.: **Comparison of experimental binding data and theoretical models in proteins containing subunits**. *Biochemistry* 1966, **5**:365–385. (2-2)

Kossiakoff, A.A. and De Vos, A.M.: **Structural basis for cytokine hormone–receptor recognition and receptor activation**. *Adv. Protein Chem.* 1998, **52**:67–108. (2-3)

Kouzarides, T.: **Histone methylation in transcriptional control**. *Curr. Opin. Genet. Dev.* 2002, **12**:198–209. (3-20)

Krause, K.L., Volz, K.W., and Lipscomb, W.N.: **2.5 Å structure of aspartate carbamoyltransferase complexed with the bisubstrate analog *N*-(phosphonoacetyl)-L-aspartate**. *J. Mol. Biol.* 1987, **193**:527–553. (2-8)

Krem, M.M., Rose, T. and Di Cera, E.: **Sequence determinants of function and evolution in serine proteases**. *Trends Cardiovasc. Med.* 2000, **10**:171–176. (4-8)

Kunin, V., Chan, B., Sitbon, E., Lithwick, G. and Pietrokovski, S.: **Consistency analysis of similarity between multiple alignments: prediction of protein function and fold structure from analysis of local sequence motifs**. *J. Mol. Biol.* 2001, **307**:939–949. (4-2)

Kuriyan, J. and Cowburn, D.: **Modular peptide recognition domains in eukaryotic signaling**. *Annu. Rev. Biophys. Biomol. Struct.* 1997, **26**:259–288. (3-1)

Kurosaki, T.: **Regulation of B-cell signal transduction by adaptor proteins**. *Nat. Rev. Immunol.* 2002, **2**:354–363. (3-2)

Kyte, J. and Doolittle, R.F.: **A simple method for displaying the hydropathic character of a protein**. *J. Mol. Biol.* 1982, **157**:105–132. (1-11)

Lahiri, S.D., Zhang, G., Dunaway-Mariano, D. and Allen, K.N.: **Caught in the act: the structure of phosphorylated beta-phosphoglucomutase from *Lactococcus lactis***. *Biochemistry* 2002, **41**:8351–8359. (2-14)

Landro, J.A., Gerlt, J.A., Kozarich, J.W., Koo, C.W., Shah, V.J., Kenyon, G.L., Neidhart, D.J., Fujita, S. and Petsko, G.A.: **The role of lysine 166 in the mechanism of mandelate racemase from *Pseudomonas putida*: mechanistic and crystallographic evidence for stereospecific alkylation by (R)-alpha-phenylglycidate**. *Biochemistry* 1994, **33**:635–643. (2-1)

Laney, J.D. and Hochstrasser, M.: **Substrate targeting in the ubiquitin system**. *Cell* 1999, **97**:427–430. (3-11)

Laskowski, R.A., Luscombe, N.M., Swindells, M.B. and Thornton, J.M.: **Protein clefts in molecular recognition and function**. *Protein Sci.* 1996, **5**:2438–2452. (4-10)

Lau, E.Y., Kahn, K., Bash, P.A., and Bruice, T.C.: **The importance of reactant positioning in enzyme catalysis: a hybrid quantum mechanics/molecular mechanics study of a haloalkane dehalogenase.** *Proc. Natl Acad. Sci. USA* 2000, **97**:9937–9942. (2-8)

Leclerc, E., Peretz, D., Ball, H., Sakurai, H., Legname, G., Serban, A., Prusiner, S.B., Burton, D.R. and Williamson, R.A.: **Immobilized prion protein undergoes spontaneous rearrangement to a conformation having features in common with the infectious form.** *EMBO J.* 2001, **20**:1547–1554. (4-15)

Lee, A.Y., Gulink, S.V. and Erickson, J.W.: **Conformational switching in an aspartic proteinase.** *Nat. Struct. Biol.* 1998, **5**:866–871. (3-3)

Lee, M. and Goodbourn, S.: **Signalling from the cell surface to the nucleus.** *Essays Biochem.* 2001, **37**:71–85. (3-12)

Leon, O. and Roth, M.: **Zinc fingers: DNA binding and protein-protein interactions.** *Biol. Res.* 2000, **33**:21–30. (1-18)

Lesk, A.M.: *Introduction to Protein Architecture* (Oxford University Press, Oxford, 2001). (4-5)

Lesk, A.M. and Chothia, C.: **Solvent accessibility, protein surfaces and protein folding.** *Biophys. J.* 1980, **32**:35–47. (1-10)

Liang, J., Edelsbrunner, H. and Woodward, C.: **Anatomy of protein pockets and cavities: measurement of binding site geometry and implications for ligand design.** *Protein Sci.* 1998, **7**:1884–1897. (4-9)

Liang, P.H. and Anderson, K.S.: **Substrate channeling and domain-domain interactions in bifunctional thymidylate synthase-dihydrofolate reductase.** *Biochemistry* 1998, **37**:12195–12205. (2-15)

Liepinsh, E. and Otting, G.: **Organic solvents identify specific ligand binding sites on protein surfaces.** *Nat. Biotechnol.* 1997, **15**:264–268. (4-9)

Liu, J., Schmitz, J.C., Lin, X., Tai, N., Yan, W., Farrell, M., Bailly, M., Chen, T. and Chu, E.: **Thymidylate synthase as a translational regulator of cellular gene expression.** *Biochim. Biophys. Acta* 2002, **1587**:174–182. (2-15)

Liu, S., Widom, J., Kemp, C.W., Crews, C.M. and Clardy, J.: **Structure of human methionine aminopeptidase-2 complexed with fumagillin.** *Science* 1998, **282**:1324–1327. (4-13)

Liu, Y., Hart, J.P., Schlunegger, M.P. and Eisenberg, D.: **The crystal structure of a 3D domain-swapped dimer of RNase A at a 2.1-Å resolution.** *Proc. Natl Acad. Sci. USA* 1998, **95**:3437–3442. (2-4)

Lockless, S.W. and Ranganathan, R.: **Evolutionarily conserved pathways of energetic connectivity in protein families.** *Science* 1999, **286**:295–299. (4-6)

Lubetsky, J.B., Dios, A., Han, J., Aljabari, B., Ruzsicska, B., Mitchell, R., Lolis, E. and Al-Abed, Y.: **The tautomerase active site of macrophage migration inhibitory factor is a potential target for discovery of novel anti-inflammatory agents.** *J. Biol. Chem.* 2002, **277**:24976–24982. (4-13)

Lundqvist, T., Rice, J., Hodge, C.N., Basarab, G.S., Pierce, J. and Lindqvist, Y.: **Crystal structure of scytalone dehydratase—a disease determinant of the rice pathogen, *Magnaporthe grisea*.** *Structure* 1994, **2**:937–944. (4-5)

Macias, M.J., Wiesner, S. and Sudol, M.: **WW and SH3 domains, two different scaffolds to recognize proline-rich ligands.** *FEBS Lett.* 2002, **513**:30–37. (3-1)

Mahadeva, R., Dafforn, T.R., Carrell, R.W. and Lomas, D.A.: **6-mer peptide selectively anneals to a pathogenic serpin conformation and blocks polymerization. Implications for the prevention of Z alpha(1)-antitrypsin-related cirrhosis.** *J. Biol. Chem.* 2002, **277**:6771–6774. (4-15)

Malcolm, B.A., Rosenberg, S., Corey, M.J, Allen, J.S., de Baetselier, A. and Kirsch, J.F.: **Site-directed mutagenesis of the catalytic residues Asp-52 and Glu-35 of chicken egg white lysozyme.** *Proc. Natl Acad. Sci. USA* 1989, **86**:133–137. (2-12)

Manning, G., Whyte, D.B., Martinez, R., Hunter, T. and Sudarsanam, S.: **The protein kinase complement of the human genome.** *Science* 2002, **298**:1912–1934. (3-12)

Markley, J.L., Ulrich, E.L., Westler W.M. and Volkman, B.F.: **Macromolecular structure determination by NMR spectroscopy.** *Methods Biochem. Anal.* 2003, **44**:89–113. (5-2)

Marmorstein, R., Carey, M., Ptashne, M. and Harrison, S.C.: **DNA recognition by GAL4: structure of a protein-DNA complex.** *Nature* 1992, **356**:408–414. (2-3)

Marti-Renom, M.A., Stuart, A.C., Fiser, A., Sanchez, R., Melo, F. and Sali, A.: **Comparative protein structure modeling of genes and genomes.** *Annu. Rev. Biophys.* 2000, **29**:291–325. (4-6)

Martin, R.B.: **Peptide bond characteristics.** *Met. Ions Biol. Syst.* 2001, **38**:1–23. (1-3)

Martin, T.F.: **PI(4,5)P(2) regulation of surface membrane traffic.** *Curr. Opin. Cell Biol.* 2001, **13**:493–499. (3-2)

Martz, E.: **Protein Explorer: easy yet powerful macromolecular visualization.** *Trends Biochem. Sci.* 2002, **27**:107–109. (5-3)

Matthews, B.W. and Bernhard, S.A.: **Structure and symmetry of oligomeric enzymes.** *Annu. Rev. Biophys. Bioeng.* 1973, **2**:257–317. (1-21)

Mattos, C., Petsko, G.A. and Karplus, M.: **Analysis of two-residue turns in proteins.** *J. Mol. Biol.* 1994, **238**:733–747. (1-5)

Mattos, C. and Ringe, D.: **Locating and characterizing binding sites on proteins.** *Nat. Biotechnol.* 1996, **14**:595–599. (4-9)

Mattos, C. and Ringe, D.: **Proteins in organic solvents.** *Curr. Opin. Struct. Biol.* 2001, **11**:761–764. (4-9)

McBride, A.E. and Silver, P. A.: **State of the Arg: protein methylation at arginine comes of age.** *Cell* 2001, **106**:5–8. (3-20)

McKessar, S.J., Berry, A.M., Bell, J.M., Turnidge, J.D. and Paton, J.C.: **Genetic characterisation of vanG, a novel vancomycin resistance locus of *Enterococcus faecalis*.** *Antimicrob. Agents Chemother.* 2000, **44**:3224–3228. (1-8)

Mesecar, A.D., Stoddard, B.L., and Koshland, D.E. Jr.: **Orbital steering in the catalytic power of enzymes: small structural changes with large catalytic consequences.** *Science* 1997, **277**:202–206. (2-8)

Mezei, M.: **Chameleon sequences in the PDB.** *Protein Eng.* 1998, **11**:411–414. (4-14)

Michal, G., ed.: *Boehringer Mannheim Biochemical Pathways Wallcharts*, Roche Diagnostics Corporation, Roche Molecular Biochemicals, P.O. Box 50414, Indianapolis, IN 46250-0414, USA. (1-0)

Miller, B.G. and Wolfenden, R.: **Catalytic proficiency: the unusual case of OMP decarboxylase.** *Annu. Rev. Biochem.* 2002, **71**:847–885. (2-6)

Miller, R.T., Jones, D.T. and Thornton, J.M.: **Protein fold recognition by sequence threading: tools and assessment techniques.** *FASEB J.* 1996, **10**:171–178. (4-7)

Milner-White, E.J.: **Description of the quaternary structure of tetrameric proteins. Forms that show either right-handed or left-handed symmetry at the subunit.** *Biochem. J.* 1980, **187**:297–302. (1-21)

Minor, D.L. Jr. and Kim, P.S.: **Context-dependent secondary structure formation of a designed protein sequence.** *Nature* 1996, **380**:730–734. (4-14)

Miranker, A. and Karplus, M.: **Functionality maps of binding sites: a multiple copy simultaneous search method.** *Proteins* 1991, **11**:29–34. (4-9)

Modis, Y., Trus, B.L. and Harrison, S.C.: **Atomic model of the papillomavirus capsid.** *EMBO J.* 2002, **21**:4754–4762. (2-4)

Morris, A.J. and Malbon, C.C.: **Physiological regulation of G protein-linked signaling.** *Physiol. Rev.* 1999, **79**:1373–1430. (1-13)

Mount, D.W.: *Bioinformatics: Sequence and Genome analysis* (Cold Spring Harbor Laboratory Press, New York, 2001). (4-1)

Murzin, A.G.: **How far divergent evolution goes in proteins.** *Curr. Opin. Struct. Biol.* 1998, **8**:380–387. (4-5)

Myers, J.K. and Oas, T.G.: **Reinterpretation of GCN4-p1 folding kinetics: partial helix formation precedes dimerization in coiled coil folding.** *J. Mol. Biol.* 1999, **289**:205–209. (1-19)

Nagano, N., Orengo, C.A. and Thornton, J.M.: **One fold with many functions: the evolutionary relationships between TIM barrel families based on their sequences, structures and functions.** *J. Mol. Biol.* 2002, **321**:741–765. (4-11)

Neidhart, D.J., Kenyon, G.L., Gelt, J.A. and Petsko, G.A.: **Mandelate racemase and muconate lactonizing enzyme are mechanistically distinct and structurally homologous.** *Nature* 1990, **347**:692–694. (4-11)

Nogales, E.: **Structural insights into microtubule function.** *Annu. Rev. Biochem.* 2000, **69**:277–302. (2-0)

Nyborg, J. and Liljas, A.: **Protein biosynthesis: structural studies of the elongation cycle.** *FEBS Lett.* 1998, **430**:95–99. (3-9)

O'Donovan, C., Apweiler, R. and Bairoch, A.: **The human proteomics initiative (HPI).** *Trends Biotechnol.* 2001, **19**:178–181. (4-0)

Oliver, S.G.: **Functional genomics: lessons from yeast.** *Philos Trans. R. Soc. Lond. B. Biol. Sci.* 2002, **357**:17–23. (4-0)

Okada, T., Ernst, O.P., Palczewski, K. and Hofmann, K.P.: **Activation of rhodopsin: new insights from structural and biochemical studies.** *Trends Biochem. Sci.* 2001, **26**:318–324. (3-8)

Ondrechen, M.J., Clifton, J.G. and Ringe, D.: **THEMATICS: a simple computational predictor of enzyme function from structure.** *Proc. Natl Acad. Sci. USA* 2001, **98**:12473–12478. (4-10)

Palade, G.E.: **Protein kinesis: the dynamics of protein trafficking and stability.** *Cold Spring Harbor Symp. Quant. Biol.* 1995, **60**:821–831. (1-13)

Pannifer, A.D., Wong, T.Y., Schwarzenbacher, R., Renatus, M., Petosa, C., Bienkowska, J., Lacy, D.B., Collier, R.J., Park, S., Leppla, S.H., Hanna, P. and Liddington, R.C.: **Crystal structure of the anthrax lethal factor.** *Nature* 2001, **414**:229–233. (2-1)

Park, H.W. and Beese, L.S.: **Protein farnesyltransferase.** *Curr. Opin. Struct. Biol.* 1997, **7**:873–880. (3-19)

Parkinson, J.S. and Kofoid, E.C.: **Communication modules in bacterial signaling proteins.** *Annu. Rev. Genet.* 1992, **26**:71–112. (3-15)

Parekh, R.B. and Rohlff, C.: **Post-translational modification of proteins and the discovery of new medicine.** *Curr. Opin. Biotechnol.* 1997, **8**:718–723. (3-18)

Patterson, S.D.: **Proteomics: the industrialization of protein chemistry.** *Curr. Opin. Biotechnol.* 2000, **11**:413–418. (4-4)

Patthy, L.: **Genome evolution and the evolution of exon-shuffling.** *Gene* 1999, **238**:103–114. (1-15)

Patthy, L.: Protein Evolution (Blackwell Science, Oxford, 1999). (4-5)

Pauling, L. and Corey, R.B.: **Configurations of polypeptide chains with favored orientations around single bonds: two new pleated sheets.** *Proc. Natl Acad. Sci. USA* 1951, **37**:729–740. (1-7)

Pauling, L.C., Corey, R.B., and Branson, H.R.: **The structure of proteins: two hydrogen-bonded helical configurations of the polypeptide chain.** *Proc. Natl Acad. Sci. USA.* 1951, **37**:205–211. (1-6)

Pauling, L.C.: *The Nature of the Chemical Bond and the Structure of Molecules and Crystals*, 3rd ed. (Cornell Univ. Press, Ithaca, New York, 1960). (1-3, 1-4)

Paulus, H.: **Protein splicing and related forms of protein autoprocessing.** *Annu. Rev. Biochem.* 2000, **69**:447–496. (3-17)

Pawson, A.J. (ed): *Protein Modules in Signal Transduction* (Springer, Berlin, New York, 1998). (3-0)

Peisach, E., Wang, J., de los Santos, T., Reich, E. and Ringe, D.: **Crystal structure of the proenzyme domain of plasminogen.** *Biochemistry* 1999, **38**:11180–11188. (3-16)

Pellegrini, M., Marcotte, E.M., Thompson, M.J., Eisenberg, D. and Yeates, T.O.: **Assigning protein functions by comparative genome analysis: protein phylogenetic profiles.** *Proc. Natl Acad. Sci. USA* 1999, **96**:4285–4288. (4-2)

Penn, R.B., Pronin, A.N., and Benovic J.L.: **Regulation of G protein-coupled receptor kinases.** *Trends Cardiovasc. Med.* 2000, **10**:81–89. (3-2)

Pereira-Leal, J.B., Hume, A.N. and Seabra, M.C.: **Prenylation of Rab GTPases: molecular mechanisms and involvement in genetic disease.** *FEBS Lett.* 2001, **498**:197–200. (3-19)

Perham, R.N.: **Self-assembly of biological macromolecules.** *Philos. Trans. R. Soc. Lond. B.* 1975, **272**:123–136. (1-19)

Perler, F.B.: **Protein splicing of inteins and hedgehog autoproteolysis: structure, function, and evolution.** *Cell* 1998, **92**:1–4. (3-17)

Perona, J.J. and Craik, C.S.: **Evolutionary divergence of substrate specificity within the chymotrypsin-like serine protease fold.** *J. Biol. Chem.* 1997, **272**:29987–29990. (4-8)

Perutz, M.F.: **Mechanisms of cooperativity and allosteric regulation in proteins.** *Q. Rev. Biophys.* 1989, **22**:139–237. (3-0)

Petrescu, A.J., Butters, T.D., Reinkensmeier, G., Petrescu, S., Platt, F.M., Dwek, R.A. and Wormald, M.R.: **The solution NMR structure of glucosylated N-glycans involved in the early stages of glycoprotein biosynthesis and folding.** *EMBO J.* 1997, **16**:4302–4310. (3-18)

Petsko, G.A.: **Structural basis of thermostability in hyperthermophilic proteins, or "there's more than one way to skin a cat".** *Methods Enzymol.* 2001, **334**:469–478. (3-11)

Petsko, G.A. and Ringe, D.: **Fluctuations in protein structure from X-ray diffraction.** *Annu. Rev. Biophys. Bioeng.* 1984, **13**:331–371. (1-22)

Petsko, G.A., Kenyon, G.L., Gerlt, J.A., Ringe, D. and Kozarich, J.W.: **On the origin of enzymatic species.** *Trends Biochem. Sci.* 1993, **18**:372–376. (4-11)

Phizicky, E.M. and Fields, S.: **Protein-protein interactions: methods for detection and analysis.** *Microbiol. Rev.* 1995, **59**:94–123. (4-4)

Ponting, C.P., Schultz, J., Copley, R.R., Andrade, M.A. and Bork, P.: **Evolution of domain families.** *Adv. Protein Chem.* 2000, **54**:185–244. (1-16)

Popot, J.L. and Engelman, D.M.: **Helical membrane protein folding, stability and evolution.** *Annu. Rev. Biochem.* 2000, **69**:881–922. (1-11)

Popot, J.L. and Engelman, D.M.: **Membrane protein folding and oligomerization: the two-stage model.** *Biochemistry* 1990, **29**:4031–4037. (1-11)

Poulos, T.L., Finzel, B.C. and Howard, A.J.: **High-resolution crystal structure of cytochrome p450 cam.** *J. Mol. Biol.* 1987, **195**:687–700. (2-3)

Quevillon-Cheruel, S., Collinet, B., Zhou, C.Z., Minard, P., Blondeau, K., Henkes, G., Aufrere, R., Coutant, J., Guittet, E., Lewit-Bentley, A., Leulliot, N., Ascone, I., Sorel, I., Savarin, P., De La Sierra Gallay, I.L., De La Torre, F., Poupon, A., Fourme, R., Janin, J. and Van Tilbeurgh, H.: **A structural genomics initiative on yeast proteins.** *J. Synchrotron. Radiat.* 2003, **10**:4–8. (4-0)

Radzicka, A. and Wolfenden, R.: **A proficient enzyme.** *Science* 1995, **267**:90–93. (2-8)

Ramachandran, G.N., Ramakrishnan, C. and Sasisekharan, V.: **Stereochemistry of polypeptide chain configurations.** *J. Mol. Biol.* 1963, **7**:95–99. (1-5)

Ramakrishnan, V.: **Ribosome structure and the mechanism of translation.** *Cell* 2002, **108**:557–572. (3-9)

Rasmussen, B.F., Stock, A.M., Ringe, D. and Petsko, G.A.: **Crystalline ribonuclease A loses function below the dynamical transition at 220 K.** *Nature* 1992, **357**:423–424. (2-2)

Rattan, S.I., Derventzi, A. and Clark, B.F.: **Protein synthesis, posttranslational modification, and aging.** *Annls N.Y. Acad. Sci.* 1992, **663**:48–62. (1-13)

Reardon, D. and Farber, G.K.: **The structure and evolution of alpha/beta barrel proteins.** *FASEB J.* 1995, **9**:497–503. (1-18)

Reva, B., Finkelstein, A. and Topiol, S.: **Threading with chemostructural restrictions method for predicting fold and functionally significant residues: application to dipeptidylpeptidase IV (DPP-IV).** *Proteins* 2002, **47**:180–193. (4-10)

Reymond, M.A., Sanchez, J.C., Hughes, G.J., Gunther, K., Riese, J., Tortola, S., Peinado, M.A., Kirchner, T., Hohenberger, W., Hochstrasser, D.F. and Kockerling, F.: **Standardized characterization of gene expression in human colorectal epithelium by two-dimensional electrophoresis.** *Electrophoresis* 1997, **18**:2842–2848. (4-4)

Rhodes, G.: *Crystallography Made Crystal Clear: A Guide to Users of Macromolecular Models* 2nd ed. (Academic Press, New York and London, 1999). (5-2)

Richards, F.M. and Richmond, T.: **Solvents, interfaces and protein structure.** *Ciba. Found. Symp.* 1997, **60**:23–45. (1-10)

Richardson, J.S. and Richardson, D.C.: **Principles and patterns of protein conformation** in *Prediction of Protein Structure and the Principles of Protein Conformation* 2nd ed. Fasman, G.D. ed. (Plenum Press, New York, 1990). (1-5, 1-15, 1-17, 1-18)

Richardson, J.S.: **Introduction: protein motifs.** *FASEB J.* 1994, **8**:1237–1239. (1-15)

Richardson, J.S.: **Looking at proteins: representation, folding, packing and design.** *Biophys. J.* 1992, **63**:1185–1209. (1-15)

Richardson, J.S.: **The anatomy and taxonomy of protein structure.** *Adv. Prot. Chem.* 1981, **34**:167–339. (1-7, 1-14)

Ringe, D.: **What makes a binding site a binding site?** *Curr. Opin. Struct. Biol.* 1995, **5**:825–829. (2-1, 2-3, 2-4)

Ringe, D. and Petsko, G.A.: **Mapping protein dynamics by X-ray diffraction.** *Prog. Biophys. Mol. Biol.* 1985, **45**:197–235. (1-22)

Rochet, J.C. and Lansbury, P.T. Jr.: **Amyloid fibrillogenesis: themes and variations.** *Curr. Opin. Struct. Biol.* 2000, **10**:60–68. (2-5)

Rockwell, N.C., Krysan, D.J., Komiyama, T. and Fuller, R.S.: **Precursor processing by kex2/furin proteases.** *Chem. Rev.* 2002, **102**:4525–4548. (3-16)

Rodbell, M.: **The complex regulation of receptor-coupled G-proteins.** *Adv. Enzyme Regul.* 1997, **37**:427–435. (3-8)

Rose, G.D. and Roy, S.: **Hydrophobic basis of packing in globular proteins.** *Proc. Natl Acad. Sci. USA* 1980, **77**:4643–4647. (1-10)

Ross, E.M. and Wilkie, T.M.: **GTPase-activating proteins for heterotrimeric G proteins: regulators of G protein signaling (RGS) and RGS-like proteins.** *Annu. Rev. Biochem.* 2000, **69**:795–827. (3-8)

Rudd, P.M. and Dwek, R.A.: **Glycosylation: heterogeneity and the 3D structure of proteins.** *Crit. Rev. Biochem. Mol. Biol.* 1997, **32**:1–100. (3-18)

Rutenber, E., Fauman, E.B., Keenan, R.J., Fong, S., Furth, P.S., Ortiz de Montellano, P.R., Meng, E., Kuntz, I.D., De Camp, D.L. and Salto, R.: **Structure of a non-peptide inhibitor complexed with HIV-1 protease. Developing a cycle of structure-based drug design.** *J. Biol. Chem.* 1993, **268**:15343–15436. (2-2)

Sablin, E.P., and Fletterick, R.J.: **Nucleotide switches in molecular motors: structural analysis of kinesins and myosins.** *Curr. Opin. Struct. Biol.* 2001, **11**:716–724. (3-6, 3-10)

Salem, G.M., Hutchinson, E.G., Orengo, C.A. and Thornton, J.M.: **Correlation of observed fold frequency with the occurrence of local structural motifs.** *J. Mol. Biol.* 1999, **287**:969–981. (1-15)

Sato, T.K., Overduin, M. and Emr, S.D.: **Location, location, location: membrane targeting directed by PX domains.** *Science* 2001, **294**:1881–1885. (3-0, 3-2)

Scheffzek, K., Ahmadian, M.R. and Wittinghofer, A.: **GTPase-activating proteins: helping hands to complement an active site.** *Trends Biochem. Sci.* 1998, **23**:257–262. (3-7)

Schena, M., Shalon, D., Davis, R.W. and Brown, P.O.: **Quantitative monitoring of gene expression patterns with a complementary DNA microarray.** *Science* 1995, **270**:467–470. (4-4)

Schmidt, A. and Lamzin, V.S.: ***Veni, vidi, vici*—atomic resolution unraveling the mysteries of protein function.** *Curr. Opin. Struct. Biol.* 2002, **12**:698–703. (5-2)

Scott, J.E.: **Molecules for strength and shape.** *Trends Biochem. Sci.* 1987, **12**:318–321. (1-6)

Sette, C., Inouye, C.J., Stroschein, S.L., Iaquinta, P.J. and Thorner, J.: **Mutational analysis suggests that activation of the yeast pheromone response mitogen-activated protein kinase pathway involves conformational changes in the Ste5 scaffold protein.** *Mol. Biol. Cell.* 2000, **11**:4033–4049. (2-5)

Sharp, K.A. and Englander, S.W.: **How much is a stabilizing bond worth?** *Trends Biochem. Sci.* 1994, **19**:526–529. (1-4, 1-12)

Shaw, J.P., Petsko, G.A. and Ringe, D.: **Determination of the structure of alanine racemase from *Bacillus stearothermophilus* at 1.9-Å resolution.** *Biochemistry* 1997, **36**:1329–1342. (4-17)

Sheinerman, F.B., Norel, R. and Honig, B.: **Electrostatic aspects of protein-protein interactions.** *Curr. Opin. Struct. Biol.* 2000, **10**:153–159. (4-10)

Sherlock, G., Hernandez-Boussard, T., Kasarskis, A., Binkley, G., Matese, J.C., Dwight, S.S., Kaloper, M., Weng, S., Jin, H., Ball, C.A., Eisen, M.B., Spellman, P.T., Brown, P.O., Botstein, D. and Cherry, J.M.: **The Stanford Microarray Database.** *Nucleic Acids Res.* 2001, **29**:152–155. (4-4)

Sherman, M.Y. and Goldberg, A.L.: **Cellular defenses against unfolded proteins: a cell biologist thinks about neurodegenerative diseases.** *Neuron* 2001, **29**:15–32. (3-11)

Shuker, S.B., Hajduk, P.J., Meadows, R.P. and Fesik, S.W.: **Discovering high-affinity ligands for proteins: SAR by NMR.** *Science* 1996, **274**:1531–1534. (4-9)

Siezen, R.J. and Leunissen, J.A.: **Subtilases: the superfamily of subtilisin-like serine proteases.** *Protein Sci.* 1997, **6**:501–523. (4-8)

Sindelar, C.V., Budny, M.J., Rice, S., Naber, N., Fletterick, R. and Cooke, R.: **Two conformations in the human kinesin power stroke defined by X-ray crystallography and EPR spectroscopy.** *Nat. Struct. Biol.* 2002, **9**:844–848. (3-10)

Silverman, R.B.: *The Organic Chemistry of Enzyme-Catalyzed Reactions* Revised ed. (Academic Press, New York, 2002). (2-6, 2-10, 2-11, 2-14)

Simons, K.T., Kooperberg, C., Huang, E. and Baker, D.: **Assembly of protein tertiary structures from fragments with similar local sequences using simulated annealing and Bayesian scoring functions.** *J. Mol. Biol.* 1997, **268**:209–225. (4-7)

Simons, K.T. Strauss, C. and Baker, D.: **Prospects for *ab initio* protein structural genomics.** *J. Mol. Biol.* 2001, **306**:1191–1199. (4-7)

Smith, C.A., Calabro, V. and Frankel, A.D.: **An RNA-binding chameleon.** *Mol. Cell* 2000, **6**:1067–1076. (4-14)

Smith, D.L., Almo, S.C., Toney, M.D. and Ringe, D.: **2.8-Å-resolution crystal structure of an active-site mutant of aspartate aminotransferase from *Escherichia coli.*** *Biochemistry* 1989, **28**:8161–8167. (4-12)

Snel, B., Bork, P. and Huynen, M.A.: **The identification of** functional modules from the genomic association of genes.** *Proc. Natl Acad. Sci. USA* 2002, **99**:5890–5895. (4-2)

Song, H., Hanlon, N., Brown, N.R., Noble, M.E., Johnson, L.N. and Barford, D.: **Phosphoprotein-protein interactions revealed by the crystal structure of kinase-associated phosphatase in complex with phosphoCDK2.** *Mol. Cell* 2001, **7**:615–626. (3-14)

Spearman, J. C.: *The Hydrogen Bond and Other Intermolecular Forces* (The Chemical Society, London, 1975). (1-4)

Spiro, R.G.: **Protein glycosylation: nature, distribution, enzymatic formation, and disease implications of glycopeptide bonds.** *Glycobiology* 2002, **12**:43R–56R. (3-19)

Spoerner, M., Herrmann, C., Vetter, I.R., Kalbitzer, H.R. and Wittinghofer, A.: **Dynamic properties of the Ras switch I region and its importance for binding to effectors.** *Proc. Natl Acad. Sci. USA* 2001, **98**:4944–4949. (3-7)

Sprang, S.R.: **G protein mechanisms: insights from structural analysis.** *Annu. Rev. Biochem.* 1997, **66**:639–678. (3-7)

Spudich, J.A.: **The myosin swinging cross-bridge model.** *Nat. Rev. Mol. Cell Biol.* 2001, **2**:387–392. (3-6, 3-10)

Stamler, J.S., Lamas, S., and Fang, F.C.: **Nitrosylation: the prototypic redox-based signaling mechanism.** *Cell* 2001, **106**:675–683. (3-20)

Steiner, D.F.: **The proprotein convertases.** *Curr. Opin. Chem. Biol.* 1998, **2**:31–39. (3-16)

Steitz, T.A.: **DNA polymerases: structural diversity and common mechanisms.** *J. Biol. Chem.* 1999, **274**:17395–17398. (4-8)

Stenmark, H., Aasland, R. and Driscoll, P.C.: **The phosphatidylinositol 3-phosphate-binding FYVE finger.** *FEBS Lett.* 2002, **513**:77–84. (3-1)

Stock, A.M., Martinez-Hackert, E., Rasmussen, B.F., West, A.H., Stock, J.B., Ringe, D. and Petsko, G.A.: **Structure of the Mg^{2+}-bound form of CheY and mechanism of phosphoryl transfer in bacterial chemotaxis.** *Biochemistry* 1993, **32**:13375–13380. (3-15)

Stock, A.M., Robinson, V.L. and Goudreau, P.N.: **Two-component signal transduction.** *Annu. Rev. Biochem.* 2000, **69**:183–215. (3-15)

Stock, D., Nederlof, P.M., Seemuller, E., Baumeister, W., Huber, R., and Lowe, J.: **Proteasome: from structure to function.** *Curr. Opin. Biotechnol.* 1996, **7**:376–385. (3-11)

Strater, N., Schnappauf, G., Braus, G., and Lipscomb, W.N.: **Mechanisms of catalysis and allosteric regulation of yeast chorismate mutase from crystal structures.** *Structure* 1997, **5**:1437–1452. (2-8)

Sudarsanam, S.: **Structural diversity of sequentially identical subsequences of proteins: identical octapeptides can have different conformations.** *Proteins* 1998, **30**:228–231. (4-14)

Sugio, S., Petsko, G.A., Manning, J.M., Soda, K. and Ringe, D.: **Crystal structure of a D-amino acid aminotransferase: how the protein controls stereoselectivity.** *Biochemistry* 1995, **34**:9661–9669. (4-12)

Sun, Y.J., Chou, C.C., Chen, W.S., Wu, R.T., Meng, M. and Hsiao, C.D.: **The crystal structure of a multifunctional protein: phosphoglucose isomerase/autocrine motility factor/neuroleukin.** *Proc. Natl Acad. Sci. USA* 1999, **96**:5412–5417. (4-13)

Swindells, M.B., MacArthur, M.W. and Thornton, J.M.: **Intrinsic phi, psi propensities of amino acids, derived from the coil regions of known structures.** *Nat. Struct. Biol.* 1995, **2**:596–603. (1-8)

Swope, M.D. and Lolis, E.: **Macrophage migration inhibitory factor: cytokine, hormone, or enzyme?** *Rev. Physiol. Biochem. Pharmacol.* 1999, **139**:1–32. (4-13)

Szwajkajzer, D. and Carey, J.: **Molecular constraints on ligand-binding affinity and specificity.** *Biopolymers* 1997, **44**:181–198. (2-4)

Tainer, J.A., Roberts, V.A. and Getzoff, E.D.: **Protein metal-binding sites.** *Curr. Opin. Biotechnol.* 1992, **3**:378–387. (1-13)

Takai, Y., Sasaki, T. and Matozaki, T.: **Small GTP-binding proteins.** *Physiol. Rev.* 2001, **81**:153–208. (3-7)

Tan, S. and Richmond T.J.: **Crystal structure of the yeast MATalpha2/MCM1/DNA ternary complex.** *Nature* 1998, **391**:660–666. (4-14)

Tefferi, A., Bolander, M., Ansell, S., Wieben, E.D. and Spelsberg, T.: **Primer on medical genomics parts I–IV.** *Mayo. Clin. Proc.* 2002, **77**:927–940. (4-0)

Templin, M.F., Stoll, D., Schrenk, M., Traub, P.C., Vohringer, C.F. and Joos, T.O.: **Protein microarray technology.** *Trends Biotechnol.* 2002, **20**:160–166. (4-17)

Theodosiou, A. and Ashworth, A.: **MAP kinase phosphatases.** *Genome Biol.* 2002, **3**:reviews3009. (3-13)

Thoden, J.B., Holden, H.M., Wesenberg, G., Raushel, F.M. and Rayment, I.: **Structural and carbamoyl phosphate synthetase: a journey of 96 Å from substrate to product.** *Biochemistry* 1997, **36**:6305–6316. (2-16)

Thornton, J.W. and DeSalle, R.: **Gene family evolution and homology: genomics meets phylogenetics.** *Annu. Rev. Genomics Hum. Genet.* 2000, **1**:41–73. (1-14)

Thornton, J.M., Orengo, C.A., Todd, A.E. and Pearl, F.M.: **Protein folds, functions and evolution.** *J. Mol. Biol.* 1999, **293**:333–342. (1-15)

Todd, A.E., Orengo, C.A., Thornton, J.M.: **Evolution of function in protein superfamilies, from a structural perspective.** *J. Mol. Biol.* 2001, **307**:1113–1143. (4-8)

Tong, A.H., Evangelista, M., Parsons, A.B., Xu, H., Bader, G.D., Page, N., Robinson, M., Raghibizadeh, S., Hogue, C.W., Bussey, H., Andrews, B., Tyers, M. and Boone, C.: **Systematic genetic analysis with ordered arrays of yeast deletion mutants.** *Science* 2001, **294**:2364–2368. (4-0)

Topham, C.M., McLeod, A., Eisenmenger, F., Overington, J.P., Johnson, M.S. and Blundell, T.L.: **Fragment ranking in modelling of protein structure. Conformationally constrained environmental amino acid substitution tables.** *J. Mol. Biol.* 1993, **229**:194–220. (1-2)

Tsou, C.L.: **Active site flexibility in enzyme catalysis.** *Annls. N.Y. Acad. Sci.* 1998, **864**:1–8. (2-2)

Tucker, C.L., Gera, J.F. and Uetz, P.: **Towards an understanding of complex protein networks.** *Trends Cell Biol.* 2001, **11**:102–106. (4-17)

Turner, M.: **Cellular memory and the histone code.** *Cell* 2002, **111**:285–291. (3-20)

Vale, R.D. and Milligan, R.A.: **The way things move: looking under the hood of molecular motor proteins.** *Science* 2000, **288**:88–95. (3-10)

van Roessel, P. and Brand, A.H.: **Imaging into the future: visualizing gene expression and protein interactions with fluorescent proteins.** *Nat. Cell Biol.* 2002, **4**:E15–E20. (4-17)

Vetter, I.R. and Wittinghofer, A.: **Nucleoside triphosphate-binding proteins: different scaffolds to achieve phosphoryl transfer.** *Q. Rev. Biophys.* 1999, **32**:1–56. (3-7)

Vetter, I.R. and Wittinghofer, A.: **The guanine nucleotide-binding switch in three dimensions.** *Science* 2001, **294**:1299–1304. (3-6)

Vocadlo, D.J., Davies, G.J., Laine, R. and Withers, S.G.: **Catalysis by hen egg-white lysozyme proceeds via a covalent intermediate.** *Nature* 2001, **412**:835–838. (2-12)

Voet, D. and Voet, G.: *Biochemistry* 2nd ed. (Wiley, New York, 1995). (1-0, 1-3, 2-12, 2-13)

Vogeley, L., Palm, G.J., Mesters, J.R. and Hilgenfeld, R.: **Conformational change of elongation factor Tu (EF-Tu) induced by antibiotic binding. Crystal structure of the complex between EF-Tu.GDP and aurodox.** *J. Biol. Chem.* 2001, **276**:17149–17155. (3-9)

von Heijne, G.: **Recent advances in the understanding of membrane protein assembly and structure.** *Q. Rev. Biophys.* 1999, **32**:285–307. (1-11)

von Mering, C., Huynen, M., Jaeggi, D., Schmidt, S., Bork, P. and Snel, B.: **STRING: a database of predicted functional associations between proteins.** *Nucleic Acids Res.* 2003, **31**:258–261. (4-2)

von Mering, C., Krause, R., Snel, B., Cornell, M., Oliver, S.G., Fields, S. and Bork, P.: **Comparative assessment of large-scale data sets of protein–protein interactions.** *Nature* 2002, **417**:399–403. (4-0, 4-17)

Wade, R.C., Gabdoulline, R.R., Ludemann, S.K. and Lounnas, V.: **Electrostatic steering and ionic tethering in enzyme-ligand binding: insights from simulations.** *Proc. Natl Acad. Sci. USA* 1998, **95**:5942–5949. (2-7)

Wall, M.E., Gallagher, S.C. and Trewhella, J.: **Large-scale shape changes in proteins and macromolecular complexes.** *Annu. Rev. Phys. Chem.* 2000, **51**:355–380. (1-22)

Wallace, A.C., Laskowski, R.A. and Thornton, J.M.: **Derivation of 3D coordinate templates for searching structural databases: application to Ser-His-Asp catalytic triads in the serine proteinases and lipases.** *Protein Sci.* 1996, **5**:1001–1013. (4-10)

Walsh, C.: *Enzymatic Reaction Mechanisms* (Freeman, San Francisco, 1979). (2-6, 2-14)

Walther, D., Eisenhaber, F. and Argos, P.: **Principles of helix-helix packing in proteins: the helical lattice superposition model.** *J. Mol. Biol.* 1996, **255**:536–553. (1-10)

Watanabe, H., Takehana, K., Date, M., Shinozaki, T. and Raz, A.: **Tumor cell autocrine motility factor is the neuroleukin/phosphohexose isomerase polypeptide.** *Cancer Res.* 1996, **56**:2960–2963. (4-13)

Weber, P.C. and Salemme, F.R.: **Structural and functional diversity in 4 alpha-helical proteins.** *Nature* 1980, **287**:82–84. (1-17)

Weiss, M.S., Blanke, S.R., Collier, R.J. and Eisenberg, D.: **Structure of the isolated catalytic domain of diphtheria toxin.** *Biochemistry* 1995, **34**:773–781. (3-3)

Wells, L., Vosseller, K. and Hart, G.W.: **Glycosylation of nucleoplasmic proteins: signal transduction and O-GlcNac.** *Science* 2001, **291**:2376–2378. (3-18)

West, A.H. and Stock, A.M.: **Histidine kinases and response regulator proteins in two-component signaling systems.** *Trends Biochem. Sci.* 2001, **26**:369–376. (3-15)

White, A., Ding, X., van der Spek, J.C., Murphy, J.R. and Ringe, D.: **Structure of the metal-ion-activated diphtheria toxin repressor/tox operator complex.** *Nature* 1998, **394**:502–506. (3-5)

Wilkinson, M.G. and Millar, J.B.: **Control of the eukaryotic cell cycle by MAP kinase signaling pathways.** *FASEB J.* 2000, **14**:2147–2157. (3-12)

Williams, R.W., Chang, A., Juretic, D. and Loughran, S.: **Secondary structure predictions and medium range interactions.** *Biochim. Biophys. Acta* 1987, **916**:200–204. (1-8)

Wilson, C., Kreychman, J. and Gerstein, M.: **Assessing annotation transfer for genomics: quantifying the relations between protein sequence, structure and function through tradition and probabilistic scores.** *J. Mol. Biol.* 2000, **297**:233–249. (4-1)

Wittinghofer, A.: **Signal transduction via Ras.** *Biol. Chem.* 1998, **379**:933–937. (3-7)

Workman, J.L. and Kingston, R.E.: **Alteration of nucleosome structure as a mechanism of transcriptional regulation.** *Annu. Rev. Biochem.* 1998, **67**:545–579. (3-20)

Wolan, D.W., Greasley, S.E., Beardsley, G.P. and Wilson, I.A.: **Structural insights into the avian AICAR transformylase mexhanism.** *Biochemistry* 2002, **41**:15505–15513. (2-15)

Wrba, A., Schweiger, A., Schultes, V., Jaenicke R. and Zavodszky, P.: **Extremely thermostable D-glyceraldehyde-3-phosphate dehydrogenase from the eubacterium *Thermotoga maritima*.** *Biochemistry* 1990, **29**:7584–7592. (2-2)

Wu, N., Mo, Y., Gao, J. and Pai, E.F.: **Electrostatic stress in catalysis: structure and mechanism of the enzyme orotidine monophosphate decarboxylase.** *Proc. Natl Acad. Sci. USA* 2000, **97**:2017–2022. (2-8)

Wurgler-Murphy, S.M. and Saito, H.: **Two-component signal transducers and MAPK cascades.** *Trends Biochem. Sci.* 1997, **22**:172–176. (3-15)

Wyrick, J.J. and Young, R.A.: **Deciphering gene expression regulatory networks.** *Curr. Opin. Genet. Dev.* 2002, **12**:130–136. (3-0)

Xia, D., Yu, C.A., Kim, H., Xia, J.Z., Kachurin, A.M., Zhang, L., Yu, L. and Deisenhofer, J.: **Crystal structure of the cytochrome bc1 complex from bovine heart mitochondria.** *Science* 1997, **277**:60–66. (1-11)

Xu, D., Xu, Y. and Uberbacher, E.C.: **Computational tools for protein modeling.** *Curr. Protein Pept. Sci.* 2000, **1**:1–21. (5-3)

Xu, W., Doshi, A., Lei, M., Eck, M.J. and Harrison, S.C.: **Crystal structures of c-Src reveal features of its autoinhibitory mechanism.** *Mol. Cell* 1999, **3**:629–638. (3-13)

Xu, W., Seiter, K., Feldman, E., Ahmed, T. and Chiao, J.W.: **The differentiation and maturation mediator for human myeloid leukemia cells shares homology with neuroleukin or phosphoglucose isomerase.** *Blood* 1996, **87**:4502–4506. (4-13)

Yaffe, M.B.: **How do 14-3-3 proteins work? Gatekeeper phosphorylation and the molecular anvil hypotheses.** *FEBS Lett.* 2002, **513**:53–57. (3-1)

Yaffe, M.B.: **Phosphotyrosine-binding domains in signal transduction.** *Nat. Rev. Mol. Cell Biol.* 2002, **3**:177–186. (3-1)

Yaffe, M.B. and Elia, A.E.: **Phosphoserine/threonine binding domains.** *Curr. Opin. Cell Biol.* 2001, **13**:131–138. (3-1)

Yang, A.S. and Honig, B.: **An integrated approach to the analysis and modeling of protein sequences and structures. I-III.** *J Mol Biol.* 2000, **301**:665–678. (4-6)

Yang, W.Z., Ko, T.P., Corselli, L., Johnson, R.C. and Yuan, H.S.: **Conversion of a beta-strand to an alpha-helix induced by a single-site mutation observed in the crystal structure of Fis mutant Pro26Ala.** *Protein Sci.* 1998, **7**:1875–1883. (4-14)

Yaswen, L., Diehl, N., Brennan, M.B. and Hochgeschwender, U.: **Obesity in the mouse model of pro-opiomelanocortin deficiency responds to peripheral melanocortin.** *Nat. Med.* 1999, **5**:1066–1070. (4-4)

Ye, S. and Goldsmith, E.J.: **Serpins and other covalent protease inhibitors.** *Curr. Opin. Struct. Biol.* 2001, **11**:740–745. (4-15)

Yennawar, N., Dunbar, J., Conway, M., Huston, S. and Farber, G.: **The structure of human mitochondrial branched-chain aminotransferase.** *Acta. Crystallogr. D. Biol. Crystallogr.* 2001, **57**:506–515. (4-12)

Yoshimura, T., Jhee, K.H. and Soda, K.: **Stereospecificity for the hydrogen transfer and molecular evolution of pyridoxal enzymes.** *Biosci. Biotechnol. Biochem.* 1996, **60**:181–187. (4-12)

Young, M.A., Gonfloni, S., Superti-Furga, G., Roux, B. and Kuriyan, J.: **Dynamic coupling between the SH2 and SH3 domains of c-Src and Hck underlies their inactivation by C-terminal tyrosine phosphorylation.** *Cell* 2001, **105**:115–126. (3-13)

Zahn, R., Liu, A., Lührs, T., Riek, R., von Schroetter, C., Lopez Garcia F., Billeter, M., Calzolai, L., Wider, G. and Wüthrich, K.: **NMR structure of the human prion protein.** *Proc. Natl Acad. Sci. USA* 2000, **97**:145–150. (4-15)

Zhang, B., Rychlewski L., Pawlowski K, Fetrow JS, Skolnick, J. and Godzik, A.: **From fold predictions to function predictions: automation of functional site conservation analysis for functional genome predictions.** *Protein Sci.* 1999, **8**:1104–1115. (4-10)

Zhang, T., Bertelsen, E. and Alber, T.: **Entropic effects of disulfide bonds on protein stability.** *Nat. Struct. Biol.* 1994, **1**:434–438. (1-13)

Zhu, Z.Y. and Blundell, T.L.: **The use of amino acid patterns of classified helices and strands in secondary structure prediction.** *J. Mol. Biol.* 1996, **260**:261–276. (1-8)

Zielenkiewicz, P. and Rabczenko, A.: **Methods of molecular modelling of protein-protein interactions.** *Biophys. Chem.* 1988, **29**:219–214. (1-19)

Zolnierowicz, S. and Bollen, M.: **Protein phosphorylation and protein phosphatases. De Panne, Belgium, September 19–24, 1999.** *EMBO J.* 2000, **19**:483–488. (3-12)

Index

References to figures are in bold,
for example **F4-33** refers to Figure 4-33.

proteasome **F1-74,** 108–109, **F3-19**
Protein Data Bank 142, **F4-28**
protein G, *Staphylococcus aureus* 159
protein kinase domains 112, **F3-24,** 153
protein kinases 87, 110–111
    activation mechanism 112–113, **F3-25**
    activation loop (segment) 112, **F3-24, F3-25,**
        F3-26, 114, **F3-27,** 115, **F3-28**
    autoregulation 88
    localization 90
    multiple alignment **F4-4**
    regulation 112–115
protein microarrays 138
protein phosphatases 110, 115
    two-component signaling systems 116–117
proteolysis
    limited 28, 29, 110, 118–119
    self (autoproteolysis) 120–121
    ubiquitin-dependent 109, **F3-20**
    see also degradation, protein; peptide bonds,
        hydrolysis
proteolytic cascade 118, 119, **F3-34**
protomer 44
protons 92
proton transfer 52–53, 74–75
proximity 66–67, **F2-25**
proximity factor 66
Pseudomonas aeruginosa 90, **F4-10**
Pseudomonas ovalis 152
pseudosymmetric proteins 44
PSI-BLAST 135
PSIPRED **F4-25**
psi torsion angle 9, **F1-9**
PTB domain 88, **F3-2**
PUA domain **F4-6**
purification, protein 139
PutA **F4-47**
pyridoxal phosphate (PLP) 73, **F2-37**
    -dependent transamination 154, **F4-43**
    in unknown gene product 164, **F4-59**
pyruvate decarboxylase 53, 73, **F2-34,** 141, **F4-20**

quantity, protein 87
quaternary structure 3, **F1-2,** 40–45
    in genetic diseases 42–43
    geometry 44, **F1-75, F1-74**
    intermolecular interfaces 42–43
    intracellular environment and 92
    ligand-induced conformational changes 47
    prediction from sequence 143
    role of complementarity 40–41

Rab 124, 125
Ramachandran plot 12, **F1-11**
RanGAP1 127
Rap-Raf oligomerization **F1-70**
Ras **F1-1,** 100–101, **F3-21**
    lipid modification 125
    mutations 101
    switching cycle 101, **F3-14**
    targeting to membrane 91
reaction sub-site, enzyme active site 64, 65
recessive mutations 43
recognition, molecular 52
recognition modules (interaction domains) 86, 88, **F3-1**
redox environment, control by 92–93
redox reactions 70–71, **F2-31**
reducing environment 10, 92
reduction 70
refinement 170
regulation of protein function
    by *N*-acetylation 126–127
    by degradation 108–109
    by location 90–91
    mechanisms 86–87

by methylation 126
by nitrosylation 127
by pH and redox environment 92–93
by phosphorylation 110–111
by proteolysis 118–121
by sumoylation 127
by switching *see* switches, molecular
regulator of G-protein signaling (RGS) proteins
    102–103
residue 4
resolution 168, **F5-1,** 169, 172
resonance 8
    peptide bonds 9, **F1-8**
response regulator protein (RR) 116–117, **F3-29,**
    F3-30
retinol-binding protein **F1-19,** 37
reverse transcriptase 147, **F4-30**
reverse turns *see* beta turns
R-factors 172
RGS proteins 102–103
rhinovirus 44, **F1-74,** 97
Rhizobium meliloti DctB protein 25, **F1-31**
rhodopsin 103, **F3-15, F4-23,** 143
ribbon diagram 172, **F5-5**
ribosome 60, **F2-16,** 104
ribozymes 62
RING finger domain **F3-2**
RNA
    binding sites 56
    catalytic (ribozymes) 62
    coenzymes and 77
    degradation 87
    editing 6, 7
    interference (RNAi) 138, 139
    messenger (mRNA) 6, **F1-5**
    regulation 87
    ribosomal 62
    transfer (tRNA) 104, **F3-16**
RNA polymerases 147
root-mean-square deviation (RMSD) 169, 172
Rosetta method 144–145, **F4-26, F4-27**
Rossmann fold 34

Saccharomyces cerevisiae (yeast) **F4-10**
    MATα2 158–159, **F4-51**
    protein functions 137, **F4-13**
    Tem1 90
    two-hybrid system 138, 139, **F4-18,** 165
    YBL036c **F4-60,** 164–165, **F4-59**
S-acylation 124, 125
SAGE (serial analysis of gene expression) 138
Salmonella typhimurium 82
salt bridges 10, **F1-10,** 42
    in phosphorylated proteins 111
Sam68 126
SAM domain **F3-2**
scaffold proteins 60, 61, **F2-18**
    protein binding to 91, **F3-4**
Schizosaccharomyces pombe 90
scorpion toxin 39, **F1-63**
scrapie 160
scytalone dehydratase 141, **F4-21**
secondary structure 3, **F1-2,** 12–13
    classifying types of folds from 12
    elements 12
        membrane proteins 18, 24–25
        see also alpha helix; beta sheet; beta turn
    prediction 18, **F1-21,** 134
    steric constraints 12, **F1-11**
selenium (Se) **F2-38**
"semi-liquid" nature of proteins 46
septin 127
sequence(s) 3, **F1-2,** 6
    alignment 132–133
        active-site residues 162

local 134–135, 137
    multiple 132, 133, **F4-4**
    pairwise 132–133, **F4-2**
analysis 132–135
chameleon 158–159, **F4-49**
comparison 132, 134
conserved 132, 133
deriving function from 136–137, **F4-11**
divergent and convergent evolution 140–141, **F4-19**
growth of information on 136, **F4-9**
identity/similarity 132–133, **F4-3,** 142
    40% rule 136–137, **F4-11**
    percent 132, 136–137
insertions and deletions 33, 132
secondary structure prediction from 18, **F1-21,** 134
signal 86, 91, **F3-4**
structure determination from 142–145
see also primary structure
Ser1 125
serine
    phosphorylation 110, 111, 112
    structure and side chain 4, **F1-3**
serine proteases
    activation 118, **F3-32**
    catalytic triad 34–35, **F1-52,** 78, 146–147
        detection methods 150, **F4-35**
    convergent evolution 140, 146–147, **F4-29,** 155
    multi-step reactions 78–79, **F2-40**
    precursors, homology modeling 143, **F4-24**
serpins (serine protease inhibitors) 161, **F4-54**
seven-transmembrane helix fold 153
SH2 domain 88, **F3-2,** 153
    in analysis of protein function **F4-6,** 135, 137
    in protein kinase regulation 112–113, **F3-26**
    in STAT signaling 58
SH3 domain 88, **F3-2**
    in analysis of protein function 134, **F4-6,** 135, 137
    in protein kinase regulation 112–113, **F3-26**
    in protein targeting 91
sickle-cell anemia 42–43, **F1-71**
side chains 4
    amino acid 4, **F1-3**
    movements 47
signal sequences 86, 91, **F3-4**
signal transduction pathways 87, 88
    protein switches 98
    regulation by phosphorylation 110, **F3-21**
    regulation by protein localization 90
signal transduction proteins
    domain arrangements 32, **F1-46**
    partner swapping 58–59, **F2-13**
    scaffold proteins 61
silk **F1-1,** 16, 37, 60
single-nucleotide polymorphism (SNP) 6, 7
site-directed mutagenesis 150
Smt3 127, **F3-49**
SNARE domain **F3-2**
sodium dodecyl sulfate (SDS) 55
solubility, glycosylation and 123
soluble proteins
    folding 20–21, **F1-23**
    hydration shell 23, **F1-26**
Sos **F3-14, F3-21**
Sp100b **F4-27**
space-filling image 172, **F5-5**
specificity
    enzyme-mediated reactions 64, **F2-24**
    ligand binding 52, 59
    sub-site, enzyme active site 64, 65
splicing, protein 120–121, **F3-35, F3-38**
sporulation, yeast 165
Src kinases 112–113, **F3-26**
Src-Lck 32, **F1-46**
S–S bridges *see* disulfide bridges
stability, protein 26–29, 108